D0787407

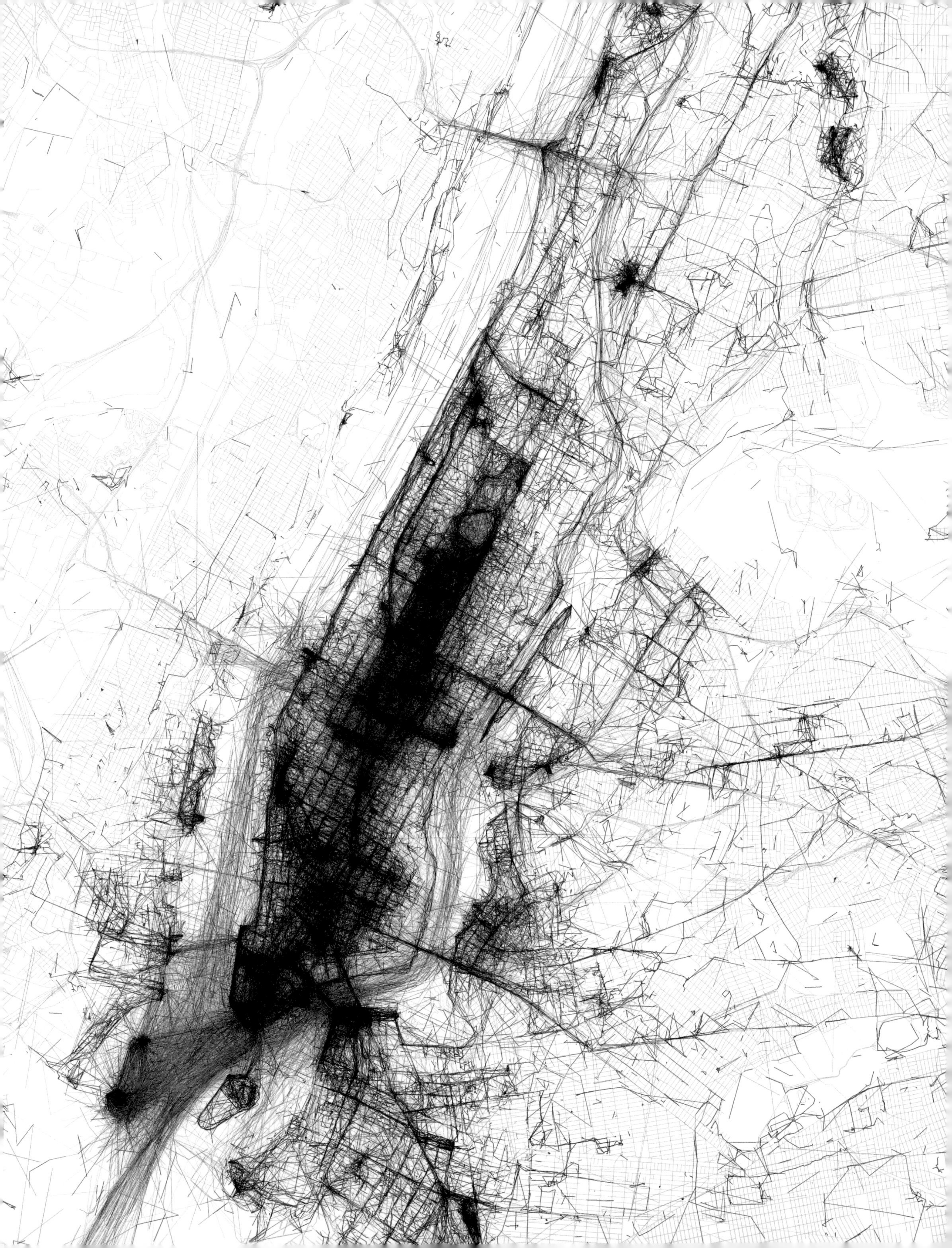

PHYS REV LETT
APPL PHYS LETT
ADV MATER
ANGEW CHEM INT EDIT
J AM CHEM SOC
CHEM REV
CHEM MATER
ACCOUNTS CHEM RES
PHYS LETT B
AM ECON REV
ECONOMETRICA
HEPATOLOGY
GASTROENTEROLOGY
CLIN INFECT DIS
J INFECT DIS
TRENDS ECOL EVOL
NEUROIMAGE
NEUROLOGY
NEURON
STROKE
IMMUNITY
J EXP MED
NAT GENET
P NATL ACAD SCI USA
SCIENCE
NATURE
J BIOL CHEM
CELL
GENE DEV
NAT MED
NAT BIOTECHNOL
BRIT MED J
LANCET
NEW ENGL J MED
CIRCULATION
JAMA-J AM MED ASSOC
J AM COLL CARDIOL
ANN INTERN MED
DIABETES CARE
GEOPHYS RES LETT
EARTH PLANET SC LETT
GEOCHIM COSMOCHIM AC
ICARUS
CLIN CANCER RES
J CLIN ONCOL
CANCER
ENVIRON SCI TECHNOL
PLANT J
PLANT PHYSIOL
PLANT CELL
APPL ENVIRON MICROB
ANAL CHEM
ASTROPHYS J SUPPL S
MON NOT R ASTRON SOC
ASTROPHYS J
ASTRON ASTROPHYS
ASTRON J
OPT LETT
REV MOD PHYS

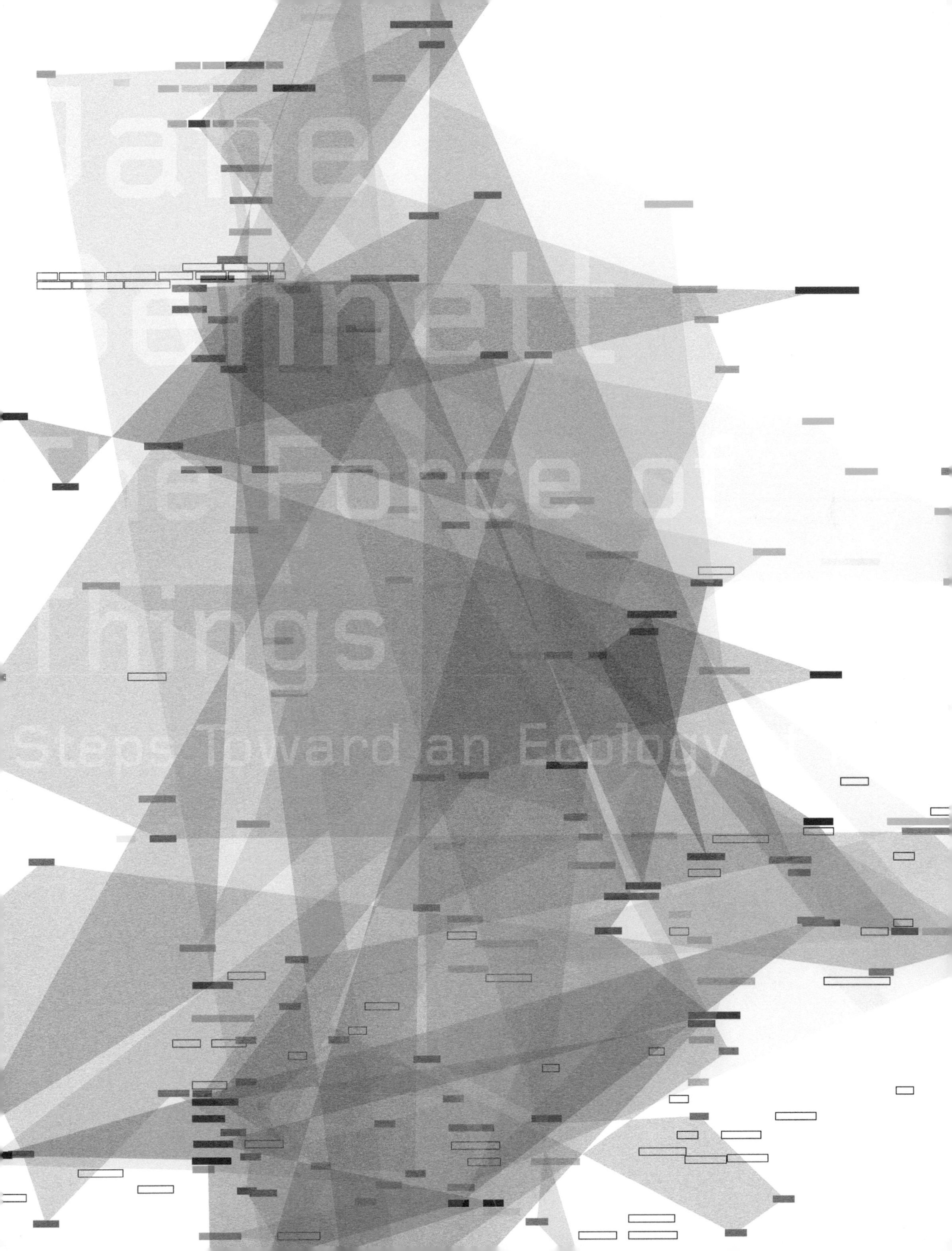
Jane
Bennett
The Force of
Things
Steps Toward an Ecology

Visual Complexity
Mapping Patterns of Information

Manuel Lima

Princeton Architectural Press
New York

Published by
Princeton Architectural Press
37 East Seventh Street
New York, New York 10003

For a free catalog of books, call 1.800.722.6657.
Visit our website at www.papress.com.

Printed and bound in China
14 13 12 11 4 3 2 First edition

Editor: Linda Lee
Designer: Jan Haux

Special thanks to: Bree Anne Apperley, Sara Bader, Nicola Bednarek Brower, Janet Behning, Megan Carey, Becca Casbon, Carina Cha, Tom Cho, Penny (Yuen Pik) Chu, Russell Fernandez, John Myers, Katharine Myers, Margaret Rogalski, Dan Simon, Andrew Stepanian, Jennifer Thompson, Paul Wagner, Joseph Weston, and Deb Wood of Princeton Architectural Press —Kevin C. Lippert, publisher

Library of Congress Cataloging-in-Publication Data
Lima, Manuel, 1978–
Visual complexity : mapping patterns of information / Manuel Lima. — 1st ed.
p. cm.
Includes bibliographical references and index.
ISBN 978-1-56898-936-5 (alk. paper)
1. Communication in science—Graphic methods. 2. Visual communication. I. Title.
II. Title: Mapping patterns of information.
Q223.L55 2011
003'.54—dc22

2010051250

Page 1: Eric Fischer, *The Geotaggers' World Atlas*, 2010. A map of geotagged photographs from Flickr and Picasa taken in New York City. The speed at which photographers traveled the urban landscape was determined by analyzing the time stamps and geotags of the images. The resulting traces were plotted on an OpenStreetMap background layer.

Page 2: Eigenfactor.org and Moritz Stefaner, *Visualizing Information Flow in Science*, 2009. A citation network of a subset of Thomson Reuters's *Journal Citation Reports* between 1997 and 2005.

Page 3: Ian Dapot, *The Force of Things*, 2005. A part of a series of posters based on mapping the relationships between cited authors and referenced ideas in Jane Bennett's essay "The Force of Things: Steps Toward an Ecology of Matter" (2005).

Page 4: Marius Watz, *Trajectories*, 2008

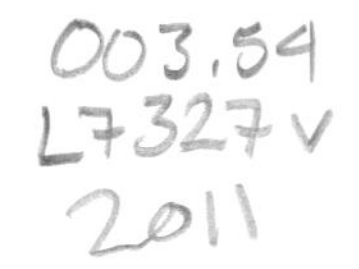

To my parents, Jorge and Maria Luisa, and my wife, Joana

Contents

Foreword

Lev Manovich

Visual Complexity looks at the intersection of two key techno-cultural phenomena of our time: networks and visualization. Both were relatively unknown only fifteen years ago but have since moved to the forefront of our social and cultural lives. While some social scientists had already started to study networks in the middle of the twentieth century, globalization and the rise of the web in the nineties and the explosion of online social networks in the last decade have drawn attention to their importance. Furthermore, although scientists had already been making graphs and charts of their data since the early nineteenth century, the ubiquity of computers, the growing programming literacy, and the wealth of data unleashed by networks at the turn of the twenty-first century democratized information visualization, making it a rapidly expanding new area of art and culture.

Author Manuel Lima is a thinker, designer, lecturer, and curator of one of the most influential online galleries that presents the best projects in information visualization: Visualcomplexity.com. However, in contrast to other important galleries that try to cover all of information visualization, Visualcomplexity.com is focused on visualizations of networks. As its ongoing curator, Lima is likely to have the best understanding of the creative impulses, exciting discoveries, and sheer range of work produced today in this area. Imbued with Lima's expertise, this book will become the essential reference for all practitioners and fans of network visualization—or, to use Lima's more evocative language, visualizations of complexity.

Many books on the recently emerging areas of software-driven design—web design, interaction design, motion graphics, and information design, for example—are simply visual portfolios. Others are how-to books that present techniques, "best practices," and practical step-by-step guides. Although we can often extract a few paragraphs or pages of important theoretical insights from both types of books, these passages usually represent only a very small percentage of a whole book.

If visualization of complex data is well on its way to becoming as important to the twenty-first century as photography and film were to the twentieth century, the time for books that encapsulate its main ideas and concepts has arrived. *Visual Complexity* is the first such book. Lima establishes conceptual coordinates and historical trajectories for both practice and appreciation of visualization. He also balances historical and theoretical discussions of larger issues with the presentation of exemplary projects in network visualization.

The rise of information visualization over the last fifteen years raises many important and interesting questions about the identity of this new medium. For example, what exactly is the difference between the hundreds of projects collected in Visualcomplexity.com's gallery and the standard graph—bar charts, pie charts, scatter plots, line charts, and so on—that many of us are routinely producing using Excel, Apple's Numbers, Google Charts, or similar software? These graph types—which were all originally invented in the first part of the nineteenth century and have therefore already been in use for about one hundred years before digital computers—share a lot in common with "digitally native" information visualization techniques developed more recently. Both depict quantified data by systematically mapping it into visual images. Both use the same graphic language: points, lines, curves, simple shapes, and other graphic primitives.

Interestingly, this language of information visualization also has many parallels with the language of geometric abstraction, which crystallized in the second decade of the twentieth century through the work of Piet Mondrian, Kasimir Malevich, Frank Kupka, and other modernist artists. However, if these artists wanted to liberate visual art from its representational function—i.e., its dependence on "visible data"—information visualization brings representation back. But this is a new kind of representation appropriate for information society: rather than representing the visible world, we now seek to represent—in order to understand—all kinds of data sets.

So what else is unique about information visualization? There are many possible answers to this question, although no single response can completely capture all the differences. Most of the projects discussed in this book are visually more dense, more complex, and more varied than the familiar charts created with graphing software. Why is this the case? First, contemporary designers, artists, and computer scientists are trying to represent considerably more data than ever before. Second, they want to represent relations between more dimensions of data than is possible with older graph types such as bar charts (one dimension) or scatter plots (two dimensions). The third reason is aesthetic and ideological: if nineteenth-century techniques for graphs fit the scientific paradigm of reduction (breaking nature down into the simplest possible elements and defining rules on how these elements interact), our current interest lies in understanding the phenomena of complexity (think chaos theory, emergence, complexity theory), which is reflected in the kinds of visualizations we find appealing.

We can also explain the visual variety of information-visualization culture. It comes from the systematic effort on the part of its practitioners to invent new visualization techniques, which is rewarded both in academia and in the cultural world. Additionally, since visualizations are now valued as artistic and cultural artifacts, we expect them to be unique—just as we expect this in fashion, product design, architecture, music, and other cultural fields.

Another question that inevitably comes up in any discussion of information visualization is whether it falls in the category of science, design, or art. Here is my proposal: rather than identifying visualization culture with a single category, let us consider it as existing in the space defined by all three.

Information visualization is widely used as a tool for understanding data—i.e., discovering patterns, connections, and structure. Since science is the area of human activity targeting the discovery of new knowledge about the world through systematic methods—such as experimentation, mathematical modeling, simulation—visualization now functions as another of these methods.

What distinguishes this new method is that it also firmly belongs to design—it involves the *visual* presentation of data in a way that facilitates the perception of patterns. Just as a graphic designer organizes information on a poster or web page to help the user navigate its layout efficiently, an information-visualization designer organizes data to help users see the patterns. At the same time, just like graphic designers, information-visualization designers do not only aim for efficiency and clarity. They chose particular visualization techniques and graphic styles in order to communicate an idea about the data and to evoke particular emotions in the viewer. For example, a network visualization may emphasize the density of the network, present it as a result of organic growth, focus on its instability and dynamism, or show the same network as a logically arranged, symmetrical, top-down, and stable structure.

A significant number of visualization projects collected on Visualcomplexity.com, and presented elsewhere online, can be considered pure art—and by *art*, we refer to the product of nonutilitarian activity as opposed to utilitarian design. The intent of these projects is not to reveal patterns or structures in data sets but to use information visualization as a technique to produce something aesthetically interesting. Old European masters created color in their paintings using a number of transparent layers; surrealists practiced automatic writings; pop artists magnified and manipulated fragments of mass media, such as comics, newspapers, and product packaging. Contemporary artists can now use the algorithms to create complex static, animated, or interactive abstractions out of data sets.

Visualizations of complex data sets can also function as art in a different sense: an activity aimed at making statements and asking questions about the world by selecting parts of it and representing these parts in particular ways. Some of the most well-known artistic visualization projects do exactly that: they make strong assertions about our world not only through the choice of visualization techniques but also through the choice of data sets.[1] As computer scientist Robert Kosara wrote in a paper presented at the 2007 Information Visualization conference: "The goal of artistic visualization is usually to communicate a concern, rather than to show data."[2] Figurative artists express their opinions about the world by choosing what they paint; writers and filmmakers do this by choosing the subjects of their narratives. Now artists can also talk about our world by choosing which data to visualize. How do you represent and ask interesting questions about society through its data traces? This is a new opportunity, responsibility, and challenge for contemporary artists.

The space defined by the disciplines of science, design, or art—or more precisely, the different goals and methods we associate with these three areas of human activity—contains lots of possibilities. A given visualization project can be situated anywhere in this space, depending on what it privileges. I think that this conceptualization can help us understand why some information-visualization projects have already achieved the status of classics within such a brief history of the field. They do not privilege one dimension of this space at the expense of others but manage to combine all three. They are functional in that they reveal interesting patterns in visualized data. They have strong design in terms of how they organize information and coordinate all the visual elements. And they embody art at its best—both pointing at and making provocative statements about important new phenomena in the world.[3] *Visual Complexity* will help to inspire practitioners to make many more exciting projects—and will help the public better understand the importance and beauty of visualizations of complexity.

Notes

1 For instance, Josh On, *They Rule*, 2004.

2 Kosara, "Visualization Criticism."

3 A few strong examples of such successful blends include Fernanda B. Viégas and Martin Wattenberg, *History Flow*, 2003; Marcos Weskamp, *Newsmap*, 2005; Mark Hansen et al., *Terre Natale (Exits 2)*, 2008; and Ben Fry, *On the Origin of Species: The Preservation of Favoured Traces*, 2009.

Introduction

In the midst of a hot, damp summer in New York City, in September 2005, I, strangely enough, found myself with too much time on my hands. I had recently graduated from Parsons School of Design and had just embarked on a much-desired full-time job in Midtown Manhattan. Even though I was now officially a member of the frantic world of digital advertising, everything else in my life seemed extremely slow paced. The weekends were the worst—two full days with nothing to do but unhurriedly relax from a mildly busy week. Just a few months before, however, I had been immersed in work, preparing for the climax of two years of obstinate dedication, my MFA thesis show. It was not surprising that after this intense period of my life, I was finding the newly rediscovered spare time to be a grueling luxury. I missed the adrenaline, the challenges, the constant array of projects and ideas. But above all, I missed a recently discovered passion for an enticing yet unfamiliar domain: the visual representation of networks.

During my time at Parsons I observed the blog phenomenon swiftly become mainstream, with many dooming predictions about the end of mass media and enthused visions about the rise of self-publishing. What I found particularly compelling about this change was the role of the blogosphere as a vibrant research environment, specifically as a venue for the study of information diffusion.

Word of mouth—the transfer of information orally from person to person—has always been a subject of interest for the social sciences, but one that has always proved difficult to scrutinize among individuals interacting within a physical environment, such as a school, company, or village. The blogosphere, on the other hand, provided an extraordinary laboratory to track and analyze how trends, ideas, and information travel through different online social groups. The possibility of bringing this vast atlas of memes to life was the driving premise of my MFA thesis, Blogviz.

Presented in May 2005, Blogviz was a visualization model for mapping the transmission and internal structure of popular links across the blogosphere. It explored the idea of meme propagation by drawing a parallel to the frequency with which the most-cited URLs appeared in daily blog entries. It was, in other words, a topological model of meme activity. But in order to properly chart these intriguing flows across varied blog entries, I needed a better understanding of how blogs were linked and how the World Wide Web was structured. During this research period, I investigated and collected dozens of projects that aimed at providing a portrait of the vast landscape of the web, as well as many network depictions from other, apparently less related domains, such as food webs, airline routes, protein chains, neural patterns, and social connections. As I quickly realized, the network is a truly ubiquitous structure present in most natural and artificial systems you can think of, from power grids to proteins, the internet, and the brain. Usually depicted by network diagrams made of nodes (a person, website, neuron, protein, or airport) and lines that connect and highlight relationships between the nodes (friendship, chemical exchange, or information flow), networks are an inherent fabric of life and a growing object of study in various scientific domains. The more of these intriguing diagrams I uncovered, the more enthralled and absorbed I became. This genuine curiosity quickly turned into a long-lasting obsession over the visual representation of networks, or more appropriately, network visualization. As I entered my new postgrad career, this passion somehow became contained, but I knew I couldn't resist it for much longer. Just a few months after graduating, in the midst of that hot, damp summer in New York City, I pulled together most of this initial body of research into one unique resource. In October 2005, VisualComplexity.com was born.

Initiated with approximately eighty projects—the result of my academic research—VisualComplexity.com quickly grew to encompass a variety of efforts, from mapping a social network of friends on Facebook to representing a global network of IP addresses. Although VisualComplexity.com has grown to over seven hundred projects, the goal remains the same: to facilitate a critical understanding of different network-visualization methods across the widest spectrum of knowledge. This vast repository—frequently referred to as a "map of maps" and depicting an assortment of systems in subject areas as diverse as biology, social networks, and the World Wide Web—is the most complete and accessible chart of the field's landscape. Some projects are rich interactive applications that go beyond the computer screen and live in large-scale multisensorial installations; others are static, meant to be experienced in print media like posters and printouts. Some require hours of rendering and complex algorithms to produce; others are simply drafted by hand or use a specific drawing software.

Making this pool of knowledge available to an even larger audience was the main impetus for this book. However, as the book gained shape, it quickly became clear that it was not just about making the pool of knowledge more accessible, but also saving it for posterity. As I reviewed projects to feature in the book, I was astounded by how many dead links and error messages I encountered. Some of these projects became completely untraceable, possibly gone forever. This disappearance is certainly not unique to network visualization—it is a widespread quandary of modern technology. Commonly referred to as the Digital Dark Age, the possibility of many present-day digital artifacts vanishing within a few decades is a considerably worrying prospect.

The reasons for this vanishing are never the same. In most instances, pieces are simply neglected over time, with authors not bothering to update the code, rendering it obsolete. In other cases, the plug-in version might become incapable of reading older formats or the application programming interface (API) from an early data-set source might change, making it extremely difficult to reuse the code that generated the original visualization. Lastly, projects are occasionally moved into different folders or domains or just taken down from the servers, simply because they highlight an outdated model that does not fit the current ambitions of their respective author or company. As I gathered many of the projects showcased in the book, I was surprised to find that it was easier to retrieve an illustration by Joachim of Fiore, produced in the twelfth century, than to attain an image of a visualization of web routers developed in 2001. Overall, this digital laissez-faire contributes to the ephemeral lifespan of most online visualizations, and consequently the whole field suffers from memory loss.

More than preserving this collective effort for the future, this book provides a broad background on the various forces shaping its development. It yields a comprehensive view of the visual representation of networks, delving into historical precedents, various contemporary methods, and a range of future prospects. It looks at the depiction of networks from a practical and functional perspective, proposing several guiding principles for current practitioners, but also explores the alluring qualities of the network schema, as a central driver for a new conception of art. This comprehensive study of network visualization should ultimately be accessible to anyone interested in the field, independent of their level of expertise or academic dexterity.

The book opens with "Tree of Life," an exploration of the sacred meaning of trees and their widespread use as a classification system over the centuries. It showcases an assortment of ancient representations—as predecessors of modern-day network diagrams—where the tree metaphor is used to visually convey a variety of topics, from theological events to an encyclopedia's table of contents.

The second chapter, "From Trees to Networks," makes the case for a new network-based outlook on the world, one that is based on diversity, decentralization, and nonlinearity. It explores several instances—from the way we envision our cities to the way we organize information and decode our brain—where an alternative network model is replacing the hierarchical tree schema.

Chapter three, "Decoding Networks," delves into the science behind network thinking and network drawing, providing a short introduction to its main precursors and early milestones. It also takes a pragmatic and utilitarian look at network visualization, acknowledging its key functions and proposing a set of guiding principles aimed at improving existing methods and techniques.

Following a series of functional recommendations for network visualization, chapter four, "Infinite Interconnectedness," presents a large number of examples divided into fourteen popular subjects. From depictions of the blogosphere to representations of terrorist networks, chapter four highlights the truly complex connectedness of modern times.

If chapter four looks at the practice primarily through its most common themes, chapter five, "The Syntax of a New Language," organizes a vast array of projects by their shared visual layouts and configurations. As designers, scientists, and researchers across the globe portray an increasing number of network structures in innovative ways, their collective effort forms the building blocks of a new network-visualization lexicon.

After presenting an abundance of network-visualization examples in chapters four and five, chapter six, "Complex Beauty," examines the alluring nature of networks, responsible for a considerable shift in our culture and society. Alternating between scientific and artistic viewpoints, this chapter explores the divide between order and complexity before culminating in a discussion of an original art movement embracing the newly discovered beauty of the network scheme.

Finally, and in the spirit of network diversity and decentralization, "Looking Ahead," the last chapter, presents different views on the influential growth of visualization, according to renowned experts, active participants, and attentive observers. The featured essays cover an array of trends and technologies shaping the progress of visualization and provide an immensely captivating perspective on what may lie ahead.

By exploring different facets of our information-driven network culture, this book ultimately unifies two rising disciplines: network science and information visualization. While network science examines the interconnections of various natural and artificial systems in areas as diverse as physics, genetics, sociology, and urban planning, information visualization aims at visually translating large volumes of data into digestible insights, creating an explicit bridge between data and knowledge. Due to its intrinsic aspiration for sense-making, information visualization is an obvious tool for network science, able to disentangle a range of complex systems and make them more comprehensible. Not only do both disciplines share a yearning for understanding, but they have also experienced a meteoric rise in the last decade, bringing together people from various fields and capturing the interest of individuals across the globe. But if this popularity is to become more than a fad, our efforts at decoding complexity need to be mastered and consolidated so that we can contribute substantially to our long journey of deciphering an increasingly interconnected and interdependent world. This book is a single step in this journey, and ultimately a testimony to the enthralling power of networks and visualization.

Acknowledgments

This endeavor would not have been possible if it had not been for the effort of dozens of individuals and institutions. First and foremost my gratitude goes to all the authors and organizations, who without exception have kindly shared their images, some spending many hours updating old code and re-creating new pieces especially for this undertaking. This book would not exist without you. The second wave of recognition goes to those who went out of their way to help me during my research and investigation, particularly Luigi Oliverio from the International Center for Joachimist Studies; Glenn Roe and Mark Olsen from the Project for American and French Research on the Treasury of the French Language, University of Chicago; Olga Pombo, author and researcher at the Center for Philosophy of Sciences of the University of Lisbon; Pablo Rodriguez Gordo, General Sub-directorate of Library Coordination (Spanish Ministry of Culture); Marcela Elgueda from Fundación Gego; and Joe Amrhein from Pierogi Gallery.

I would also like to thank the contributing writers of the last chapter—Christopher Grant Kirwan, David McConville, Andrew Vande Moere, and Nathan Yau—whose essays have enriched the book in a remarkable way. A big thanks to my editor, Linda Lee, for all the advice and support, and to Alexandre Nakonechnyj, Lev Manovich, and Fernanda Viegas for their feedback and patient review of the manuscript. Finally, and most warmly, my caring gratitude goes to my wife, Joana, for her understanding, encouragement, and immense patience during my occasional ramblings about networks and visualization and the long nights spent in front of the computer.

Tom Beddard, *Fractal Tree*, 2009

A tree representation generated by a Glynn fractal—a type of Julia Set fractal—which itself is derived from a simple mathematical function that produces a complex pattern by repeating itself

01 | The Tree of Life

As buds give rise by growth to fresh buds, and these, if vigorous, branch out and overtop on all sides many a feebler branch, so by generation I believe it has been with the great Tree of Life, which fills with its dead and broken branches the crust of the earth, and covers the surface with its ever branching and beautiful ramifications.

—Charles Darwin

The distributions and partitions of knowledge are not like several lines that meet in one angle, and so touch but in a point; but are like branches of a tree, that meet in a stem, which hath a dimension and quantity of entireness and continuance, before it comes to discontinue and break itself into arms and boughs.

—Francis Bacon

Trees are among the earliest representations of systems of thought and have been invaluable in organizing, rationalizing, and illustrating various information patterns through the ages. As the early precursors of modern-day network diagrams, tree models have been an important instrument in interpreting the evolving complexities of human understanding, from theological beliefs to the intersections of scientific subjects. This favored scheme, usually highlighting a hierarchical ordering in which all divisions branch out from a central foundational trunk, is ultimately a universal metaphor for the way we organize and classify ourselves and the world around us.

The two epigraphs to this chapter are drawn from Darwin, *The Origin of Species*, 172; and Bacon, *Francis Bacon*, 189.

Sacred Trees

For thousands of years trees have been the subject of worship, esteem, and mythology. They are a common motif in world religions and a central theme in the art and culture of many ancient civilizations, from Babylon to the Aztecs. As symbols of prosperity, fertility, strength, and growth, trees have been considered sacred by, or have had an astral meaning to, numerous societies over the ages. With their roots firmly entrenched in the ground and branches reaching toward the skies, they embody a link between heaven, Earth, and the underworld—a unifying symbol of all elements, physical and metaphysical. "A tree that reaches into heaven," says Rachel Pollack, a science fiction writer and tarot expert, "is a very vivid and enticing metaphor, and so has proved useful to humans the world over as a way to formulate our desire to encounter the divine."[1]

For thousands of years, forests have had an impact on humans, not only as a symbol of the mysteries of nature but also as tangible providers of shelter and resources. In *The Real Middle Earth: Exploring the Magic and Mystery of the Middle Ages* (2002), Brian Bates explains how forests were considered places of magic and power, "like a great spirit which had to be befriended."[2] It is not surprising, notes Bates, that so many ancient folktales are set in the woods, as "forests seem to be a natural template for the human imagination."[3] Even though the West has lost its connection to nature as a divine revelation, many of these ancient myths still bear a considerable influence in contemporary society. "The folklore of the modern European peasant, and the observances with which Christmas, May Day, and the gathering of the harvest are still celebrated in civilized countries," explains J. H. Philpot, "are all permeated by the primitive idea that there was a spiritual essence embodied in vegetation, that trees, like men, had spirits, passing in and out amongst them, which possessed a mysterious and potent influence over human affairs."[4]

The romances of the Middle Ages contain innumerous fables of enchanted forests and gallant knights, with the woods serving as the perfect backdrop. The magic, however, started a long time ago. In his captivating *The Forest in Folklore and Mythology* (1928), Alexander Porteous explains the meaning of this primordial fascination:

> In the early strivings of the mind of primitive man to account for the scheme of creation, the tree took a foremost place, and the sky, with its clouds and luminaries, became likened to an enormous Cosmogonic Tree of which the fruits were the sun, moon, and stars. Many races of the earth evolved their own conception of a World Tree, vast as the world itself. They looked upon this tree as the cradle of their being, and it bore different names among different nations, and possessed different attributes.[5]

The tree of life, or the world tree, is "an image of the whole universe, or at least of our planet, that embodies the notion that all life is interrelated and sacred."[6] This mystical concept has been frequently associated with actual trees in the real world, adopting distinct shapes and traits depending on the era and area of the globe. Lotus trees, pomegranate trees, almond trees, and olive trees are among the many varieties that have embodied this myth. But the tree of life is ultimately a symbol of all trees. Behind their multifaceted physical manifestations, elucidates renowned anthropologist Edwin Oliver James, "lies the basic themes of creation, redemption, and resurrection, resting upon the conception of an ultimate source of ever-renewing life at the centre of the cosmos, manifest and operative in the universe, in nature, and in the human order."[7] Author and tree mythology researcher Fred Hageneder further explains the universal nature of the tree:

> According to many of the teachings of ancient wisdom, the universe comprises a spiral or circular movement around a central axis, the *axis mundi*. And this centre pole has often been depicted as the *Tree of Life*, or *Universal Tree*.... It portrays the universe as much more than a lifeless, clockwork mechanism that blindly follows the laws of physics; rather, it presents our world as a living, evolving organism, imbued with divine spirit.[8]

Throughout the ages, different cultures developed their own concept of a tree of life, with all major religions around the world containing tales and legends of sacred trees. While pre-Christian Scandinavia had its Ash Yggdrasil, early Hinduism had its tree of Jiva and Atman, and later its Ashvastha, or Sacred Fig tree—also called Bodhi tree by Buddhists, and under which Gautama Buddha is believed to have meditated and attained enlightenment.

As for Christians, Jews, and Muslims, they all share the mystical tale of the tree of knowledge of good and evil, which originally came from ancient Sumer—a pre-Babylonian civilization spanning over three thousand years in Mesopotamia, modern-day Iraq. "There is, indeed, scarcely a country in the world where the tree has not at one time or another been approached with reverence or with fear, as being closely connected with some spiritual potency," affirms J. H. Philpot in *The Sacred Tree in Religion and Myth* (1897).[9] The Bible itself has several references to mystical trees, with the most popular being the tree of knowledge from the Book of Genesis—a tree situated in the center of the Garden of Eden and from which Adam was forbidden to eat. fig. 1 But perhaps one of the most bewildering religious manifestations of the tree of life is the Sephirotic tree—a mystical symbol, central to esoteric Judaism. fig. 2

Kabbalah (*aytz chayim* in Hebrew) is a Jewish mystical tradition, which translates as "received," an allusion to the teachings passed through generations or directly from God. A pivotal element of the Kabbalah wisdom is the Sephirotic tree: a diagram of ten circles symbolizing ten pulses, or emanations, of divine energy, called *sephirot* in Hebrew—the derivation of *sapphire*. fig. 3 The ten nodes of this schema, reading zigzaggedly from top to bottom, are deciphered in the following manner: (1) *Kether*—crown, (2) *Hokhmah*—wisdom, (3) *Binah*—understanding, (4), *Chesed*—mercy, (5) *Gevurah*—power, (6) *Tiferet*—beauty, (7) *Netzach*—eternity, (8) *Hod*—glory, (9) *Yesod*—foundation, (10) *Malkuth*—kingdom. Even though there are several interpretations to the Sephirotic tree, the diagram ultimately depicts the different stages of divine creation, indicating that the "Creator sent the energy down in a specific pattern from Kether to Malkuth,"[10] or in other words, from a sublime and intangible presence to a physical and earthly existence. Using the tree metaphor to represent the emanation of the Universe, a map of all existence, the Sephirotic tree has remained a powerful symbol over many centuries and still bears a great significance in the mystic study of the Torah.

fig. 1

Albrecht Dürer, *The Fall of Man*, 1509

In this engraving depicting the famous biblical tale from the Book of Genesis, Adam and Eve stand in the Garden of Eden, with the tree of knowledge of good and evil in the background.

fig. 2

Sephirotic tree, from Paulus Riccius, *Portae lucis* (Doors of light), 1516

The pen-and-ink illustration depicts a Jewish Kabbalist meditating while holding the tree of life.

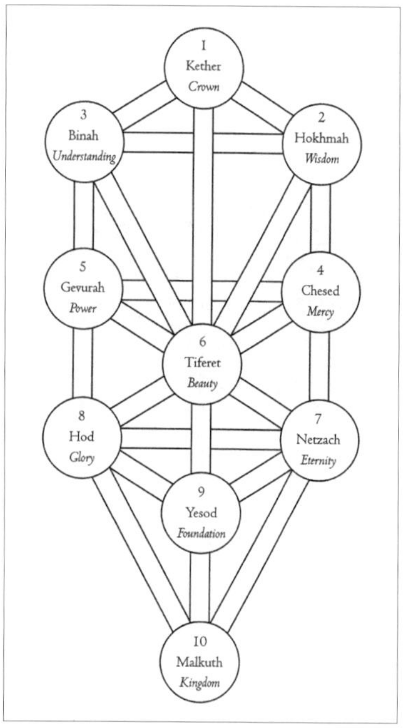

fig. 3

Sephirotic-tree diagram, from Rachel Pollack, *The Kabbalah Tree: A Journey of Balance & Growth*, 2004.

Trees of Knowledge

Our primeval connection with nature and the tree might explain why its branched schema has not only been a symbol with sacred and pagan meanings but also an important metaphor for the classification of the natural world and the meanders of human understanding. Used to address social stratification, domains of human understanding, family ties, or evolutionary relationships between species, the tree has been a ubiquitous model since it can pragmatically express multiplicity (represented by its boughs, branches, twigs, and leaves) from unity (its central foundational trunk). fig. 4 Its arrangement implies a succession of subgroups from larger groups, which are in turn connected to a common root, or starting point. Because of this expressive quality, the metaphorical structure of a tree has been used for thousands of years, from early Sumerian times to modern-day science and operating systems. Currently the scheme still finds relevance in genetics, linguistics, archeology, epistemology, philosophy, genealogy, computer science, and library and information science, among many other areas. fig. 5 As Pollack eloquently puts it: "As the traditions of Western and Christian Kabbalah clearly demonstrate, the tree operates very well as a symbol for many systems of belief. It really has grown into a kind of organizing principle for our human efforts to understand the world."[11]

Most of us are familiar with the tree metaphor. You have probably seen an archetypal organizational chart of your company, a genealogical tree of your family, or perhaps a map of musical influences. While the metaphor is truly widespread, it is still possible to distill the use of trees, as an epistemological model, into two major domains: genealogy (in its broad philosophical sense, tracing the development of ideas, subjects, people, and society through history) and classification (a systematic taxonomy of values and subvalues). Whereas genealogy incorporates the tree to illustrate growth and subdivision over time, classification applies the hierarchical model to show our desire for order, symmetry, and regularity.

Portuguese scholar Olga Pombo, who has thoroughly investigated the classification of science, points to German philosopher Alwin Diemer as the progenitor of a fundamental framework for classification. In *Conceptual Basis of the Classification of Knowledge* (1974), Diemer divided the conventions of classification into four main domains: ontological (classification of species), informational (classification of information), biblioteconomical (classification of books), and gnosiological (classification of knowledge). (See also chapter 2, "Classifying Information," pages 61–64, and "Ordering Nature," pages 64–69.) Even though the ontological, informational, and biblioteconomical domains have been greatly marked by the tree model, it was in the gnosiological domain that the recursive metaphor of the tree had one of the most striking manifestations.

The idea of capturing the entirety of human knowledge and classifying it by means of a tree is an aged aspiration, a meme hundreds of years old. The biblical tree of knowledge, for instance, represented the collective knowledge of humanity—everything we have learned as a species, embodied in a tree. The idea of an arboreal organizational scheme is so ingrained in our minds that we employ it figuratively in a variety of daily circumstances, which in turn conditions the way we understand things and express them to others. When we say "the root of a problem" or "the root of scientific research," we are alluding to some sort of hierarchical model with a defined

fig. 4

Genealogy of Henry II (973–1024), the Holy Roman Emperor, from Hartmann Schedel, *Nuremberg Chronicle*, 1493

In this genealogical scheme, the tree metaphor is taken literally, with the woman's womb depicted as the source of offshoots.

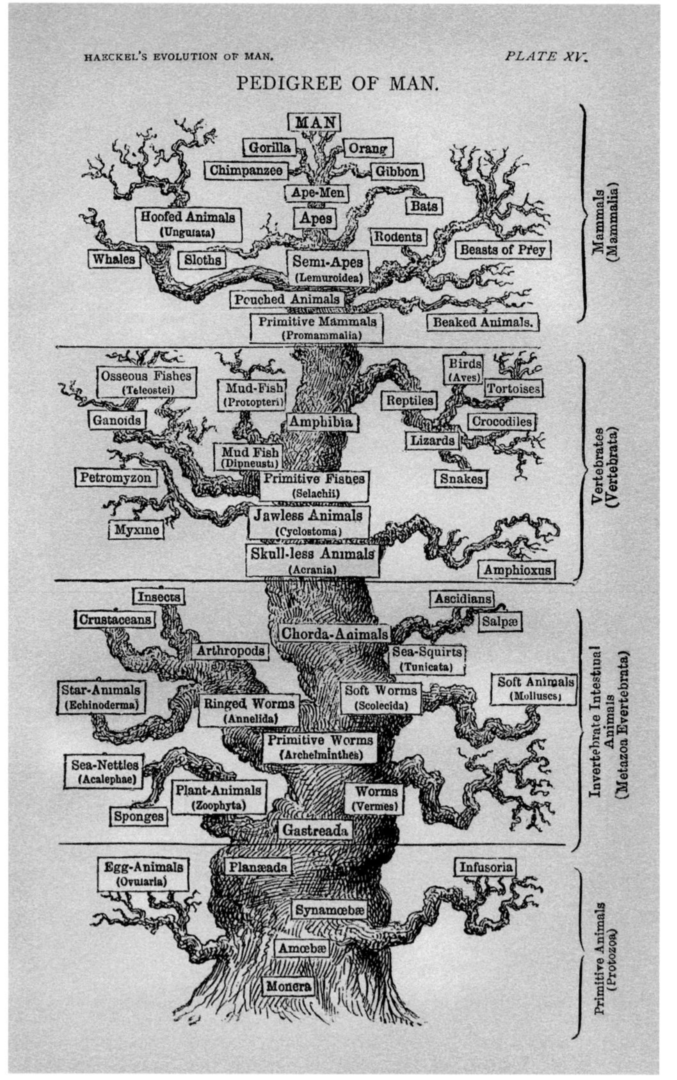

fig. 5

Ernst Haeckel, *Pedigree of Man*, 1879

Like many other early evolutionists, Haeckel believed that humans were the pinnacle of evolution—the highest form of life—as shown in this depiction of the tree of life, with man at the zenith.

foundation, a unifying basis. We also use it to convey the distinct areas of human knowledge, as in "the branches of science" or, more specifically, "genetics is a branch of science." The origin of the word *knowledge* itself is strongly tied to trees. "In the Germanic languages, most terms for learning, knowledge, wisdom, and so on are derived from the words for tree or wood," says Hageneder. "In Anglo-Saxon we have *witan* (mind, consciousness) and *witige* (wisdom); in English, 'wits,' 'witch,' and 'wizard'; and in modern German, *Witz* (wits, joke). These words all stem from the ancient Scandinavian root word *vid*, which means 'wood' (as in forest, not timber)."[12]

Early Pioneers

The earliest known concept for a hierarchical organization of knowledge comes to us from ancient Greece, through the work of one of its main characters: Aristotle. In *Categories*—the first of six works on logic, collectively called *Organon* (ca. 40 BCE)—Aristotle (384–322 BCE) delivers a fundamental vision on classification. He starts by exploring a series of semantic relationships, as in the equivocal, unequivocal, and derivative naming of things. He then presents the notion of the predicable, used in different forms of speech and the division of beings, before organizing every entity of human apprehension according to ten categories: substance, quantity, quality, relation, place, time, position, state, action, and affection. In the remaining text, Aristotle discusses in detail the definitions of all given categories and concludes with the different types of movement in nature (e.g., generation, destruction, increase). As professor of philosophy Anthony Preus explains, Aristotle's structure is not simply based in a tenfold classification but "suggests that each category serves as the genus for a group of immediately subordinate kinds, or species, which in turn serve as genera for further species subordinate to them, and so on until one reaches a level at which no further division is possible."[13]

This outstanding work is one of the most important philosophical treatises of all time and has been a long-lasting influence in Western culture. It grabbed the attention of innumerous philosophers over the centuries, such as Porphyry, René Descartes, Gottfried Leibniz, Immanuel Kant, and Martin Heidegger, all of whom variously defended, opposed, or modified Aristotle's original ideas. The cornerstone of Aristotle's philosophical theorizing, *Categories* laid the foundation for all subsequent classification efforts in a variety of scientific areas and still remains a subject of study and encouragement in the pursuit of a comprehensible universal categorization.

Tree of Porphyry

Porphyry (234–ca. 305 CE) was a Greek philosopher born in the city of Tyre, modern-day Lebanon. He is mostly renowned for his contribution to *The Six Enneads* (ca. 270 CE)—the only collection of writings by Porphyry's teacher and founder of Neoplatonism, the Greek philosopher Plotinus. But it was in his short introduction, or *Isagoge* (ca. 270 CE), to Aristotle's *Categories* that Porphyry made one of the most striking contributions to knowledge classification. In this highly influential introduction, translated into Latin by Anicius Manlius Severinus Boethius and disseminated across medieval Europe, Porphyry reframes Aristotle's original predicables into a decisive list of five classes: genus (*genos*), species (*eidos*), difference (*diaphora*), property (*idion*), and accident (*sumbebekos*). Most importantly, he introduces a hierarchical, finite structure of classification, in what became known as the tree of Porphyry, or simply the Porphyrian tree. fig. 6

Expanding on Aristotle's *Categories* and visually alluding to a tree's trunk, Porphyry's structure reveals the idea of layered assembly in logic. It is made of three columns of words, where the central column contains a series of dichotomous divisions between genus and species, which derive from the supreme genus, Substance. Even though Porphyry himself never drew such an illustration—his original tree was purely textual in nature—the symbolic tree of Porphyry was frequently represented in medieval and Renaissance works on logic and set the stage for theological and philosophical developments by scholars throughout the ages. It was also, as far as we know, the earliest metaphorical tree of knowledge.

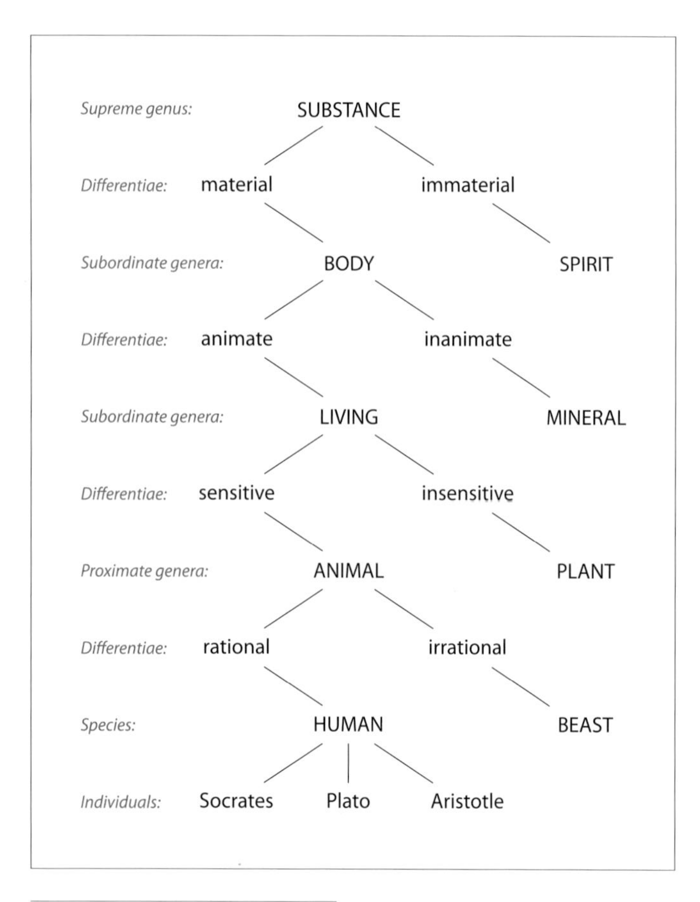

fig. 6

The porphyrian tree, the oldest known type of a classificatory tree diagram, was conceived by the Greek philosopher Porphyry in the third century AD. This figure shows a Porphyrian tree as it was originally drawn by the thirteenth-century logician Peter of Spain.

Liber figurarum (Book of figures)

Joachim of Fiore (ca. 1135–1202) was a twelfth-century Italian abbot and the founder of the monastic order of San Giovanni in Fiore, whose followers are called Joachimites. Very little is known with certainty about this extraordinary man, and most of his life accounts came to us from a biography published by a later monk of the monastery of Fiore, Jacobus Græcus Syllanæus, in 1612. Joachim opposed many religious dogmas and was a firm believer in a more liberal Church. He envisioned a new age in which mankind would reach total freedom and the hierarchy of the Church would become unnecessary under the rule of the Order of the Just, an alliance between Christians, Jews, and Muslims.

Some see Joachim of Fiore as a visionary and a prophet, others as a mere dissident. Still considered a heretic by the Vatican, Joachim of Fiore left behind a number of his writings and treatises that attest to his productive intellect. Among them is the extraordinary *Liber figurarum*, one of the most important and stunning collections of symbolic theology from the Middle Ages. fig. 7 The illustrations shown in the manuscript were conceived by Joachim in different stages of his life and published posthumously in 1202. They depict a variety of characters and institutions from the Old and New Testaments, and many employ an organic arboreal schema to highlight the centrality of Christ, the gradation of biblical protagonists, and the links with the past—as in the recurrent use of branches to symbolize the twelve tribes of Israel. fig. 8, fig. 9, fig. 10

fig. 7

The Tree of the Two Advents, from Joachim of Fiore, *Liber figurarum*, 1202

This remarkable figure presents the main characters and institutions of the Christian salvation history. From bottom to top: Adam, Jacob the Patriarch, Ozias the Prophet, and Jesus Christ (repeated twice). The figure of Christ dominates the center of the genealogical tree (representing the first coming, or Redemption), as well as the very top (the place of the second coming, or Resurrection). The lower branches, originating from the figure of Jacob the Patriarch, correspond to the twelve tribes of Israel, and the top branches, radiating from the image of Jesus Christ, symbolize the twelve Christian churches.

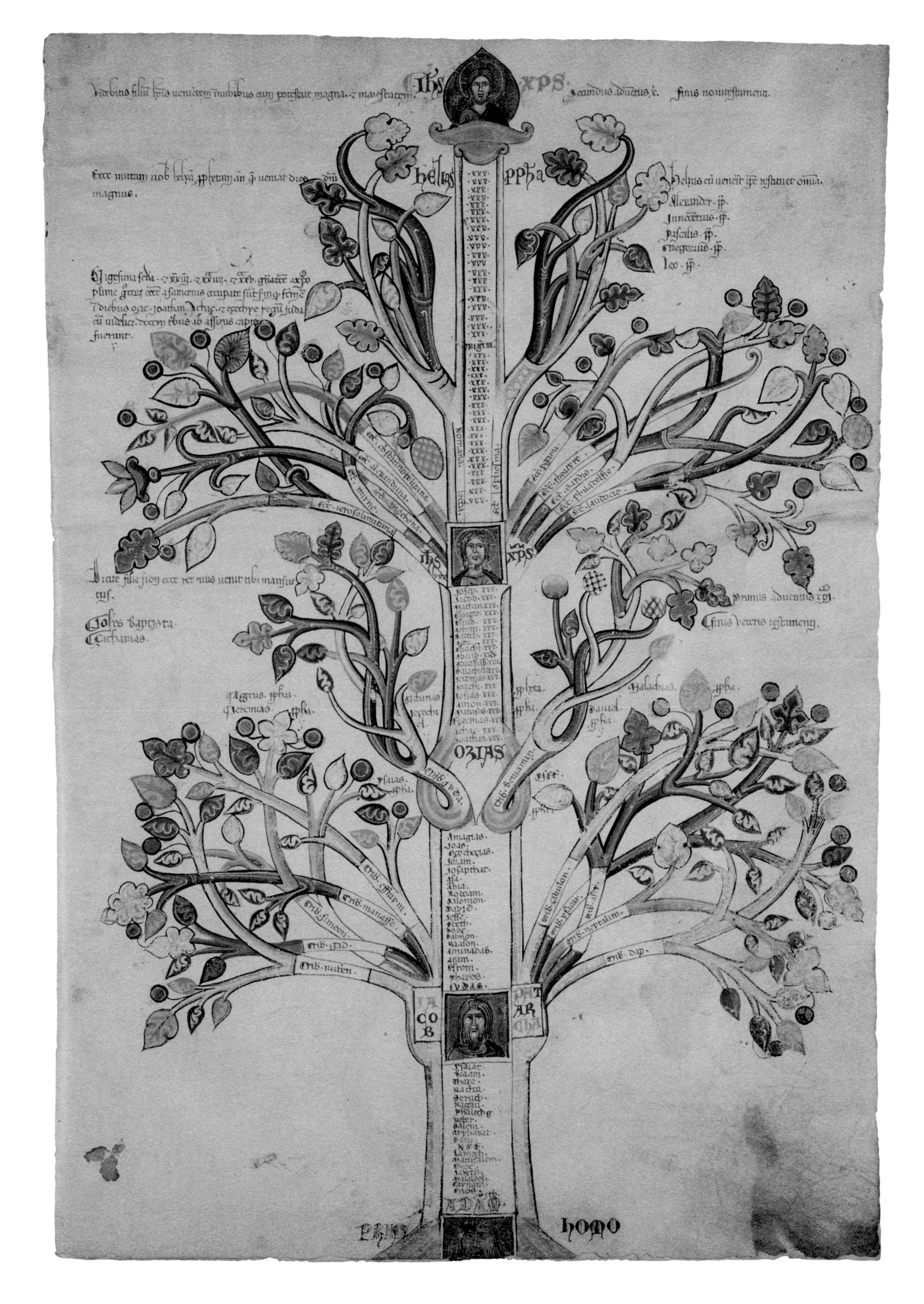

IHS XPS
secundus adventus xi.
finis novi testamenti
helias ppha
Alexander pp
Innocentius pp
Pascalis pp
Gregorius pp
Leo pp
IHS XPS
primus adventus xpi
Finis veteris testamenti
Johes baptista
Zacharias
OZIAS
IA COB
PAT AR CHA
HOMO

fig. 8

The Trinitarian Tree Circles, from Joachim of Fiore, *Liber figurarum*

This tree represents the development of the history of Christianity, divided into three circles, or states, of the world. At the bottom, Noah's tree (rooted in his three sons) gives origin to the first circle (age of the Father), leading to the second (age of the Son), and ultimately the third (age of the Holy Spirit). The amount of foliage increases in density toward the top, culminating in lush vegetation, symbolizing the glory of the universal church.

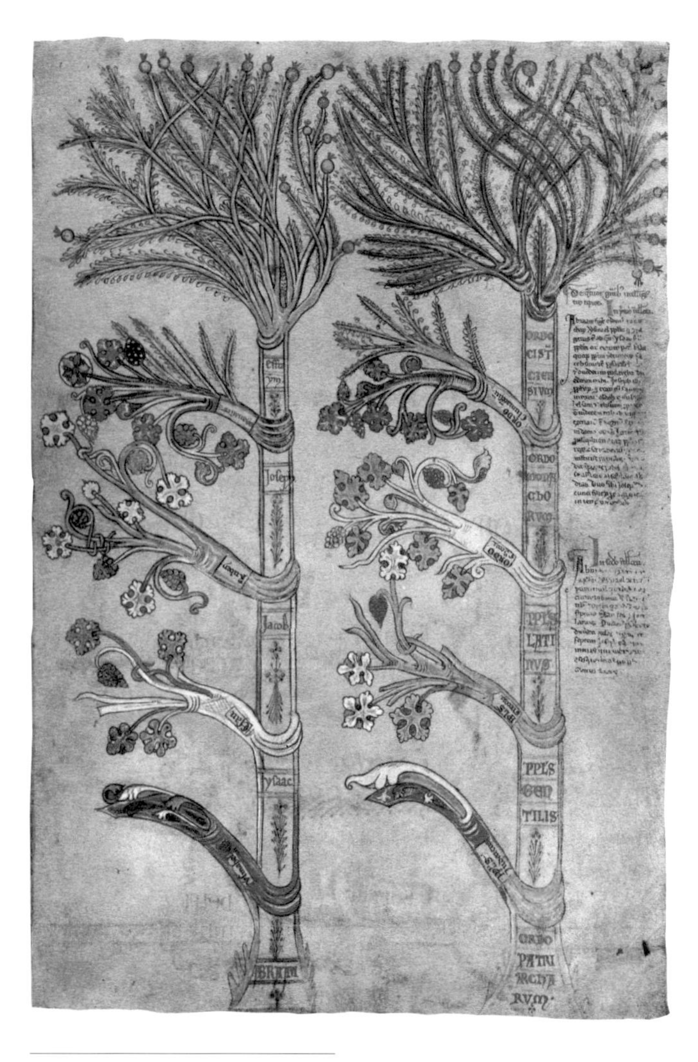

fig. 9

A Pair of Trees with Side-shoots, from Joachim of Fiore, *Liber figurarum*

A depiction of the Christian salvation history, with the names of its main protagonists stacked up along the two tree trunks. The first tree traces the Old Testament—Abraham, Isaac, Jacob, Joseph, Ephraim; and the second, the New Testament.

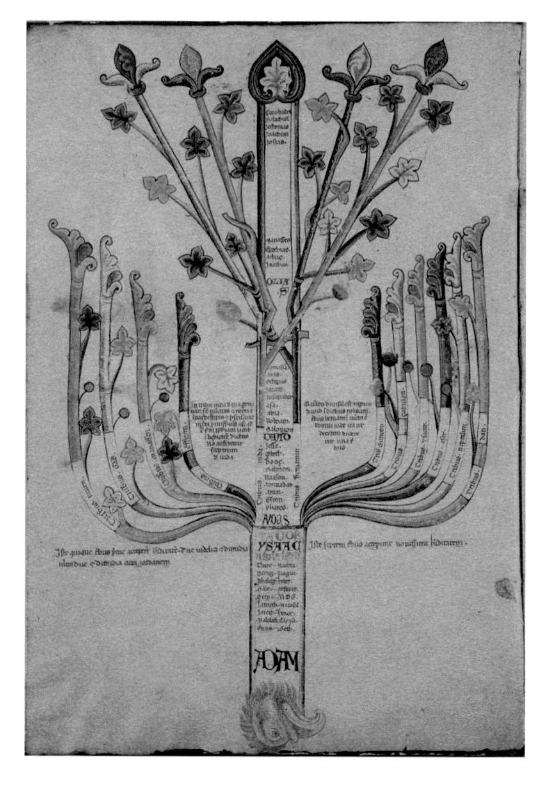

Editor Peter L. Heyworth sees this tree allegory as a teaching tool for the rich biblical heritage. "In this period, the art of preaching itself came to be likened to a tree. *Praedicare est arborizare* [to preach is to plant a tree]: the well-grown sermon must be rooted in a theme, that flourishes in the trunk of a biblical *auctoritas*, and thence grows into its branches and twigs: the divisions and subdivisions whereby the preacher extends his subject-matter."[14] The sermon *Praedicare est arborizare* is an example of a medieval formulation in which trees were often allegorized in relation to discursive phenomena, which in turn might explain why the tree metaphor is so prevalent in the illustrations featured in *Liber figurarum*. With a strong theological nature, *Liber figurarum* represents a remarkable effort of systematization of historical accounts, events, and social ties by means of trees and is a seminal work in this period of history.

Arbor scientiae (Tree of science)

Born in the Spanish Balearic island of Mallorca in 1232, Ramon Llull (1232–1315) was one of the most astonishing figures of medieval Europe. A poet, mystic, philosopher, and devout Christian, he wrote one of the earliest, if not the first, European novel, *Blanquerna* (1283), and was the spearhead of the consolidation of the Catalan language. Llull left behind more than 250 works in Catalan, Latin, and Arabic, with many more subsequent translations into French, Spanish, and Italian. His most renowned piece is *Ars magna* (The great art), first published in 1271 as *Ars magna primitiva* (The first great art) and recurrently iterated by Llull in subsequent editions over thirty years.

But one of Llull's most well-known works, pertaining to knowledge representation, was his wonderful *Arbor scientiae* from 1296, which includes a magnificent compilation of sixteen trees of scientific domains following a leading tree, itself called the arbor scientiae. fig. 11 An expression of his mystical universalism, this encyclopedic work concentrates on the central image of a tree of science, able to sustain all areas of knowledge.[15] Appearing in the very beginning of the book, the illustration of the tree of science works as an introduction to his beguiling concept and a sort of arborescent table of contents. This great tree comprises eighteen roots, which relate to nine transcendent principles (not detailed) and nine art principles: difference, concord, contrariety, beginning, middle, end, majority, equality, and minority. The top of the tree is made of sixteen branches, each bearing a fruit and a label, representing the different domains of science, which are then depicted as individual trees in the remaining pages of the work.

The first set of trees relates to profane knowledge and includes the *arbor elementalis* (physics, metaphysics, and cosmology), the *arbor vegetalis* (botany and medicine), the *arbor sensualis* (animals and sensible beings), the *arbor imaginalis* (mental entities, psychology), the *arbor humanalis* (anthropology and the studies of man), the *arbor moralis* (ethics, moral vices, and virtues), and the *arbor imperialis* (government and politics of a prince). fig. 12 The second group covers the entirety of religious knowledge and comprises the *arbor apostolocalis* (ecclesiastical studies and the organization of the church), the *arbor celestialis* (astronomy and astrology), the *arbor angelicalis* (angels), the *arbor eviternalis* (immortality, the afterlife, hell and paradise), the *arbor maternalis* (the study of Virgin Mary), the *arbor christinalis* (christology), and the *arbor divinalis* (theology). There are two additional trees, *arbor exemplificalis* and *arbor quaestionalis*. The first contains a series of metaphorical examples pertaining to the four

fig. 10

The Tree-Eagle (Old Dispensation), from Joachim of Fiore, *Liber figurarum*

The eagle, a powerful symbol of spiritual enlightenment and contemplation, is prominently featured in this tree that depicts the advent of the age of the Holy Spirit. The central trunk lists various generations in the history of Christianity, from Adam to Zorobabel, while the lower branches symbolize the twelve tribes of Israel, separated by the tribes that first entered the promised land (left) and the tribes that arrived later (right).

fig. 11 (left)

Arbor scientiae (Tree of science), from Ramon Llull, *Arbor scientiae venerabilis et caelitus illuminati Patris Raymundi Lullii maioricensis Liber ad omnes scientias utilissimus*, 1515

fig. 12

Arbor moralis (Moral tree), from Llull, *Arbor scientiae venerabilis*

fig. 13

Diagrams of Llull's intriguing combinatory-logic concepts, from Llull, *Ars Brevis V.M.B. Raymundi Lullij Tertij Ord. S. Francisci, Doc. Illu.: mendis castigata, Capitibus Divisa, atque scholiis locupletata*, 1669

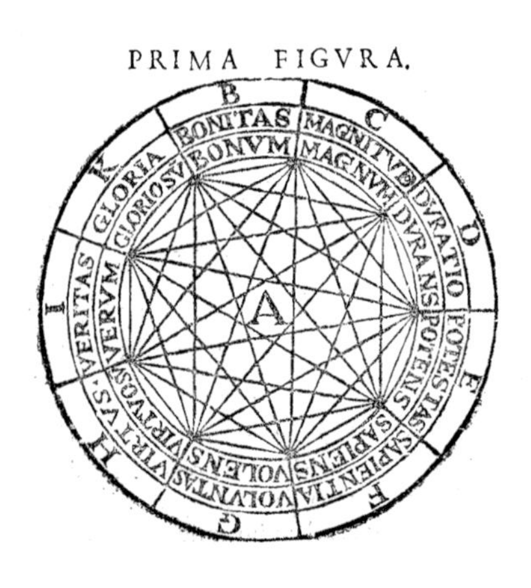

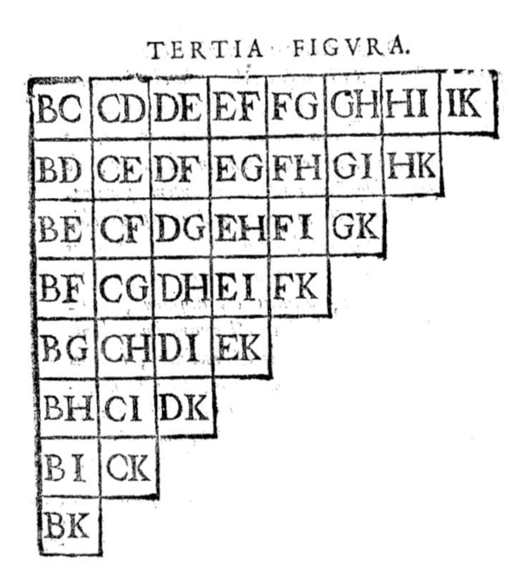

natural elements (fire, water, earth, and air), with the goal of transforming science into an accessible narrative, while the second features a large body of four thousand questions related to the preceding trees.

Despite Llull's magnificent arrangement of the trees of science, which centuries later influenced the classification efforts of Francis Bacon and René Descartes, his most recognized contribution to European thinking was the pursuit of an "organic and unitary corpus of knowledge and a systematic classification of reality," which included a series of diagrams, symbolic notations, and mechanical apparatuses.[16] fig. 13 This approach was instrumental in the research of German philosopher and mathematician Gottfried Leibniz (1646–1716), particularly in his conception of an imaginary universal language capable of expressing the most sophisticated mathematical, scientific, and metaphysical concepts—the famous *characteristica universalis* (universal characteristic). Leibniz first mentioned this lexicon made of pictographic characters, which reduces all debate to calculation, in his *Dissertatio de arte combinatoria* (Dissertation on the combinatorial art), published when Leibniz was only nineteen. This vocabulary created the seed for the later development of the binary system—the foundation of all modern computers—that Leibniz eventually presented in his ingenious *Explication de l'arithmétique binaire* (Explanation of binary arithmetic), published in 1705.

Encyclopedism

The desire for gathering the sum of human knowledge in a comprehensive compendium is as old as our desire to organize it. The earliest known attempt to do so appeared in the form of *Naturalis historia* (Natural history), published in 80 AD, a thirty-seven-chapter encyclopedia describing different aspects of the natural world and human developments in art, architecture, and medicine, among other domains, written by Pliny the Elder, a Roman statesman. Leading up to the Middle Ages, the seminal *Etymologiae* (ca. 630 AD), by the prolific scholar Saint Isidore of Seville, is one of the most significant encyclopedic ventures. The work comprised 449 chapters in twenty volumes and encapsulated much of the knowledge of the age. Later on Bartholomeus Anglicus's *De proprietatibus rerum* (1240) became one of the most read encyclopedias of the time, while Vincent of Beauvais's *Speculum majus* (1260) was the most comprehensive—with over three million words. But it was in the midst of the French Renaissance that one of the most consequential efforts at rationalizing knowledge took place, by the hands of French scholar Christophe de Savigny.

Tableaux accomplis de tous les arts libéraux (Complete tables of all liberal arts)

In 1587 de Savigny published in Paris the magnificent *Tableaux accomplis de tous les arts libéraux*. This pivotal work contains sixteen beautifully decorated tables covering the following arts and sciences (in the order they appear in the book): grammar, rhetoric, dialectic, arithmetic, geometry, optics, music, cosmography, astrology, geography, physics, medicine, ethics, jurisprudence, history, and theology. The book is solely composed of the sixteen tables, each one accompanied by a corresponding one-page description. In every table, a tree of interrelated topics takes center stage, surrounded by an oval ornamental piece containing various graphic elements pertaining to the depicted discipline. fig. 14 *Tableaux accomplis de tous les arts libéraux*, one of the most enticing medieval pieces on the rationalization and visual representation of knowledge, became an important influence on the subsequent work of Bacon, ultimately consolidating the widespread use of the tree metaphor.

The consolidation: Francis Bacon and René Descartes

In 1605 English philosopher and fervent promoter of the Scientific Revolution Bacon (1561–1626) published one of the major landmarks in the history of science, and arguably the most significant philosophical work in English until then. In *The Advancement of Learning*, Bacon not only suggests a new science of observation and experimentation, as a substitute to secular Aristotelian science, but also explores with great minutia the wide arrangement of all human knowledge, from the general to the particular. He starts by dividing man's understanding into three main parts: "History to his Memory, Poesy [poetry] to his Imagination, and Philosophy to his Reason."[17] He then suggests various subdivisions to the three main categories and drills down to its key disciplines, such as physics, mathematics, and anatomy, describing and contextualizing them in great detail. During his expositions he alludes to the tree of knowledge: "The distributions and partitions of knowledge are not like several lines that meet in one angle, and so touch but in a point; but are like branches of a tree, that meet in a stem, which hath a dimension and quantity of entireness and continuance, before it comes to discontinue and break itself into arms and boughs."[18]

While this effort on the classification of knowledge is thought to have been inspired by de Savigny's pictorial encyclopedia, produced a few years before, Bacon's essay fostered much of the subsequent thinking in scholarly circles, making him the great precursor of modern encyclopedism and a key influence in Descartes's conception of the tree of knowledge.

Descartes (1596–1650), often called the father of modern philosophy, continued exploring Bacon's ideas on the arboreal scheme of science in many of his works, including *The World* (1629–33), *Dioptrics* (1637), *Meteorology* (1637), and *Geometry* (1637). But it was in his *Principia philosophiae* (*Principles of Philosophy*) (1644), his longest and most ambitious piece, that Descartes delved further into the topic. This exceptional work was meant to have six parts (although he only concluded the first four): I—The Principles of Human Knowledge, II—The Principles of Material Things, III—The Visible Universe, IV—The Earth, V—Living Things, VI—Human Beings. The 207 completed principles are normally short (one paragraph each) and resemble a list of small knowledgeable units, a synthesis of most of his theories in philosophy and physics, dealing with everything from geometry to the perception of the senses.

In a letter to the French translator of the work, while explaining the rationale behind the principles, Descartes describes his image of the tree of knowledge:

> Thus, all Philosophy is like a tree, of which Metaphysics is the root, Physics the trunk, and all the other sciences the branches that grow out of this trunk, which are reduced to three principles, namely, Medicine, Mechanics, and Ethics. . . . But as it is not from the roots or the

fig. 14

Geometry, from Christophe de Savigny, *Tableaux accomplis*, 1587

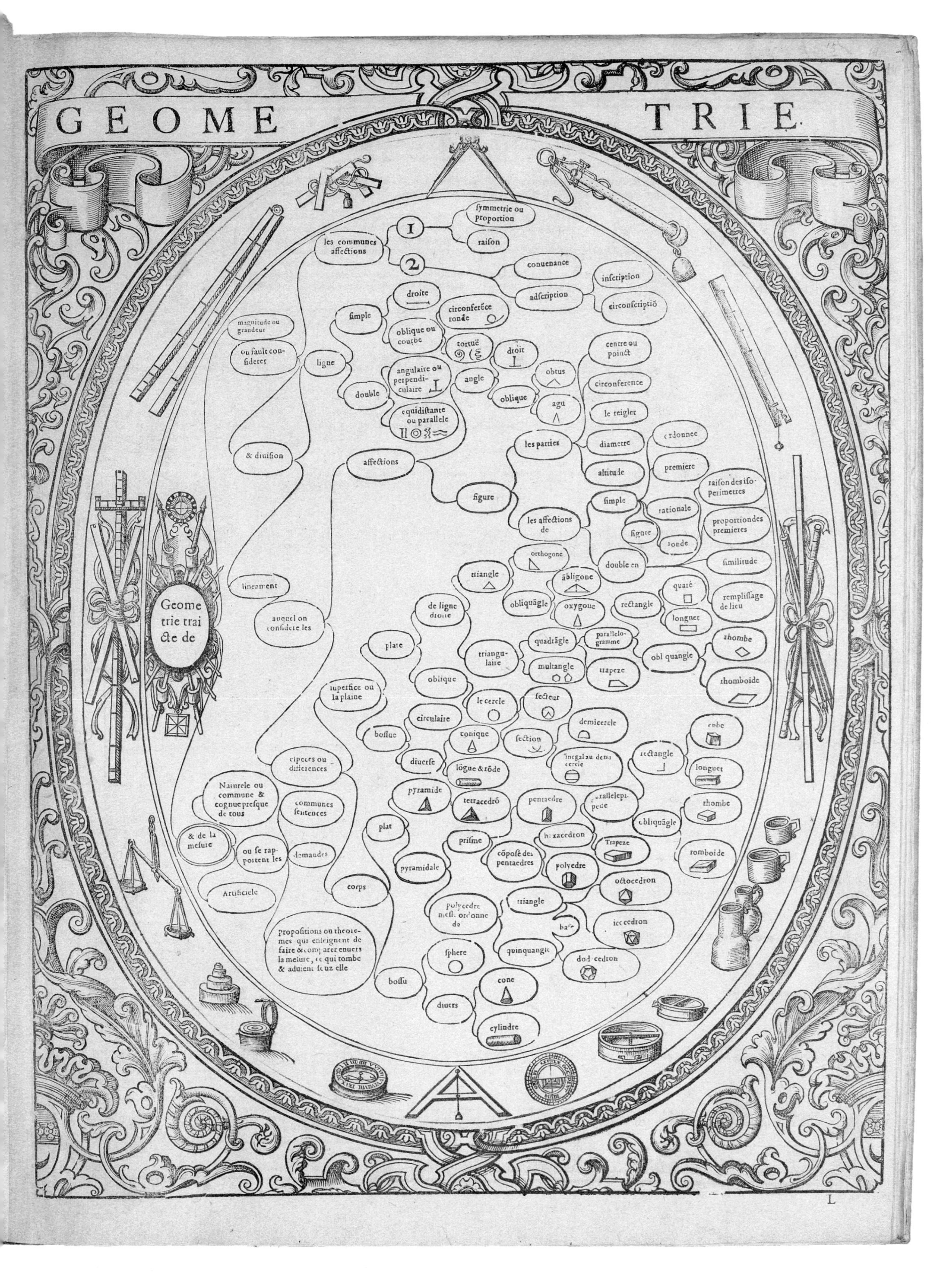
GEOME TRIE.
Geome
trie trai
cte de
lineament
auquel on
confidere les
magnitude ou
grandeur
ou fault con-
fideres
& diuifion
ligne
fimple
droite
oblique ou
courbe
circonferêce
ronde
tortuë
double
angulaire ou
perpendi-
culaire
equidiftante
ou parallele
angle
droit
oblique
obtus
agu
les communes
affections
1
fymmetrie ou
proportion
raifon
2
conuenance
adfcription
infcription
circonfcriptiõ
affections
figure
les parties
centre ou
poinct
circonference
le reigler
diametre
altitude
ordonnee
premiere
les affections
de
fimple
figure
rationale
ronde
raifon des ifo-
perimetres
proportion des
premieres
double en
fimilitude
orthogone
triangle
ãbligone
obliquãgle
oxygone
rectangle
quaré
longuet
remplissage
de lieu
de ligne
droite
plate
triangu-
laire
quadrãgle
multangle
parallelo-
gramme
trapeze
obl quangle
rhombe
rhomboide
oblique
fuperfice ou
la plaine
le cercle
fecteur
circulaire
bossue
conique
fection
demicercle
diuerfe
lõgue & rõde
inegal au demi
cercle
rectangle
cube
longuet
efpeces ou
differences
pyramide
tetraedrõ
pentaedre
parallelepi-
pede
rhombe
obliquãgle
romboide
Naturele ou
commune &
cognue prefque
de tous
communes
fentences
plat
prifme
hexaedron
cõpofé des
pentaedres
Trapeze
polyedre
& de la
mefure
ou fe rap-
portent les
demandes
pyramidale
octoedron
Artificiele
corps
polyedre
triangle
icocedron
quinquangle
dodecedron
Propofitions ou theore-
mes qui enfeignent de
faire & comparer enuers
la mefure, ce qui tombe
& aduient fur elle
fphere
boffu
cone
diuers
cylindre

trunks of trees that we gather the fruit, but only from the extremities of their branches, so the principal utility of philosophy depends on the separate uses of its parts, which we can only learn last of all.[19]

While neither Bacon nor Descartes developed a visual representation of the tree of knowledge, it is through their words that we can ascertain the construction of such a hierarchical classification scheme, which contributed decisively to the establishment of the general metaphor of the tree as the underlying epistemological model of all sciences.

Cyclopædia

Published in 1728 and composed of two volumes, this work by Ephraim Chambers was one of the earliest general encyclopedias written in English. The noticeably long full title of the piece describes its holistic aim: *Cyclopædia, or, An universal dictionary of arts and sciences: containing the definitions of the terms, and accounts of the things signify'd thereby, in the several arts, both liberal and mechanical, and the several sciences, human and divine: the figures, kinds, properties, productions, preparations, and uses, of things natural and artificial: the rise, progress, and state of things ecclesiastical, civil, military, and commercial: with the several systems, sects, opinions, etc: among philosophers, divines, mathematicians, physicians, antiquaries, criticks, etc: the whole intended as a course of ancient and modern learning.*

One of the most significant achievements of *Cyclopædia*, with respect to knowledge classification, was the introduction of a horizontal tree diagram, in which the hierarchical ordering of subjects reads from left to right, instead of the common top-down or bottom-up arrangement. fig. 15, fig. 16 The leftmost branch of knowledge has a series of sub-branches (e.g., artificial, external, real) before reaching the final, rightmost branches (e.g., astronomy, geography, sculpture), which represent particular sections in the book and fulfill the goal of the chart to serve as a table of contents. Bearing a strong resemblance to the successive forking of the Porphyrian tree, the diagram maps forty-seven different disciplines in the book, including meteorology, geometry, alchemy, architecture, commerce, medicine, and poetry. We can once more perceive the tree metaphor, not only to express the various relations between the topics, but also as a unifying element, connecting all areas of knowledge under the same foundation.

Encyclopédie

During the mid-eighteenth century, in the height of French Enlightenment, one of the most astounding encyclopedic efforts took place by Denis Diderot and Jean le Rond d'Alembert. First published in 1751, the *Encyclopédie, ou dictionnaire raisonné des sciences, des arts et des métiers* (Encyclopedia, or a systematic dictionary of the sciences, arts, and crafts) was one of the largest encyclopedias produced until then, accounting for 20 million words in 71,818 articles and 3,129 illustrations over thirty-five volumes. Inspired by a French translation of Chambers's *Cyclopædia, Encyclopédie* became an important drive for the subsequent launch of *Encyclopedia Britannica* seventeen years later, and a precursor to many modern encyclopedias.

This innovative encyclopedia paid special attention to the mechanical arts, and it was the first to include contributions from well-known authors, many of the great names of French Enlightenment among them, such as

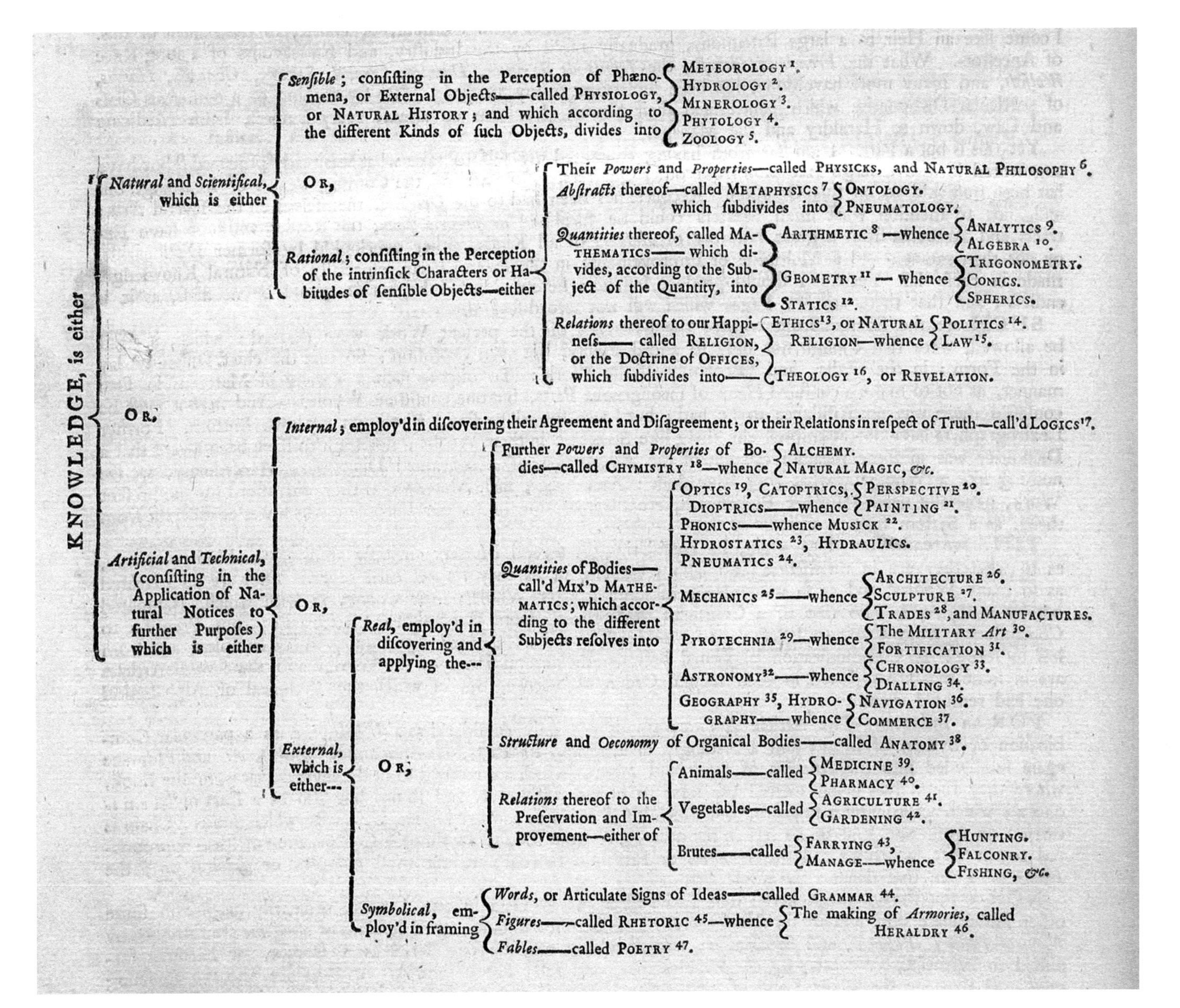

KNOWLEDGE, is either

- *Natural* and *Scientifical*, which is either
  - *Sensible*; consisting in the Perception of Phænomena, or External Objects——called PHYSIOLOGY, or NATURAL HISTORY; and which according to the different Kinds of such Objects, divides into
    - METEOROLOGY [1].
    - HYDROLOGY [2].
    - MINEROLOGY [3].
    - PHYTOLOGY [4].
    - ZOOLOGY [5].
  - OR, *Rational*; consisting in the Perception of the intrinsick Characters or Habitudes of sensible Objects—either
    - Their *Powers* and *Properties*—called PHYSICKS, and NATURAL PHILOSOPHY [6].
    - *Abstracts* thereof—called METAPHYSICS [7] which subdivides into
      - ONTOLOGY.
      - PNEUMATOLOGY.
    - *Quantities* thereof, called MATHEMATICS—— which divides, according to the Subject of the Quantity, into
      - ARITHMETIC [8] —whence
        - ANALYTICS [9].
        - ALGEBRA [10].
      - GEOMETRY [11] — whence
        - TRIGONOMETRY.
        - CONICS.
        - SPHERICS.
      - STATICS [12].
    - *Relations* thereof to our Happiness—— called RELIGION, or the Doctrine of OFFICES, which subdivides into——
      - ETHICS [13], or NATURAL RELIGION—whence
        - POLITICS [14].
        - LAW [15].
      - THEOLOGY [16], or REVELATION.
- OR, *Artificial* and *Technical*, (consisting in the Application of Natural Notices to further Purposes) which is either
  - *Internal*; employ'd in discovering their Agreement and Disagreement; or their Relations in respect of Truth—call'd LOGICS [17].
  - OR, *External*, which is either---
    - *Real*, employ'd in discovering and applying the---
      - Further *Powers* and *Properties* of Bodies—called CHYMISTRY [18]—whence
        - ALCHEMY.
        - NATURAL MAGIC, &c.
      - *Quantities* of Bodies—call'd MIX'D MATHEMATICS; which according to the different Subjects resolves into
        - OPTICS [19], CATOPTRICS, DIOPTRICS——whence
          - PERSPECTIVE [20].
          - PAINTING [21].
        - PHONICS——whence MUSICK [22].
        - HYDROSTATICS [23], HYDRAULICS.
        - PNEUMATICS [24].
        - MECHANICS [25]——whence
          - ARCHITECTURE [26].
          - SCULPTURE [27].
          - TRADES [28], and MANUFACTURES.
        - PYROTECHNIA [29]—whence
          - The MILITARY *Art* [30].
          - FORTIFICATION [31].
        - ASTRONOMY [32]——whence
          - CHRONOLOGY [33].
          - DIALLING [34].
        - GEOGRAPHY [35], HYDROGRAPHY—— whence
          - NAVIGATION [36].
          - COMMERCE [37].
      - *Structure* and *Oeconomy* of Organical Bodies——called ANATOMY [38].
      - *Relations* thereof to the Preservation and Improvement—either of
        - Animals——called
          - MEDICINE [39].
          - PHARMACY [40].
        - Vegetables—called
          - AGRICULTURE [41].
          - GARDENING [42].
        - Brutes——called
          - FARRYING [43].
          - MANAGE——whence
            - HUNTING.
            - FALCONRY.
            - FISHING, &c.
    - OR, *Symbolical*, employ'd in framing
      - *Words*, or Articulate Signs of Ideas——called GRAMMAR [44].
      - *Figures*——called RHETORIC [45]—whence The making of *Armories*, called HERALDRY [46].
      - *Fables*——called POETRY [47].

fig. 16

Table of contents, from Chambers, *Cyclopaedia*

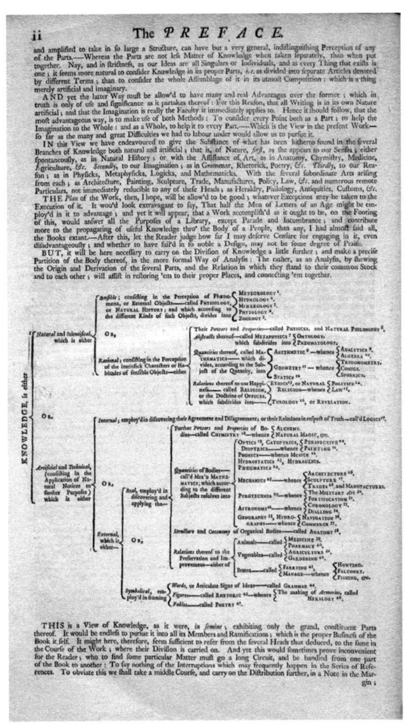

ii The PREFACE.

fig. 15

Table of contents as it appears in the preface of the original publication, from Ephraim Chambers, *Cyclopædia, or, An universal dictionary of arts and sciences*, 1728

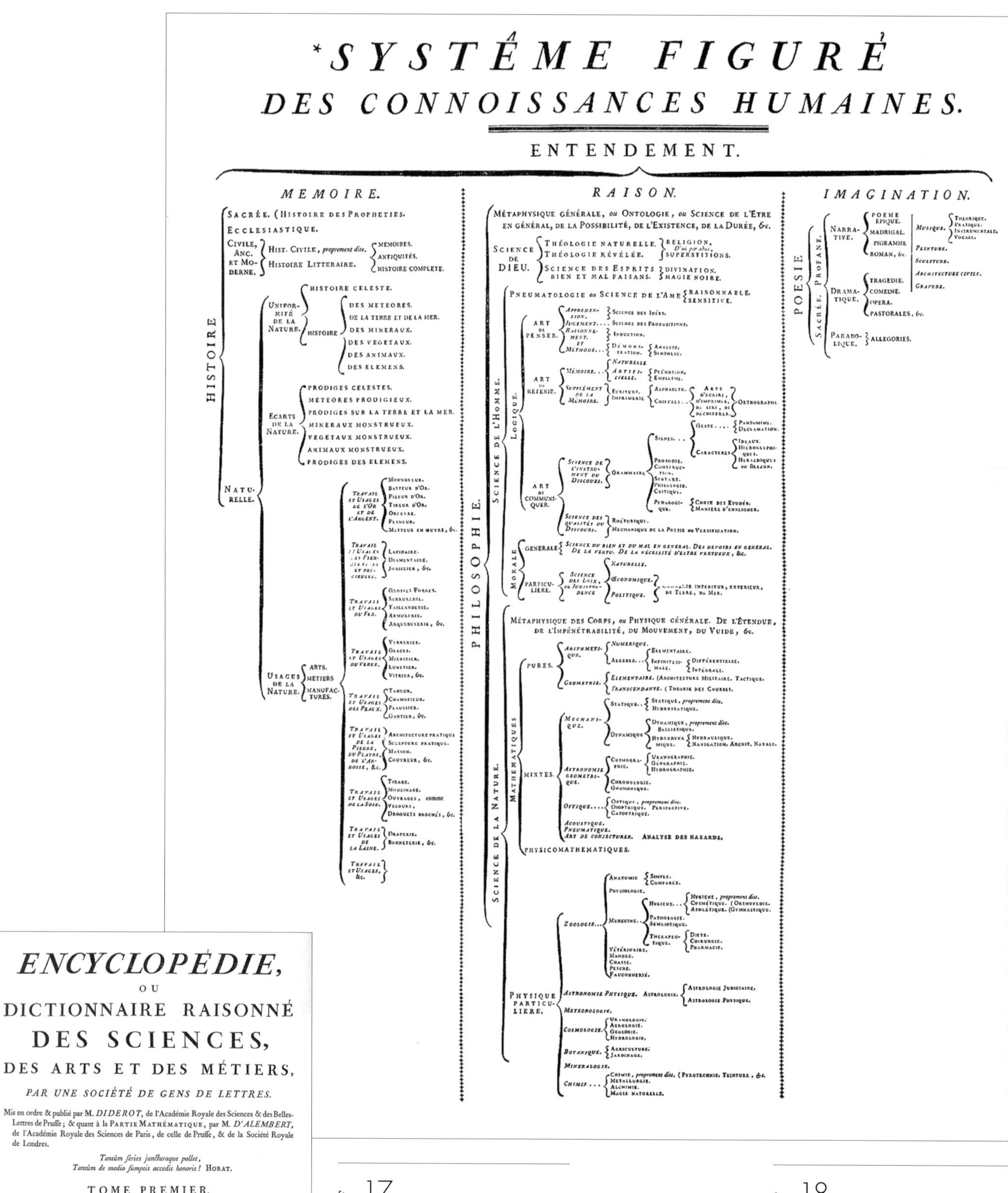

ENCYCLOPÉDIE,

OU

DICTIONNAIRE RAISONNÉ

DES SCIENCES,

DES ARTS ET DES MÉTIERS,

PAR UNE SOCIÉTÉ DE GENS DE LETTRES.

Mis en ordre & publié par M. *DIDEROT*, de l'Académie Royale des Sciences & des Belles-Lettres de Pruffe; & quant à la PARTIE MATHÉMATIQUE, par M. *D'ALEMBERT*, de l'Académie Royale des Sciences de Paris, de celle de Pruffe, & de la Société Royale de Londres.

Tantùm feries junčturaque pollet,
Tantùm de medio fumptis accedit honoris! HORAT.

TOME PREMIER.

A PARIS,

Chez BRIASSON, *rue Saint Jacques, à la Science.*
DAVID l'aîné, *rue Saint Jacques, à la Plume d'or.*
LE BRETON, Imprimeur ordinaire du Roy, *rue de la Harpe.*
DURAND, *rue Saint Jacques, à Saint Landry, & au Griffon.*

M. DCC. LI.

AVEC APPROBATION ET PRIVILEGE DU ROY.

fig. 17

Title page, from Denis Diderot and Jean le Rond D'Alembert, *Encyclopédie, ou dictionnaire raisonné des sciences, des arts et des métiers* (Encyclopedia, or a systematic dictionary of the sciences, arts, and crafts), 1751

fig. 18

Système figuré des connaissances humaines (Figurative system of human knowledge), from Diderot and D'Alembert, *Encyclopédie*

Voltaire, Montesquieu, and Rousseau. Diderot's article "Encyclopedia" indicates his main motivation behind this pursuit: "Indeed, the purpose of an encyclopedia is to collect knowledge disseminated around the globe; to set forth its general system to the men with whom we live, and transmit it to those who will come after us, so that the work of preceding centuries will not become useless to the centuries to come."[20]

Diderot believed an encyclopedia to be, above all, a directory of associations, where the connections between the different areas of science could be exposed and further pursued by each individual reader. "Every science overlaps with others: they are two continuous branches off a single trunk,"[21] asserts Diderot. The branching analogy appears once again. In the same article, Diderot expresses an intriguing vision of the future, reminiscent of our now ubiquitous hypertext:

> Thanks to encyclopedic ordering, the universality of knowledge, and the frequency of references, the connections grow, the links go out in all directions, the demonstrative power is increased, the word list is complemented, fields of knowledge are drawn closer together and strengthened; we perceive either the continuity or the gaps in our system, its weak sides, its strong points, and at a glance on which objects it is important to work for one's own glory, or for the greater utility to humankind. If our dictionary is good, how many still better works it will produce![22]

This conception of an encyclopedia as a growing organism with many possible directions, as a map of scientific domains, explains why Diderot and d'Alembert included an illustration of the collective knowledge of humankind in the *Encyclopédie*. The piece entitled *Système Figuré des Connaissances Humaines* (Figurative system of human knowledge), and later called the tree of Diderot and d'Alembert, was first featured in the original 1751 edition and executed by French designer and engraver Charles-Nicolas Cochin. fig. 17, fig. 18 The scheme organizes all areas of science (knowledge) under three main branches: memory (or history), reason (or philosophy), and imagination (or poetry). If conceptually the scheme is inspired by Bacon's classification, graphically it has a clear similitude to Chambers's tree diagram in *Cyclopædia*, showcasing an analogous succession of curly brackets from higher to lower categories.

In 1780 an alternative version of the original illustration was made, this time as a much more literal arboreal metaphor. *Essai d'une distribution généalogique des sciences et des arts principaux* (Genealogical distribution of arts and sciences) was featured as a fold-out frontispiece in the *Table analytique et raisonnée des matieres contenues dans les XXXIII volumes in-folio du Dictionnaire des sciences, des arts et des métiers, et dans son supplement.* figs. 19–20 This tree depicts a genealogical distribution of knowledge, with its three prominent branches matching the early diagram: memory/history (left), reason/philosophy (center), and imagination/poetry (right). The heavy tree bears fruits of different sizes, representing the different domains of science, in an intricate branching configuration. Its slightly unbalanced look is caused by the dominance of the central bough of philosophy, which holds most of the tree's branches and shadows the withered ones of history and poetry.

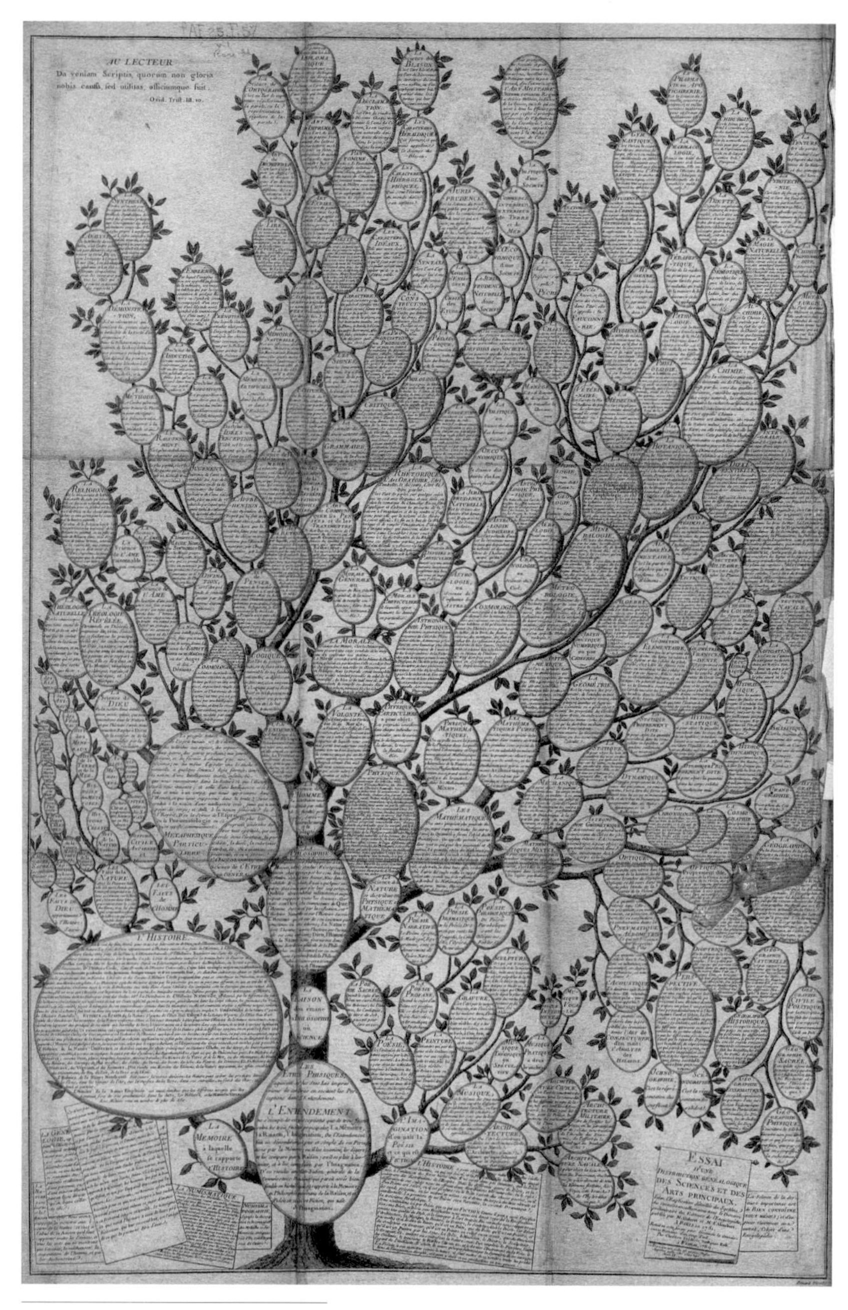

fig. 19

Chrétien Frederic Guillaume Roth, *Essai d'une distribution généalogique des sciences et des arts principaux* (Genealogical distribution of arts and sciences), from Diderot and D'Alembert, *Encyclopédie*

fig. 20

Detail of *Essai d'une distribution généalogique*

The End of an Era

Various types of depictions of trees mapping an incredible array of topics have surfaced throughout the decades. The descendants of these ancient tree diagrams are still an integral part of the structure and navigation of most modern computer systems, allowing one to browse, filter, and organize files in a nested hierarchy. Nonetheless, the *Essai d'une distribution généalogique des sciences et des arts principaux* marked the end of the golden age of embellishment, in which trees were seen as powerful figures embedded with loftier connotations. Over time, tree diagrams acquired a generic nonfigurative design and became utilitarian tools rigorously studied by those in computer science and the mathematical field of graph theory. Even though they have lost most of their allegorical symbolism, contemporary tree models still use many labels of the past (e.g., root, branches, leaves). Today trees are used in the representation of taxonomic knowledge in a variety of subject areas; and, as an exceptionally suitable scheme in the modeling of hierarchical structures, they will most certainly continue their widespread sphere of influence well into the future.

Notes

1 Pollack, *The Kabbalah Tree*, 2.
2 Bates, *The Real Middle Earth*, 44.
3 Ibid.
4 Philpot, *The Sacred Tree in Religion and Myth*, 3.
5 Porteous, *The Forest in Folklore and Mythology*, 191.
6 Hageneder, *The Living Wisdom of Trees*, 8.
7 James, *The Tree of Life*, 1.
8 Hageneder, *The Living Wisdom of Trees*, 8.
9 Philpot, *The Sacred Tree in Religion and Myth*, 1.
10 Pollack, *The Kabbalah Tree*, xvi.
11 Ibid., xvii.
12 Hageneder, *The Living Wisdom of Trees*, 8.
13 Preus and Anton, *Essays in Ancient Greek Philosophy V*, 19.
14 Heyworth, *Medieval Studies for J. A. W. Bennett*, 216.
15 Pombo, "Combinatória e Enciclopédia em Rámon Lull."
16 Rossi, *Logic and the Art of Memory*, 38.
17 Bacon, *Francis Bacon*, 175.
18 Ibid., 189.
19 Ibid.
20 Diderot, "Encyclopedia."
21 Ibid.
22 Ibid.

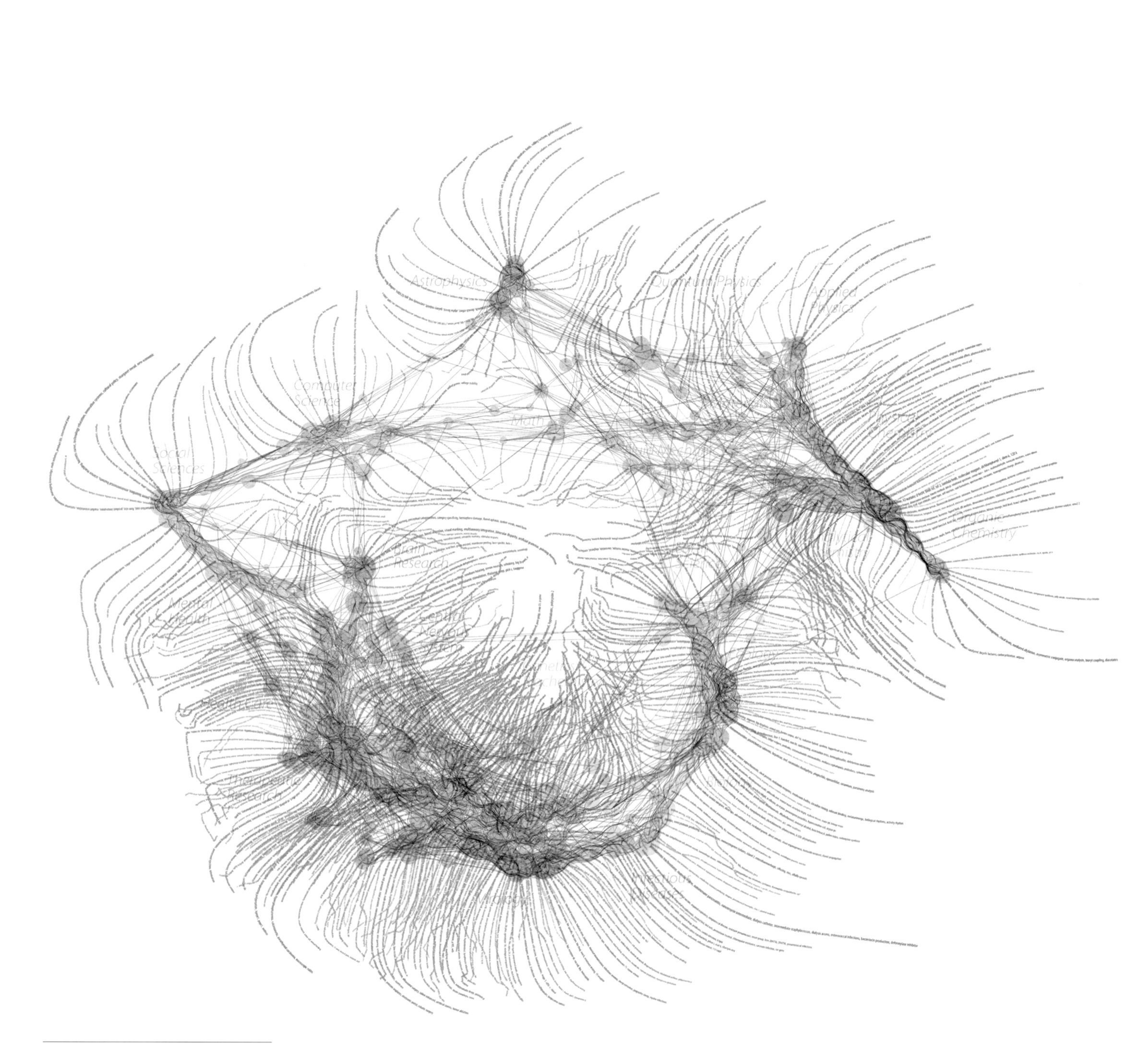

W. Bradford Paley, Dick Klavans, and Kevin Boyack, *Mapping Scientific Paradigms*, 2006

A map of science constructed by sorting roughly 800,000 scientific papers into 776 different scientific paradigms (colored circles) based on how often papers were cited together by authors of other papers. Links were made between the paradigms that shared common members (authors).

02 | From Trees to Networks

In simplicity of structure the tree is comparable to the compulsive desire for neatness and order that insists the candlesticks on a mantelpiece be perfectly straight and perfectly symmetrical about the centre. The semilattice, by comparison, is the structure of a complex fabric; it is the structure of living things, of great paintings and symphonies.

—Christopher Alexander

We're tired of trees. We should stop believing in trees, roots, and radicles. They've made us suffer too much. All of arborescent culture is founded on them, from biology to linguistics. Nothing is beautiful or loving or political aside from underground stems and aerial root, adventitious growths and rhizomes.

—Gilles Deleuze and Félix Guattari

Besides its evocative symbolism and undeniable applicability in different types of organizational contexts, the hierarchical tree model has innumerous connotations that have been occasionally criticized. Many opposing voices, usually highlighting its most pernicious attributes, have refuted the centralized, top-down tree metaphor, advocating for flexible alternative models able to accommodate the complex connectedness of modern society. The first disapproving view associates trees with the notion of centralism, or centralization, which expresses either an unequivocal concentration of power and authority in a central person or group of people, or a particular

The two epigraphs to this chapter are drawn from Alexander, "A City is Not a Tree," 58–62; and Deleuze and Guattari, *A Thousand Plateaus*, 15.

system in which most communications are routed through one central hub. Centralism is also linked to other less reputable concepts, such as totalitarianism, authoritarianism, and absolutism—typical of severely oppressive hierarchical systems.

In *The Tree of Life* (2007), Guillaume Lecointre and Hervé Le Guyader provide two additional views associated with the widespread concept of trees: finalism and essentialism.[1] Finalism, as the name implies, envisages a world where everything flows toward a predetermined final goal. Essentialism has an absolute understanding of the nature of being, in which every entity has a set of properties belonging to a precisely defined kind or group. It sees the essence of things as permanent, immutable characteristics—a fundamental rule for the enduring tree organization. If finalism describes the unidirectional courses of trees, essentialism alludes to their inert branches, which never shift or interact. Given that centralism, finalism, and essentialism form the basis of a common tree arrangement, we might also describe it as an authoritarian, unidirectional, and stagnant model.

In part due to its aforementioned traits, the tree model has been attacked, most notably by French philosophers Gilles Deleuze and Félix Guattari, who, in response, developed an antagonistic philosophical theory. Deleuze and Guattari oppose trees due to their forced totalitarianism and despotism—they are always dependent on a central authority. The authors argue in *A Thousand Plateaus* (1987) that trees are a condition of theoretical rigidness and unidirectional progress, where everything returns to a central trunk through vertical and linear connections. Trees therefore embody an organization that has never truly embraced multiplicity.

In opposition to this authoritarian model, Deleuze and Guattari in their *Capitalism and Schizophrenia* (1972–80) propose the concept of *rhizome*, aimed at acknowledging multiplicities and multilinearities: "In contrast to centered systems with hierarchical modes of communication and pre-established paths, the rhizome is an acentered, nonhierarchical, nonsignifying system without a General and without an organizing memory or central automaton, defined solely by a circulation of states."[2] Distinct from a tree topology and its individual branches, the rhizome connects any point to any other point, in a transverse and autonomous way, allowing for a flexible network of intercommunicability to emerge. "The rhizome pertains to a map that must be produced, constructed, a map that is always detachable, connectable, reversible, modifiable, and has multiple entryways and exits and its own lines of flight."[3]

The rhizomatic model is a significant influence in postmodern thinking, particularly in areas like communication theory, cyberspace theory, complex systems, nonlinear narrative, and hypermedia. But perhaps one of the most famous demonstrations of the principle's applicability is hypertext. Pertaining to any text with references (hyperlinks) to other texts, hypertext is the fundamental building block of the World Wide Web—arguably the largest rhizomatic system ever created by man. But rhizomatic theory also helps us apprehend the intricacies of the world: the rhizome is not a self-imposed conjectural view on our existence, but a fundamental topology of nature and an underlying element to the complex fabric of life. Perhaps for this reason, the rhizome has become a philosophical mentor in the ongoing struggle of modern science to cope with increasingly complex challenges.

fig. 1

Warren Weaver's concept of the three stages of modern science, according to the type of problems it tried to solve—problems of simplicity, problems of disorganized complexity, and problems of organized complexity—as discussed in Weaver, "Science and Complexity" (1948)

A few decades before Deleuze and Guattari's conception of the rhizome, American scientist Warren Weaver was already aware of the inherent complexities of nature and the hurdles anticipated by the scientific community in deciphering them. In 1948 in an article entitled "Science and Complexity," Weaver divided the history of modern science into three distinct stages: The first period, covering most of the seventeenth, eighteenth, and nineteenth centuries, encapsulated what he denominated as "problems of simplicity." Most scientists during this period were fundamentally trying to understand the influence of one variable over another. The second phase, taking place during the first half of the twentieth century, involved "problems of disorganized complexity." This was a period of time when researchers started conceiving systems with a substantial number of variables, but the way many of these variables interacted was thought to be random and sometimes chaotic. The last stage defined by Weaver, initiated in the second half of the twentieth century and continuing to this day, is critically shaped by "problems of organized complexity." Not only have we recognized the presence of exceedingly complex systems, with a large number of variables, but we have also recognized the notion that these variables are highly interconnected and interdependent. fig. 1

In reference to the last stage, Weaver wrote in 1948:

> These problems [such as commodity price fluctuation, currency stabilization, war strategies, or the behavioral patterns of social groups]—and a wide range of similar problems in the biological, medical, psychological, economic, and political sciences—are just too complicated to yield to the old nineteenth century techniques which were so dramatically successful on two-, three-, or four-variable problems of simplicity....These new problems, and the future of the world depends on many of them, requires science to make a third great advance, an advance that must be even greater than the nineteenth century conquest of problems of simplicity or the twentieth century victory over problems of disorganized complexity. Science must, over the next 50 years, learn to deal with these problems of organized complexity.[4]

Weaver's paper has been a great influence on contemporary thinking about complexity and the emergence of recent fields like complexity science and network theory. The conjecture he made for the second half of the twentieth century, and the advancement of science in dealing with increasingly complex challenges, is as true now as it has ever been. As we will see throughout this chapter, many of our contemporary hurdles, from the way we organize our cities to the way we decode our brain, concern problems of organized complexity that cannot be portrayed, analyzed, or understood by employing a centralized tree metaphor. In opposition to top-down hierarchies, these new challenges deal primarily with rhizomatic properties such as decentralization, emergence, mutability, nonlinearity, and ultimately, diversity.

The complex connectedness of modern times requires new tools of analysis and exploration, but above all, it demands a new way of thinking. It demands a pluralistic understanding of the world that is able to envision the wider structural plan and at the same time examine the

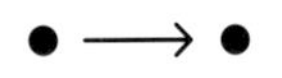

17th, 18th, and 19th Centuries
Problems of Simplicity

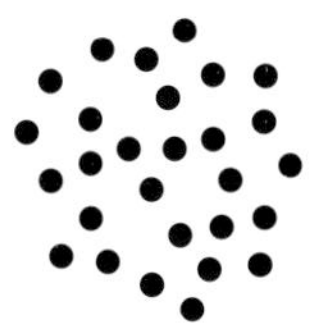

First half of the 20th Century
Problems of Disorganized Complexity

Post -1950
Problems of Organized Complexity

intricate mesh of connections among its smallest elements. It ultimately calls for a holistic systems approach; it calls for network thinking.

Planning a City

In 1965 the influential architect Christopher Alexander, most famously known for his book *A Pattern Language: Towns, Buildings, Construction* (1977), wrote one of the most influential pieces of postmodern criticisms on architecture, a short essay entitled "A City is Not a Tree." In it, Alexander refutes the hierarchical and centralized organization of the urban landscape, characteristic of model cities such as Brasilia, in favor of organic cities like London and New York. He declares that many of these artificial cities have failed due to the rigid and insipid thinking of their creators, who planned areas of activity (e.g., residential, industrial, commercial) as independent and incommunicable modules, in a typical tree structure of separate branches.

Although we have become accustomed to this type of stringent urban layout, it is a fairly recent conception. Before the Industrial Revolution, most people lived in residential spaces located above their work environment, and the line between personal and professional spheres was very thin. The effects of industrialization meant that people lived in one area of town, worked in another, and probably shopped in yet another area. Anyone who has ever lived in a large city with its sprawling suburban areas knows how this segregation translates to lengthy daily commuting time. fig. 2 Alexander is extremely forthright about the consequences of this fragmentation: "In any organized object, extreme compartmentalization and the dissociation of internal elements are the first signs of coming destruction. In a society, dissociation is anarchy. In a person, dissociation is the mark of schizophrenia and impending suicide."[5] As Alexander clearly implies, human beings do not naturally comply with this highly compartmentalized modus operandi. Our connections, among ourselves and with the surrounding environment, do not follow this type of conceptual order and simplicity. We are ambiguous, complex, and idiosyncratic. "The reality of today's social structure is thick with overlap—the systems of friends and acquaintances form a semilattice, not a tree,"[6] states Alexander on the convergent nature of social groups. He is convinced that the reductionist conception of urban spaces, typical of a tree organization, blinds our judgment of the city and limits the problem-solving abilities of many planners and system analysts.

Alexander understands well the invisible mesh of interconnections that overlays the urban environment and suggests a semilattice organization, similar to a network, that can better suit the complexities of human relationships. Due to its structural intricacy, the semilattice is a source of rich variety. Within the semilattice the "idea of overlap, ambiguity, multiplicity of aspect [are not] less orderly than the rigid tree, but more so....They represent a thicker, tougher, more subtle and more complex view of structure."[7] Alexander finishes the article with the unequivocal assertion that "the city is not, cannot and must not be a tree.... The city is a receptacle for life."[8] fig. 3

Published four years before Alexander's piece, Jane Jacobs, in the classic *The Death and Life of Great American Cities* (1961), delivers a biting criticism of the urban calamities perpetrated in the United States during the 1950s, which caused social alienation by isolating a large number of communities and urban spaces. The book also offers

fig. 2

Brandon Martin-Anderson, *Shortest Path Tree*, 2006

Street and transit information are inputted into software that computes the shortest routes between one specific location and the remaining areas of town (San Francisco). The width of a branch (route) is proportional to the sum of branches reachable by that branch.

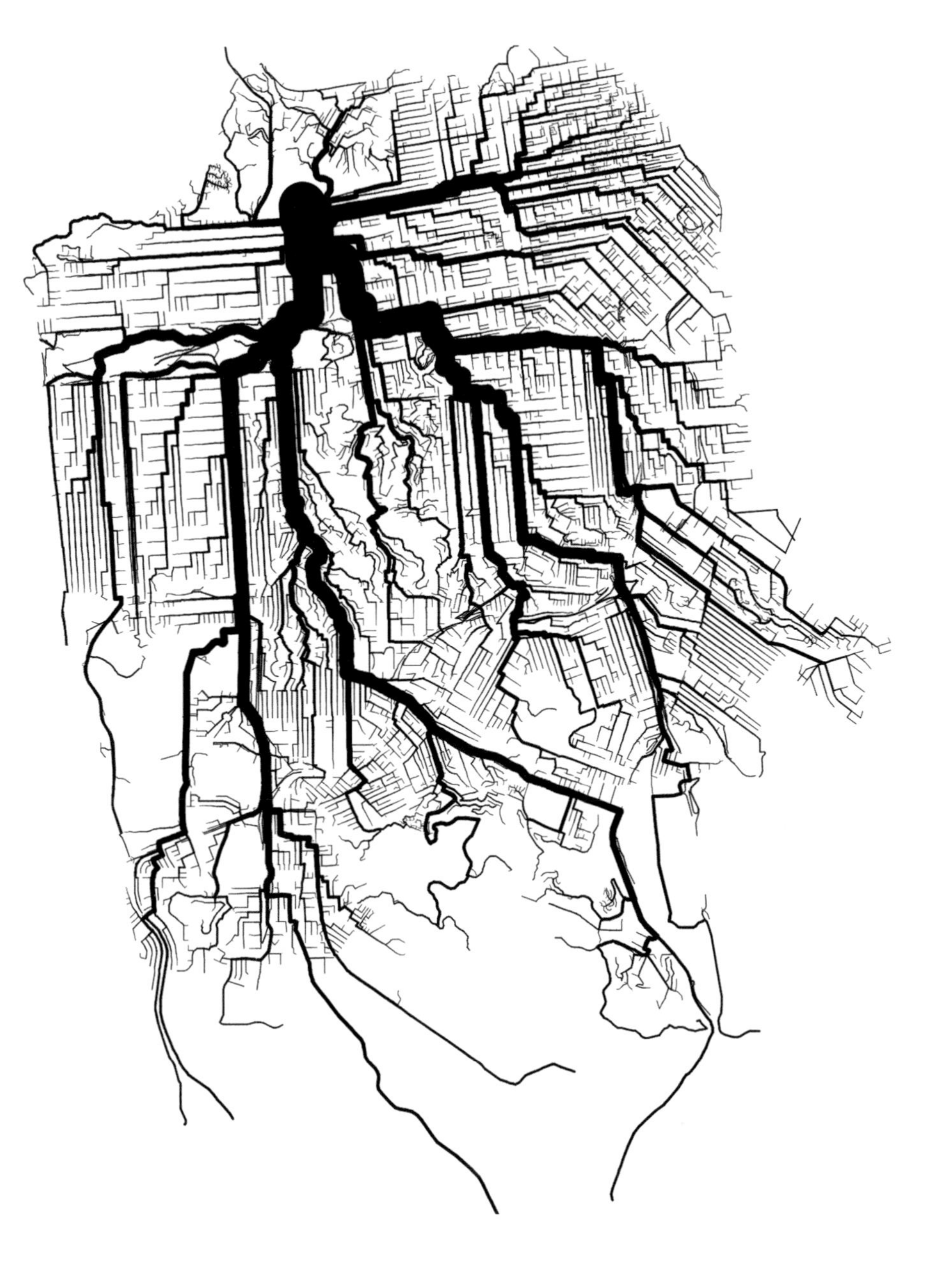

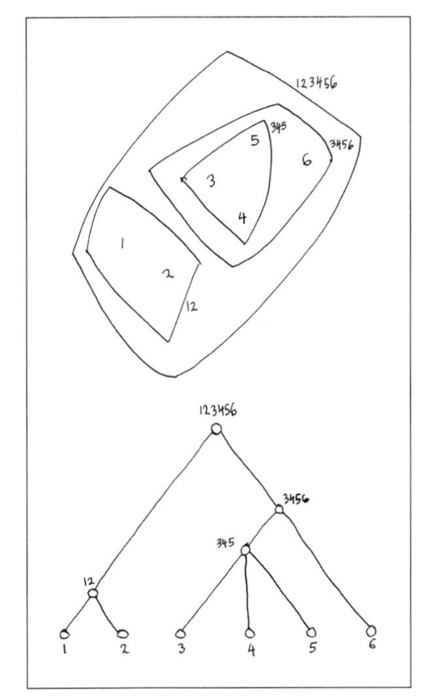

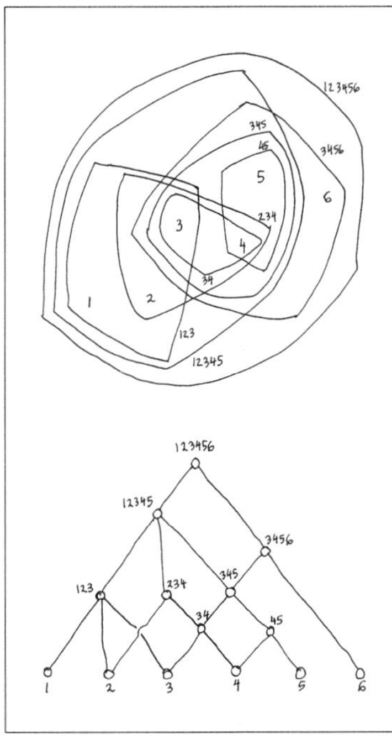

fig. 3

Nikos Salingaros, Tree versus semilattice, 1965

A set of diagrams that appeared in Christopher Alexander's essay "A City is Not a Tree" (1977), which complemented his discourse on the different ways to organize a city. It shows the scheme of a simplified tree model (left), which excludes the likelihood of overlapping areas, and a more tolerant semilattice (right), which allows for different urban layers to coexist.

a convincing exaltation for organicism—a philosophical notion that proposes that reality is best understood as an organic whole. Jacobs eloquently describes the aim of her book in the very first introductory paragraph: "This book is an attack on current city planning and rebuilding. It is also, and mostly, an attempt to introduce new principles of city planning and rebuilding, different and even opposite from those now taught in everything from schools of architecture and planning to the Sunday supplements and women's magazines."[9] She opposes any simplified or ingenuous plan of a city, and in reference to Weaver's original thesis, Jacobs states, "Cities happen to be problems in organized complexity, like the life sciences. They present situations in which a half-dozen or even several dozen quantities are all varying simultaneously and in subtly interconnected ways."[10] Jacobs reiterates that contemporary urban planning has obliterated the city, because it has rejected its main actors—human beings.

The work of Jacobs and Alexander has been fundamental in the emergence of New Urbanism, a movement started in the 1980s with the goal of investigating a more efficient integration between the primary uses of a city. Even though this movement has caused various new developments to emerge in the last decades across the globe—promoting pedestrian-friendly neighborhoods, green-building construction, historic preservation, and a balanced development between jobs and housing—we still have a long way to go. This much needed transformation will have to acknowledge the city as a living organism in constant mutation, a highly complex network involving a vast number of variables. It will have to conceive the city as an open space bursting with overlap and spontaneity, where the natural conditions for creativity, recreation, and cooperation can easily prosper. This will only happen if we stop imposing artificial barriers on our spaces and truly embrace the diverse social nature of man.

Neural Landscape

The study of the brain has truly come a long way. Ancient accounts, dating back to Greek philosophy, indicate an early belief in the assessment of one's character or personality (e.g., criminality) from one's facial features (e.g., distance between the eyes), in what became commonly referred to as "physiognomy." It is not surprising that, as social beings, people would associate outer appearance with inner character. This became a widespread conviction that was held by many prominent figures of antiquity, including Aristotle.

Toward the eighteenth century, researchers abandoned the study of facial features to pursue the study of the skull itself. And by 1796, by the hands of German physician Franz Joseph Gall, the pseudoscience of phrenology was born. Phrenologists believed that the shape of the skull, with its bumps and hollows, exposed the thoughts within. Believing that the mind was made up of different mental faculties represented in distinct areas of the brain, phrenologists measured the cranial bone to find the engorged or shrunken areas of the skull that corresponded to the area of the brain responsible for a particular personality trait, character, or behavior. Gall's list of the "brain organs" contained twenty-seven different regions, including "the love of one's offspring," "the carnivorous instinct; the tendency to murder," "the memory of words," "the sense of sounds; the gift of music," "the poetical talent," and "the organ of religion." fig. 4, fig. 5

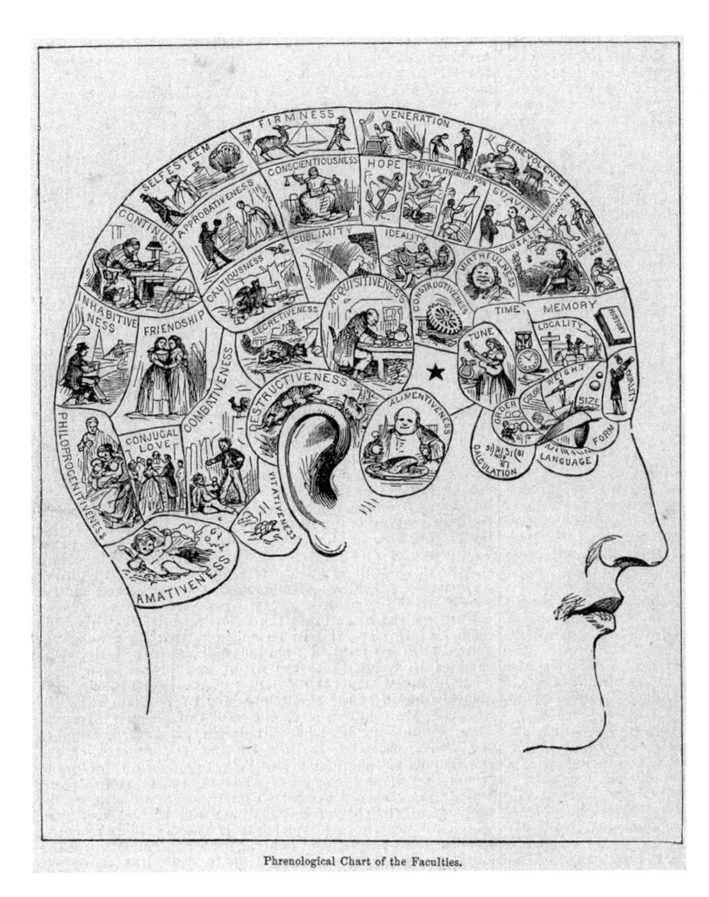

Phrenological Chart of the Faculties.

fig. 4

Phrenology diagram, from W. H. De Puy, *People's Cyclopedia of Universal Knowledge*, 1883

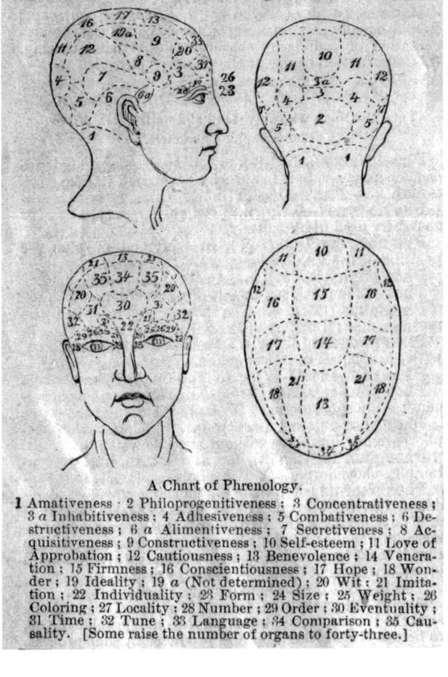

Phre-nol′o-gy (-nŏl′ō-jȳ), *n.* [Gr. φρήν, φρενός + *-logy.*] **1.** Science of the special functions of the several parts of the brain, or of the supposed connection between the faculties of the mind and organs in the brain. **2.** Physiological hypothesis that mental faculties, and traits of character, are shown on the surface of the head or skull; craniology. — **Phre-nol′o-gist**, *n.* — **Phren′o-log′ic** (frĕn′ō-lŏj′ĭk), **Phren′o-log′ic-al**, *a.*

A Chart of Phrenology.

1 Amativeness; 2 Philoprogenitiveness; 3 Concentrativeness; 3 *a* Inhabitiveness; 4 Adhesiveness; 5 Combativeness; 6 Destructiveness; 6 *a* Alimentiveness; 7 Secretiveness; 8 Acquisitiveness; 9 Constructiveness; 10 Self-esteem; 11 Love of Approbation; 12 Cautiousness; 13 Benevolence; 14 Veneration; 15 Firmness; 16 Conscientiousness; 17 Hope; 18 Wonder; 19 Ideality; 19 *a* (Not determined); 20 Wit; 21 Imitation; 22 Individuality; 23 Form; 24 Size; 25 Weight; 26 Coloring; 27 Locality; 28 Number; 29 Order; 30 Eventuality; 31 Time; 32 Tune; 33 Language; 34 Comparison; 35 Causality. [Some raise the number of organs to forty-three.]

fig. 5

The definition of phrenology with corresponding diagram, from Noah Webster, *Webster's Academic Dictionary*, 1895

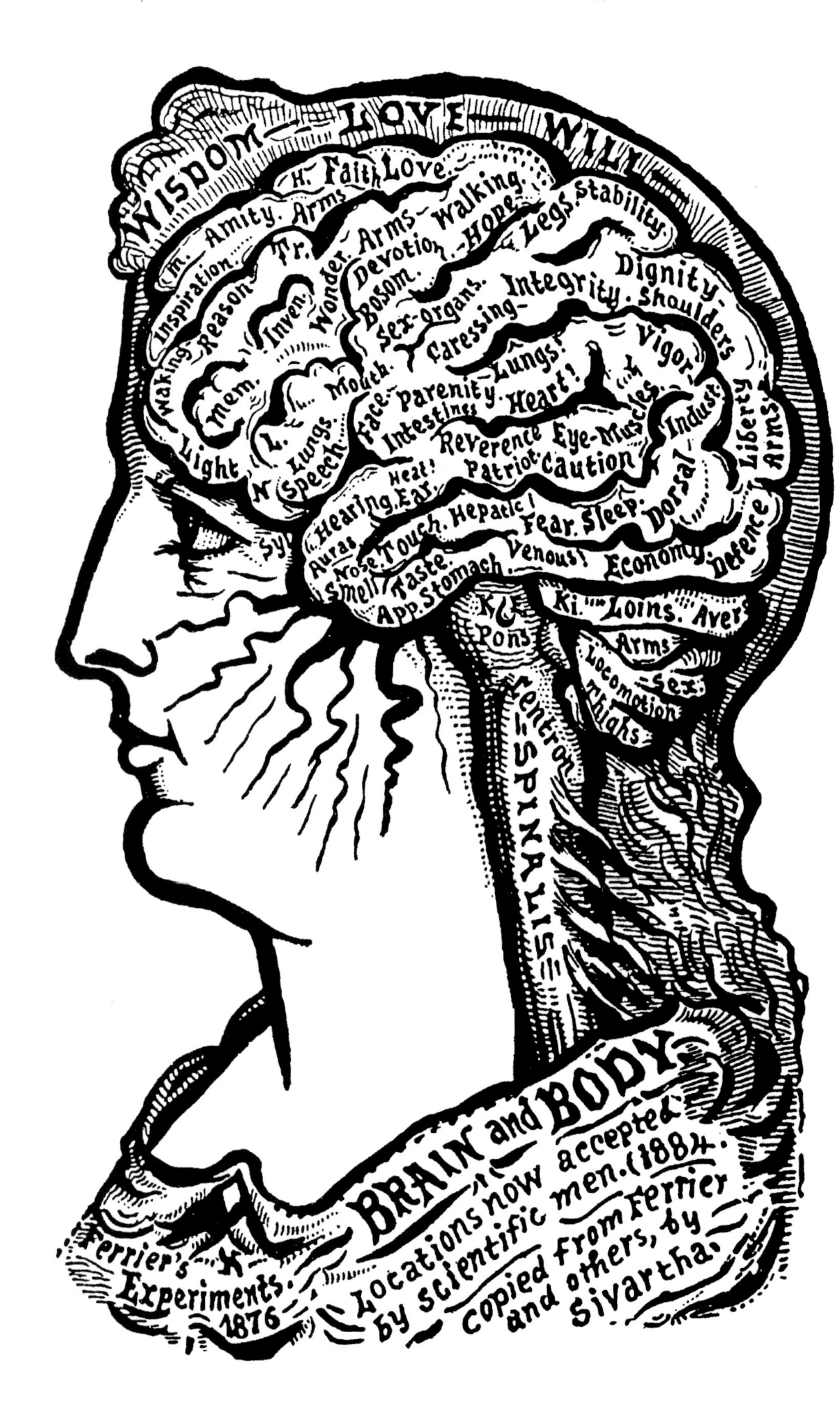

fig. 6

Brain and Body, from Alesha Sivartha, *The Book of Life: The Spiritual and Physical Constitution of Man*, 1912

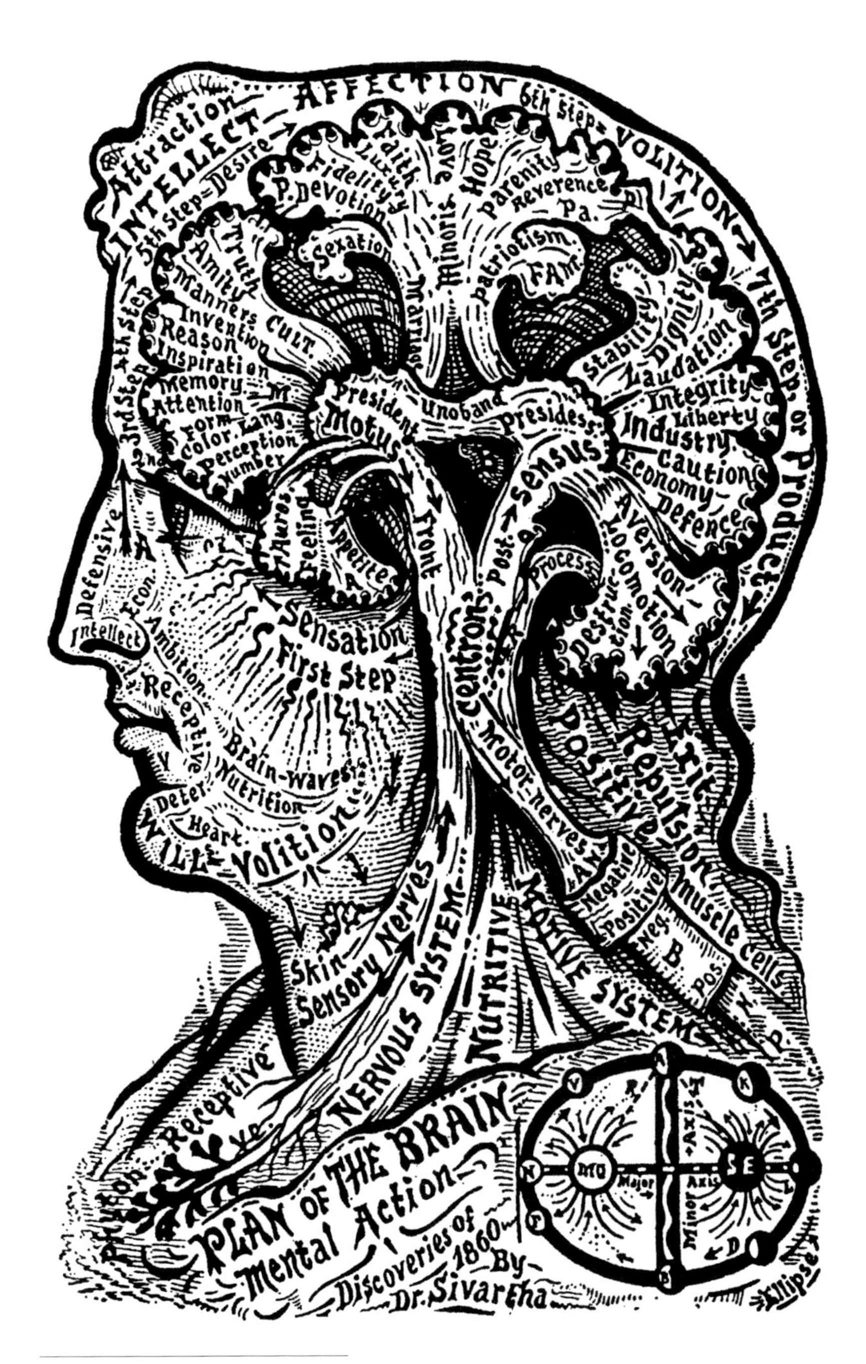

fig. 7

Brain Structure, from Sivartha, *The Book of Life*

fig. 

Locations, from Sivartha, *The Book of Life*

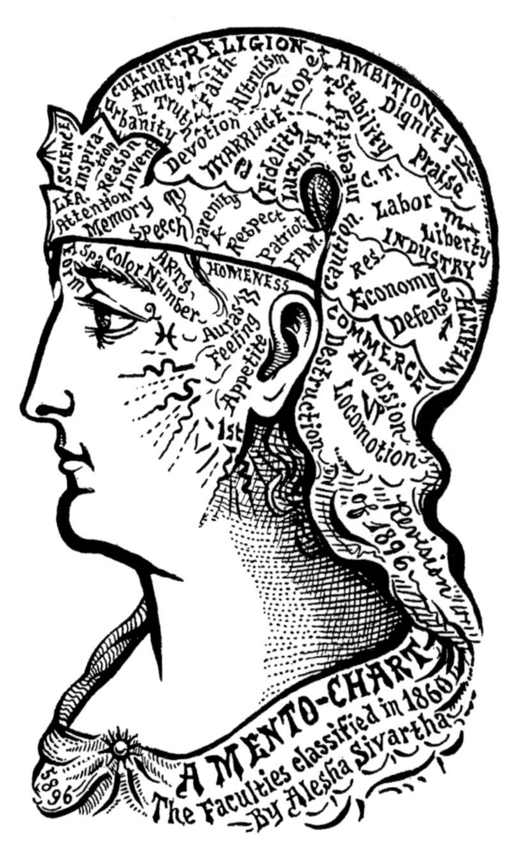

A century later, as phrenology was already in decline, the work of philosopher Alesha Sivartha captured the pinnacle of this ideology in *The Book of Life: The Spiritual and Physical Constitution of Man* (1912). In this magnificent collection of brain maps and pseudoscientific illustrations, Sivartha explores many of the ideas associated with phrenology, with the brain as the setting for the structuring of all types of social, political, ethical, and cultural concepts. fig. 6, fig. 7, fig. 8

Phrenology was eventually recognized as an extremely flawed system that simply went down in history as a divergent scientific pursuit. It did, however, contribute to a long-lasting meme regarding the modularity of the mind. Many people still think of the brain in terms of left and right, front and back modules, or believe in the presence of a unique centralized control responsible for the great diversity of cognitive tasks and human behaviors. "The idea of a 'centre' for different functions in the brain is so intuitively appealing, it is hard to relinquish," explains renowned neuroscientist Susan Greenfield.[11] But it is an inadequate model. Instead of a centralized control, our voluntary and involuntary actions are triggered by a series of electrochemical impulses percolating through millions of neurons, like a multitude of musical instruments coming together for a symphony.

It is estimated that an adult human brain has around one hundred billion neurons, with each neuron being biologically wired to thousands of its neighbors by dendrites and axons. The brain's staggering complexity represents one of the toughest puzzles, challenging modern neuroscientists to constantly reassess their assumptions. One of the men leading this quest is Henry Markram, director of the Center for Neuroscience and Technology at École Polytechnique Fédérale de Lausanne (EPFL), in Switzerland, and project director of the remarkably impressive Blue Brain project.

Markram and his team of neuroscientists, paired with IBM's sophisticated supercomputer, Blue Gene, are creating a blueprint of the neocortex—a part of the cerebral cortex that accounts for nearly 80 percent of the human brain. fig. 9 In this active region, made up from a dense network of neurons and fibers—commonly known as gray matter due to its gray color in preserved brains—many higher cognitive functions such as conscious thought, memory, and communication occur.

The goal of the Blue Brain project—to generate a holistic model of such an intricate structure—is not a straightforward task and conceivably requires an astounding amount of computing power. This is how writer Jonah Lehrer describes the technical setup behind the project:

> In the basement of a university in Lausanne, Switzerland, sit four black boxes, each about the size of a refrigerator, and filled with 2,000 IBM microchips stacked in repeating rows. Together they form the processing core of a machine that can handle 22.8 trillion operations per second. It contains no moving parts and is eerily silent. When the computer is turned on, the only thing you can hear is the continuous sigh of the massive air conditioner. This is Blue Brain.[12]

Blue Brain is an incredible initiative, only comparable in greatness to the Human Genome Project—an international scientific initiative with the goal of mapping and sequencing the entire human genome. The immense

fig. 9

École Polytechnique Fédérale de Lausanne, *Blue Brain Project*, 2008

A computer-generated model produced with IBM's Blue Gene supercomputer, part of the Blue Brain project, shows the thirty million connections between ten thousand neurons in a single neocortical column—arguably the most complex part of a mammal's brain. The different colors indicate distinctive levels of electrical activity.

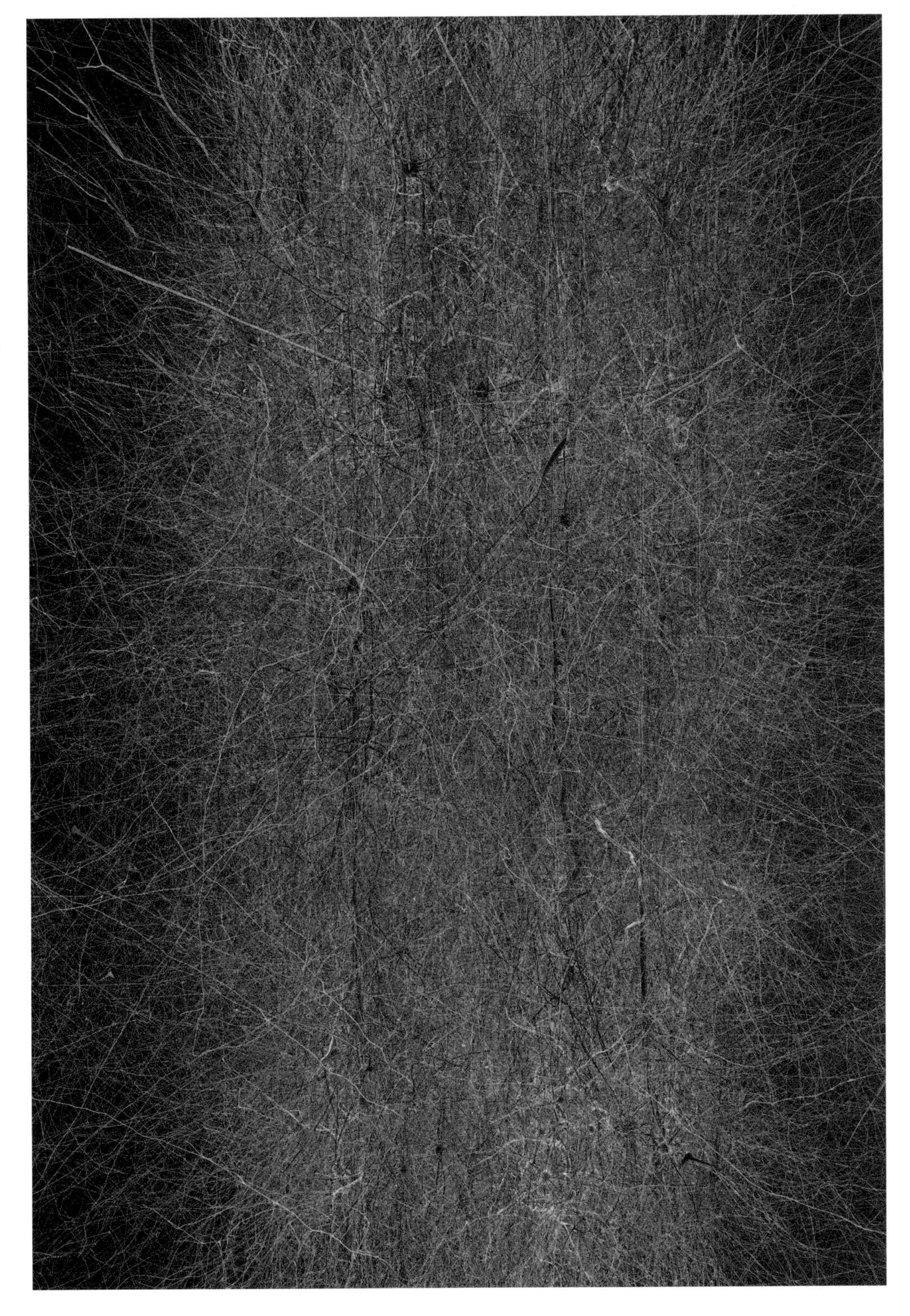

knowledge that could arise from this breakthrough is still hard to grasp, and so are its consequences. Once we have the entire map of the neural circuitry, the possibility of re-creating it in a functional simulation would be the logical next step. In an ambitious statement at the annual TEDGlobal conference in Oxford, England, in July 2009, with the theme "The Substance of Things Not Seen," Markram declared, "It is not impossible to build a human brain and we can do it in ten years."[13]

The first significant milestone of the project has already been achieved: the replication of the neocortical column—a small slice of the brain cortex containing approximately ten thousand neurons, and about thirty million synaptic connections between them. "Now we just have to scale it up," says Markram in his usual optimistic attitude, referring to the replication of the rest of the brain.[14] His encouraging view is founded in the conviction that Blue Brain is conveying a different way of looking at ourselves, and ultimately at science. Markram considers that merely looking at the isolated parts is not providing us the whole picture, and the reductionist approach of neuroscience, successful as it was until now, has exhausted itself. "This doesn't mean we've completed the reductionist project, far from it. There is still so much that we don't know about the brain. But now we have a different, and perhaps even harder, problem. We're literally drowning in data. We have lots of scientists who spend their life working out important details, but we have virtually no idea how all these details connect together. Blue Brain is about showing people the whole."[15]

If Markram is right in his conjecture regarding a holistic approach to replace an outmoded reductionist view, then we might see many other areas of science following this systemic modeling method. What this new reasoning might represent for the advancement of science might be even more significant than Blue Brain itself.

Ubiquitous Datasphere

The internet is one of the most extraordinary and complex systems ever built by man. It has become so influential in our lives that it is easy to forget its relatively young age. In the middle of the Cold War and while working for the U.S. military intelligence agency RAND Corporation to develop a new communication system that could survive a nuclear attack, the then thirty-year-old Paul Baran produced several documents that attested the vulnerabilities of the communication infrastructure of the time. His proposition for a safer alternative would become a central driving force for the subsequent development of the internet. In 1964 Baran suggested three possible models for the novel system: centralized (with a single decision center), decentralized (more than one decision center), and distributed (made by uniformly distributed nodes with no decision center). fig. 10 Baran recommended the very last one, a model with a noticeable mesh structure, more resilient to an eventual attack. The distributed topology Baran proposed, which he published in a series of reports entitled *On Distributed Communications* (1964), would then be further developed and implemented by another American agency: the Advanced Research Projects Agency, commonly known as ARPA. fig. 11, fig. 12

Since the first message sent across two computers in October 1969 as a part of the early ARPANET, the internet has grown at an astounding pace. Contemporary maps of the convoluted landscape of routers, servers, and

fig. 10

Paul Baran, Network models, 1964

The three architectures—centralized, decentralized, and distributed—possible for a novel communication system. From Baran, "On Distributed Communications: Introduction to Distributed Communications Networks," 1964 (paper published internally within Rand Corporation).

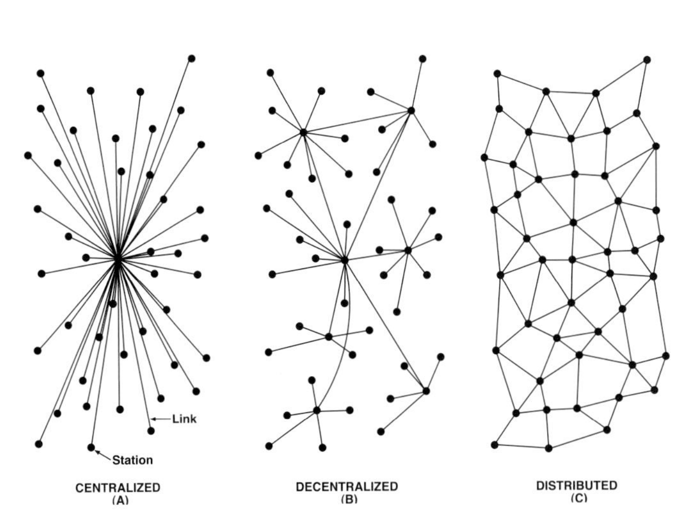

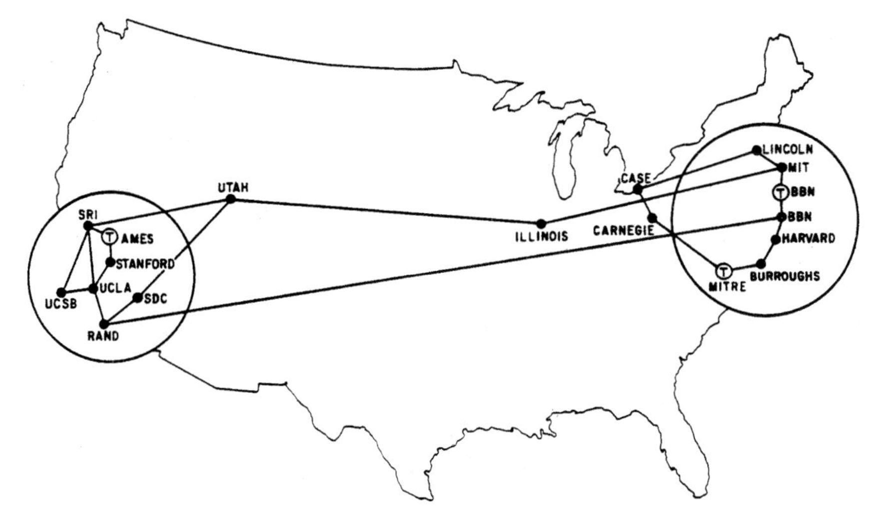

fig. 11

A map of the Advanced Research Projects Agency Network (ARPANET) from September 1971, showing some of its earliest nodes: at University of California, Los Angeles; University of California, Santa Barbara; Rand Corporation; Massachusetts Institute of Technology; and Harvard University. From F. Heart, A. McKenzie, J. McQuillian, and D. Walden, *ARPANET Completion Report*, January 4, 1978.

fig. 12

A map of an expanding ARPANET from March 1977. Heart, McKenzie, McQuillian, and Walden, *ARPANET Completion Report*.

ARPANET LOGICAL MAP, MARCH 1977

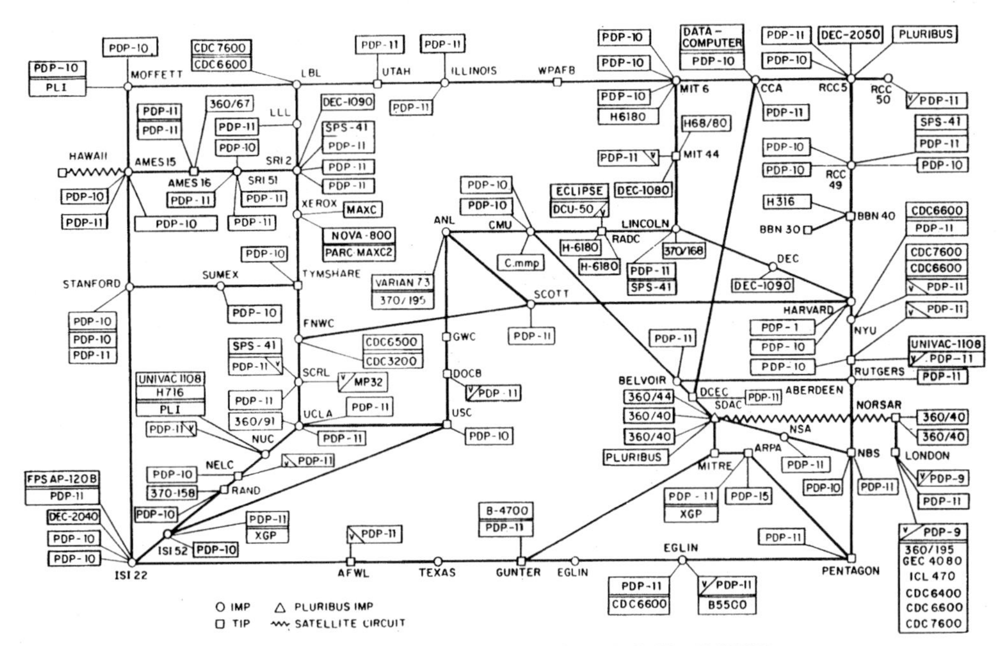

(PLEASE NOTE THAT WHILE THIS MAP SHOWS THE HOST POPULATION OF THE NETWORK ACCORDING TO THE BEST INFORMATION OBTAINABLE, NO CLAIM CAN BE MADE FOR ITS ACCURACY)

NAMES SHOWN ARE IMP NAMES, NOT (NECESSARILY) HOST NAMES

connections are astonishingly complex, highlighting one of the most intricate man-made structures. But one of the most interesting attributes of the internet is that it sustains another equally tangled network of nodes and links, embodying an enormous volume of data: the World Wide Web.

Twenty years after the famous ARPANET experiment, in 1990, English physicist Tim Berners-Lee and Belgian scientist Robert Cailliau, while working for the European Organization for Nuclear Research (CERN) in Geneva, proposed the construction of a "web of nodes" that could store "hypertext pages" visualized by "browsers" in a certain network. The original name for this global hypertext system was "mesh," but in December of the same year, it was launched with its brand new designation: World Wide Web. After the first hypertext pages were added to the web, it was just a matter of time until other websites followed in a vertiginous expansion. In June 1993 there were only about 130 websites in the whole world. By June of 1998 the web had grown to 2,410,067, and by June 2003 there were 40,936,076 indexed websites. Currently the number is greater than two hundred million, and it might be considerably higher, given that many websites include other websites under the same domain, and a remaining number is undocumented and far from the reach of legal estimates.

This proliferation of websites denotes the vitality of the World Wide Web, mostly due to its underlying democratic distribution of information. According to professors of Communication and Creative Arts at Purdue University Calumet Lee Artz and Yahya R. Kamalipour, this massively complex network constitutes an "information biosphere . . . a single, interconnected information organism of free expression and free trade."[16] fig. 13 But if currently the autonomous web is mostly characterized by the enhanced influence of common citizens, through the use of empowering online tools and services, the next web revolution will be even more powerful.

While there are millions of information-embedded web pages online, they are, in most cases, unable to automatically extract knowledge from their inherent interconnections. A single neuron is insignificant, but as it communicates with thousands of neighbors through synapses, it suddenly becomes part of a whole, much bigger than the sum of its parts. This whole keeps changing over time by the addition and deletion of nodes, increasing or decreasing the strength of their connections, in order to constantly adapt to human experience and new learning requirements. It is through this process that the brain retrieves an old memory, analyzes the thread of a sudden event, or composes an argument for a particular idea. This is the level of malleability, commonly called neuroplasticity, that the web is expected to develop in the next few years or decades.

It might certainly take longer than that, but there are several signs attesting to this "neurological" transformation of the web. Not only is data becoming more widely accessible—as companies, institutions, and governments open their data sets to the general public—it is also becoming enriched with prolific metadata, allowing new sets of comparison and interweaving. In a March 2009 talk at Technology Entertainment and Design (TED) conference, Berners-Lee made a vehement exaltation for linked data. One year later, in February 2010, he came back to the renowned conference to corroborate his vision with various practical examples, stating that "if people put data on the web—government data, scientific data, community data—whatever it is, it will be used by other people to do wonderful things in ways they never could have imagined."[17]

Initiatives such as the U.S. government's launch of Data.gov—an advanced online portal that combines hundreds of databases from a variety of public agencies, institutions, and departments across the country—are a brilliant indicator of change and one that is being replicated the world over. Launched in May 2009, Data.gov opens the door to political transparency and public scrutiny, aiming at a broadly informed democracy, but even more importantly, it adds a very reliable set of building blocks to the growing mesh of online knowledge.

Increasingly distant from any centralized communication model of the past, the significant impending shift of the web will command an even more detailed and tangled layer of complexity, where data becomes widely interrelated with and detached from constraining documents. The early documentcentric web will give way to a pulsating ecosystem of data, a truly ubiquitous datasphere, the main challenge of which—apart from privacy concerns—will be interoperability—universal standards and formats that enable an effortless data interlace. The last thing we want is a large collection of indiscernible data points lying in servers across the world. It is particularly interesting to see how the internet's fundamental model of autonomy has been replicated over time, just like looking into a convoluted fractal representation, where the same underlying principle of complexity and interconnectedness is applied to ever more tiny parts of the structure, from routers to servers, web pages, and now data.

Social Collaboration

The idea of social stratification is one of the most ubiquitous and oldest sociological constructs in the world. From feudalism to capitalism, there has always existed rigid layers of society based on social differentiation, with a strong emphasis on the division of labor. It is not only society at large that succumbs to this unyielding model; businesses, armies, churches, governments, and many other entities also follow a defined ranking among their members, always lead by a central commanding figure.

Although it is difficult to define the origins of social stratification, Professor Anthony J. McMichael considers one of the central seeds for its development to have occurred around 10,000 BC in the Fertile Crescent, a bountiful, crescent-shaped area covering parts of the Middle East. The development of agriculture drove modern humans to slowly abandon their nomadic lifestyles for more reliable, confined harvests. This transition from hunting and gathering to settled farming and agriculture gave birth to the first agricultural revolution, known as the Neolithic Revolution. This transformation brought prosperity, the enlargement of rural settlements, and the surge of many new types of labor, which in turn caused the emergence of social class, status, and power—central elements to the emergence of hierarchical systems of domination.

But it was during the Industrial Revolution that many of our hierarchical conceptions of society were widely put into practice. This critical period caused not only a major change in the way cities were planned but also a key transformation in the way businesses and society at large were organized. The fundamental process of rationalization that occurred during this stage became a central impetus for the solidification of bureaucracy, corporate ranking, and

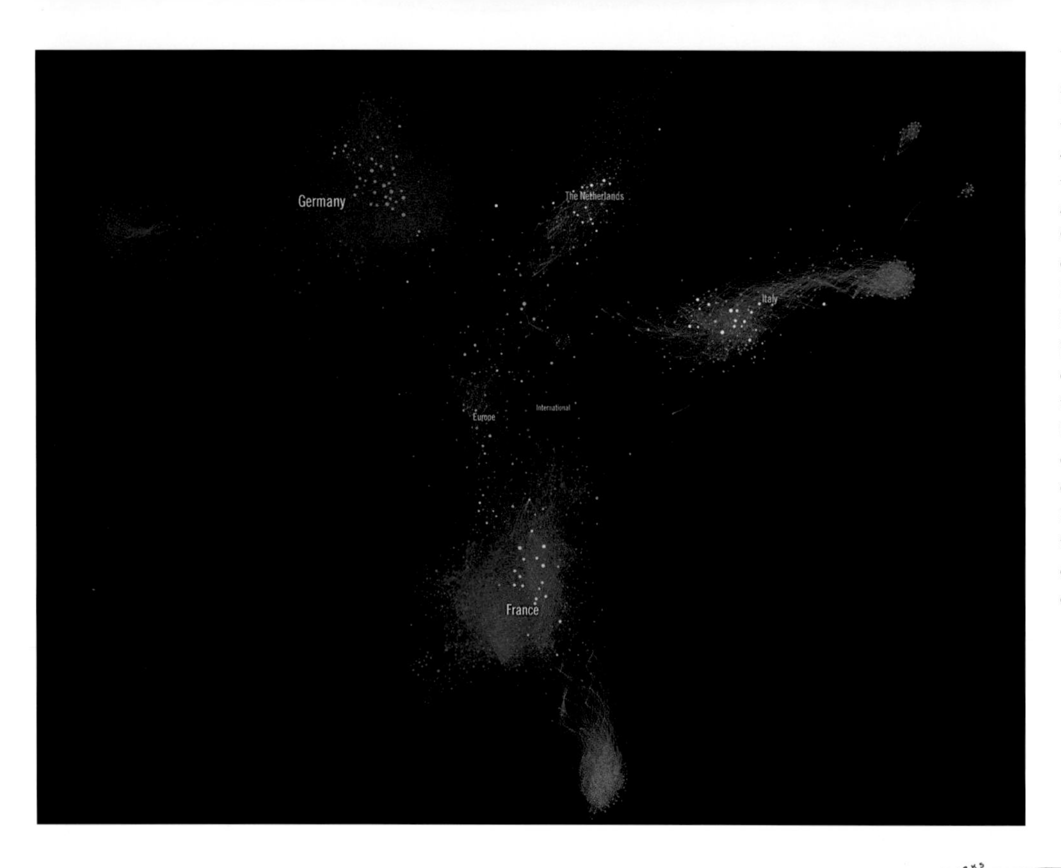

fig. 13

Antonin Rohmer, *Eurosphere*, 2009

A map of the political blogosphere in Central Europe (France, Germany, Italy, and the Netherlands), showing shared links between communities of political bloggers and portals, communities of journalists and experts, communities of political pundits, media websites, trade unions, think tanks, public institutions, NGOs, and activists. The distance between websites, within a community and between communities, reveals the amount of interactions: the closer websites are, the more they engage with each other. The data was collected over a one-year period.

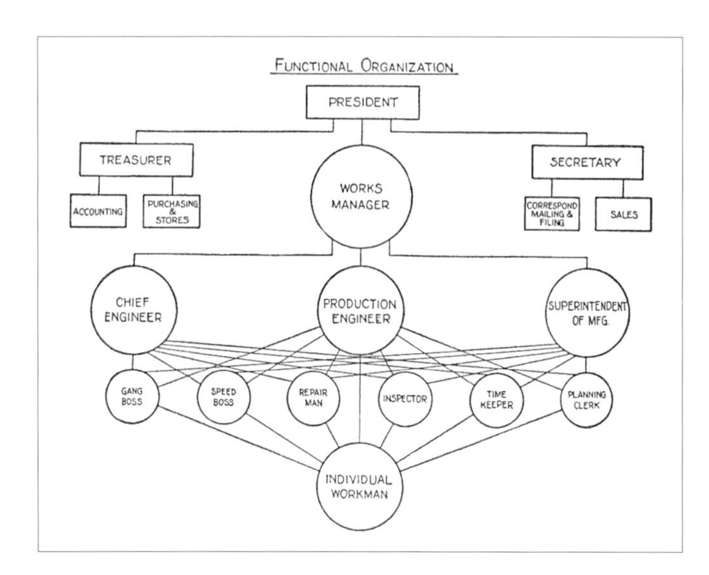

fig. 14

A typical corporate organizational chart, showing the hierarchical structure from the president to the individual workman. From Smith, *Graphic Statistics in Management.*

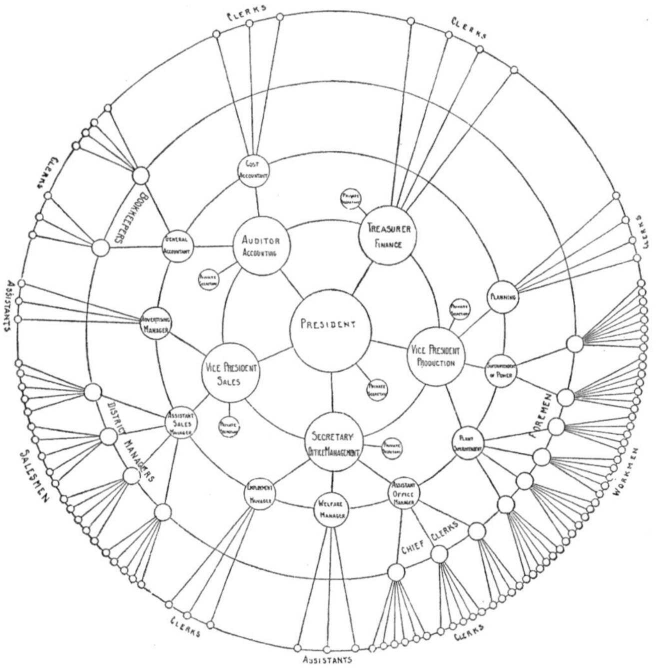

fig. 15

This radial organization chart highlights the centralized decision-making power structure of most companies, with the president at the very core, followed by successive degrees of dependency. From Smith, *Graphic Statistics in Management.*

management models based on centralized control. Many of these ideas are still so ingrained in our modern existence that it is hard to relinquish them. fig. 14, fig. 15

We tend to consider social hierarchy as the norm across societies and, in some cases, as a necessity or sociological predisposition. But in fact, many past and present-day hunter-gatherer societies have little or no concept of economic or political status, class, or permanent leadership. Many indigenous communities in the Americas and Australia today follow the same type of decentralized and egalitarian structure, since stratification is broadly seen as a cause of conflict and instability.

In *The Starfish and the Spider* (2006), Ori Brafman and Road Beckstrom describe the difficulties encountered by the Spanish conquistadors when facing the leaderless Apache tribes in the sixteenth century in present-day New Mexico. The Spanish tried to convert them to Christianity and to a sedentary agrarian lifestyle, but the Apache resisted and fought back and held them off for two centuries. "It wasn't that the Apaches had some secret weapon that was unknown to the Incas and the Aztecs. Nor had the Spanish army lost its might," explain the authors. "No, the Apache defeat of the Spanish was all about the way the Apaches were organized as a society"—a simple, yet complex flattened layout, typical of a decentralized configuration.[18]

Decentralized social structures do not have a leader in control or defined ranks, and most importantly, they have no headquarters. Additionally, when attacked, centralized systems tend to become even more centralized, while decentralized organizations become more open and decentralized. Even when the Spanish went after and started killing Nant'ans (a type of spiritual leader), new Nant'ans would immediately emerge. The further the Spanish charged, the more decentralized and hard to conquer the Apaches became.

This particular chapter of history is not entirely distant from present-day news headlines announcing the state of affairs between armies and terrorist cells. Although with an entirely distinct nature and political motivation, the main tool of terrorist resistance today is the same as the one perpetrated by the Apache four hundred years ago: decentralization.

Decentralized social groups are not only common in hunter-gatherers and terrorist networks, but continue to grow today, more than in any other period since the Industrial Revolution, challenging many secular theories and also becoming intrinsically associated with our technological progress. As Don Tapscott and Anthony D. Williams affirm in their bestselling *Wikinomics* (2006), "Profound changes in the nature of technology, demographics, and the global economy are giving rise to powerful new models of production based on community, collaboration, and self-organization rather than on hierarchy and control."[19] *Wikinomics* is just one of many books published in the last few years exploring the topic of decentralization and its effects on modern society. Much has been written, blogged, and discussed on the evolution of this new nonstratified scheme, usually coupled with encouraging democratized ideas such as the wisdom of crowds, democratic journalism, local decision, self-empowerment, independence, and self-organization, and repeatedly illustrated by its main frontrunners: Wikipedia, the open-source movement, Craigslist, and Digg.

As the web is a prime case of a leaderless structure, embodied by its overlaying mesh of self-governing information

fig. 16

Linkfluence, *CPAN Explorer*, 2009

A map showing the collaborations between the most active Perl programming language developers within the large online community CPAN (Comprehensive Perl Archive Network). The size of the node denotes the number of Perl modules (discrete components of software for the Perl programming language) an author has released on CPAN, and connections represent shared modules between authors.

nodes, it is not surprising that the most well-known cases of decentralization are happening on the web. This growing autonomy is also fomenting a critical social behavior that sustains many online services and initiatives. In Thomas L. Friedman's international bestselling book *The World is Flat* (2005), he describes ten "flatteners"—key drivers for transforming the world into a level playing field. Among them is what Friedman considers to be the most subversive force of all: online collaboration. fig. 16

When geography and the speed of communication are no longer obstacles, collaboration is virtually unlimited. The web is not only a data biosphere but also a social biosphere with access to an increasingly diverse set of online communication tools that are environments for globalized collaboration and community building. As professor of new media Geert Lovink eloquently states, "What defines the Internet is its social architecture. It's the living environment that counts, the live interaction, not just the storage and retrieval procedure."[20]

As the intricacy of online social network services and their inherent collaboration increases, so does the prospect of a human collective intelligence, a sphere of human thought, or "noosphere," as it was famously called by Russian mineralogist and geochemist Vladimir Vernadsky. In fact, the image of someone closing their laptop and turning off their online status is quickly becoming an illustration of the past, as the virtual web becomes more and more intertwined with real life. As it continues to expand and pervade every level of human activity, the web also diffuses many of its collaborative social models based on decentralization and democratization, and the effects of this conceptual permeation might create an everlasting shift in the stratification of society.

Classifying Information

The need to store and organize information has been with us since early Sumerian times, but it is from medieval Europe that we have the most vivid ancient accounts. In the twelfth century, the emergence of universities marked the decline of the great cloister libraries and the rise of a new form of knowledge archive: the university library. As these new centers of scholarly teaching and research became increasingly independent from religion, they expanded their curricula to include secular disciplines, propelling an exuberant intellectual activity and an intense eagerness for even more knowledge—in the form of books. This fervor was soon met with an increase in book production, aided by the invention of printing and cheaper paper. fig. 17

The insatiability for information led to challenges in indexing many of the newly discovered works. Professor Alfred W. Crosby explains in *The Measure of Reality* (1996) how medieval schoolmen "were at loss for a principle by which to arrange masses of information for easy retrieval."[21] Alternative categorization systems had to be devised, with many new indices featuring Arabic numerals, which had just started to disseminate across Europe. Strangely enough, one of the solutions to the dilemma, just before the use of alphabetization became standard in the twelfth century, was quite distinct from most modern objective taxonomies. Looking at the arrangement of information through a subjective lens, medieval schoolmen organized library catalogs by order of prestige, beginning with the Bible, followed by books on and by the church fathers (influential theologians, Christian teachers, and bishops), and so on, with books on liberal arts coming last.[22]

Many generations later, on the other side of the ocean, American librarian Melvil Dewey (1851–1931) devised one of the most relevant contributions to knowledge organization. Created in 1876, the Dewey Decimal Classification (DDC) method is a widely used library classification system based on decimal numbers. Having undergone twenty-two major revisions, the last one in 2004, its uncomplicated structure has been adopted by innumerous libraries, both in the United States and abroad.

As the founder of the American Metric Bureau—an organization established in Boston in 1876 with the goal of promoting the metric system in the United States—Dewey was a passionate supporter of the decimal system of measurement first adopted by France, which in part explains his preference for the inherent simplicity of powers of ten. The system he created organizes all knowledge into ten classes (e.g., Religion [100], Social Sciences [200], and Literature [800]). These classes are subdivided into ten divisions, and each division into ten sections. fig. 18 To locate a particular book on ecology, one would start on Science (500), then Life Sciences (570), and finally Ecology (577). In comparison, that same book on ecology would have a call number QH540-549.5 under DDC's main competitor, the Library of Congress Classification (LCC) system. The unclutteredness of the purely numerical method was an important reason for its widespread adoption, but even though the simplest of its kind, DDC is still based on an absolute tree arrangement of fixed positions, allowing for few, if any, interlinks. Other methods have challenged Dewey's structure, some using the notion of faceted classification, which is based on calling out clearly defined properties of a book (e.g., title, author, subject, year). But "both trees and faceted systems specify the categories, or facets, ahead of time," explains technologist, writer, and philosopher David Weinberger. "They both present users with tree-like structures for navigation, letting us climb down branches to get to the leaf we're looking for."[23] More recently a new method has emerged that promises to cut across the branches and reach forthwith for the leaves.

Initially used in online services such as Flickr, Del.icio.us, and YouTube, folksonomy is a method employed by thousands of websites, services, and applications. A term coined by information architect Thomas Vander Wal in July 2004, *folksonomy* is a portmanteau of *folk* and *taxonomy*, and is also known as social classification or social tagging. Folksonomy is an alternative system for categorizing content by means of informal tags—specific keywords assigned to a piece of information (e.g., a web page, video, image, computer file)—which describe the item and facilitate its retrieval during browsing or searching. This emergent bottom-up classification is intrinsically distinct from top-down hierarchies like the Dewey system. In DDC, each book or document has a unique reference in a single immutable hierarchical structure. In contrast, any digital object created by folksonomy is defined by different tags (metadata), allowing it to be ordered and located in multiple ways. It is also a highly adaptable method, since it ultimately relies on the natural language of the community or individual using it. fig. 19

In an interesting reference to this phenomenon, David Weinberger, who has been following this change up close, states: "Autumn has come to the forest of knowledge, thanks to the digital revolution. The leaves are falling and the trees are looking bare. We are discovering that traditional knowledge hierarchies that have served us so well are unnecessarily restricted when it comes to organizing information in the digital world."[24] According to Weinberger, while the old method conceived immutable

fig. 17

Jan Cornelius Woudanus, *Leiden University Library*, 1610

This renowned print depicts the organization of shelves at the library of the University of Leiden, the oldest university in the Netherlands, founded in 1575.

500	**Science**	**600**	**Technology**	**900**	**History and Geography**
570	Life Sciences	610	Medical Sciences	970	General history of North America
572	Biochemistry	613	Promotion of Health	973	General history of United States
572.8	Biochemical Genetics	613.2	Dietetics	973.9	Twentieth Century
572.86	DNA (Deoxyribonucleic acid)	613.24	Weight-gaining diet	973.91	1901-1953

fig. 18 (above)

The hierarchical structure of the Dewey Decimal Classification system, showing three individual paths within the Sciences (500), Technology (600), and History and Geography (900) categories

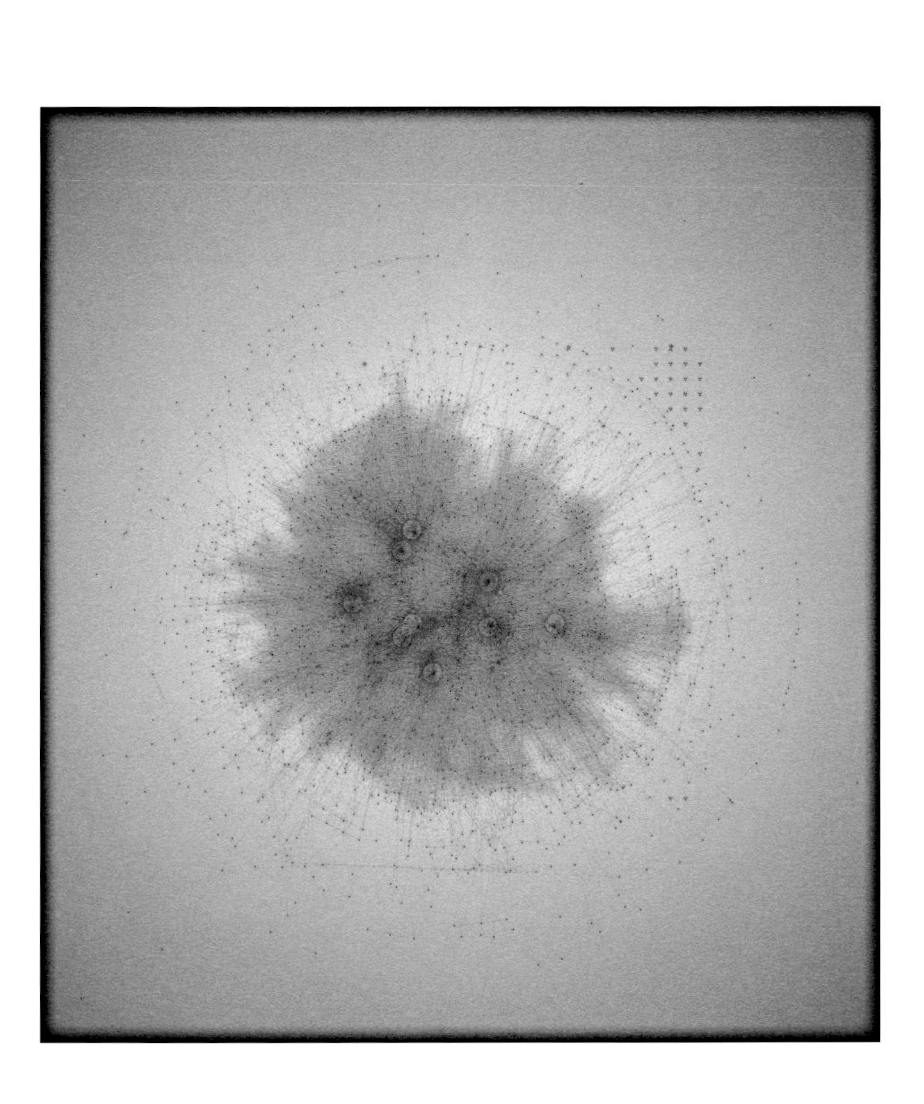

fig. 19 (left)

Kunal Anand, *Looks del.icio.us*, 2008

An intricate graph of tag relationships in an individual's Del.icio.us account, a pioneer social-bookmarking system and one of the precursors of folksonomy

trees, the new creates "piles of leaves" in an idiosyncratic, flexible way.

The adoption of folksonomy would certainly have been much less likely if not for the significant advances in modern computer science, particularly in search algorithms and data mining. In the past it would have been unthinkable to conceive a folksonomic model without an advanced computerized system that could facilitate the simultaneous search and retrieval of thousands of items. Recent progress has also placed folksonomy in the midst of a much larger technological movement—influenced by parallel advances in tracking and navigation satellite systems, microchip implants, and radio-frequency identifiers—that aims at converting a variety of objects (physical or digital) into intelligible, reachable, and indexable entities.

It is still too early to speculate on the role of folksonomy in the next few years or decades, and some might even say that it will never be as dominant as it promises. One of these hushing voices is renowned information architect Peter Morville, who considers the revolutionary rhetoric of the free-tagging movement behind folksonomy to be an overhyped exaggeration. Although acknowledging the benefits of folksonomy as a trendspotting and personal bookmarking tool, Morville says in his book *Ambient Findability* (2005) that when it comes to findability, folksonomies' "inability to handle equivalence, hierarchy, and other semantic relationships causes them to fail miserably at any significant scale.... If forced to choose between the old and new, I'll take the ancient tree of knowledge over the transient leaves of popularity any day."[25]

Whatever your view on the topic, one thing is indubitable: this innovative system has brought a new agility and malleability to the way we index and access information, possibly causing an important shift in the long run. Will folksonomy be a point of departure for profound structural change or a mere fad portrayed in history as a transitory alternative to the hierarchical tree model? Only time will tell.

Ordering Nature

Our long obsession for orderly arrangement can be best observed in the keystone of science: the classification of the natural world. As in the establishment of knowledge classification, we turn to Aristotle for the first proposal on the division of living things as a linear gradation principle. The most well-known student of Plato and the teacher of Alexander the Great, Aristotle was highly dedicated to the tangible world. His views on the physical sciences, published as a collection of treatises in his seminal *Physica* (Physics) (ca. 350 BC), have been a key influence in medieval erudition and an impetus for many radical views leading to the Scientific Revolution. Aristotle's conception of a "natural philosophy," dedicated to the uncovering of natural phenomena, was the precursor of the natural sciences and the dominant term referring to scientific inquiry before the emergence of modern science.

Both Plato and Aristotle believed in universalism (a school of thought that believes in universal facts or properties), but while Plato contemplated the universal apart from particular things—since the particular was merely a prototype or imitation of an ideal form—Aristotle saw the universal in particular things, in what he considered to be the essence of things. His belief in essentialism—the presence of an immutable essence in every object—led to the desire for an absolute ordering of nature, a perfect universe, where all species are hierarchically arranged in a natural ladder from lowest to highest. Even though some of the classes and divisions he set forth in *Historia Animalium* (History of animals), published circa 350 BC, quickly became obsolete, others, such as genus, species, and substance, have endured for a long time and still resonate in modern natural taxonomies.

Throughout the Middle Ages, many of Aristotle's ideas were revived, expanded, and sometimes misrepresented. *Scala naturae* (natural ladder), commonly referred to as the Great Chain of Being, was a medieval philosophical concept that viewed the world as a hierarchical tree of virtues, which assessed the most heavenly virtues of all matter and life on the planet. Different versions of this partisan system have been developed over the centuries, highlighting particular beliefs or ideologies of the time. Some placed God in the very pinnacle of the hierarchy, with angels immediately below, followed by members of the church, laymen, animals, and plants, until the very last foundational layer, earth itself. Other variations, based on the Divine Right of Kings—a doctrine of royal absolutism in which the monarch is subject to no earthly authority—positioned the king at the very top of the ladder, above the aristocratic lords, with the peasants below them. Applicable in limitless circumstances, scala naturae was always portrayed as an immutable linear gradation of perfection, or heavenly virtue.

The Renaissance gave rise to serious efforts to categorize a growing number of unidentified species, particularly through the work of naturalists and physicians, including Conrad von Gesner, Andrea Cesalpino, Robert Hooke, John Ray, Augustus Rivinus, and Joseph Pitton de Tournefort. But one of the greatest contributions to biological classification and nomenclature came a bit later, by Swedish physician and zoologist Carl Linnaeus (1707–1778). The Linnaean taxonomy, as it was later called, was set forth in his *Systema naturae* (1735) and organized nature into three main kingdoms: *Regnum animale* (animal), *Regnum vegetabile* (vegetable), and *Regnum lapideum* (mineral). Based on a nested hierarchy of successive categories, or ranks, kingdoms were divided into classes, and classes were divided into orders, and so on, until the very last rank. Organisms were essentially grouped by their shared physical traits, and many were kept in similar arrangements until modern times. fig. 20

With the recognition of Darwin's theory of evolution as the central principle for species formation, Linnaeus's classification somewhat fell out of favor. This is where phylogenetics came in. Seen as "the history of the descendants of living beings," or the study of evolutionary association across different groups of organisms, phylogenetics is a contemporary by-product of Darwin's drastic biological revolution, which greatly informed the naming and classification of species.[26] Central to its study is the Phylogenetic tree—a hierarchical representation of evolutionary relationships between ranges of biological species that share a common ancestor. Darwin himself suggested such an evolutionary tree, in a rough sketch back in 1837. fig. 21 Years later his tree concept would be materialized in his seminal *On the Origin of Species by Means of Natural Selection* (1859). The only illustration in this masterpiece appears in the fourth chapter, "Natural Selection," in what he denominated the *Tree of Life*. Further in the chapter's summary, Darwin expands on the tree metaphor to better explain his views on evolution:

fig. 20

The taxonomic ranks of the fruit fly, the modern human, and the pea, in a hierarchical tree according to Linnaean taxonomy

	Fruit fly	Human	Pea
Domain	Eukarya	Eukarya	Eukarya
Kingdom	Animalia	Animalia	Plantae
Phylum	Arthropoda	Chordata	Magnoliophyta
Subphylum	Hexapoda	Vertebrata	Magnoliophytina
Class	Insecta	Mammalia	Magnoliopsida
Subclass	Pterygota	Theria	Rosidae
Order	Diptera	Primates	Fabales
Suborder	Brachycera	Haplorrhini	Fabineae
Family	Drosophilidae	Hominidae	Fabaceae
Subfamily	Drosophilinae	Homininae	Faboideae
Genus	Drosophila	Homo	Pisum
Species	D. melanogaster	H. sapiens	P. sativum

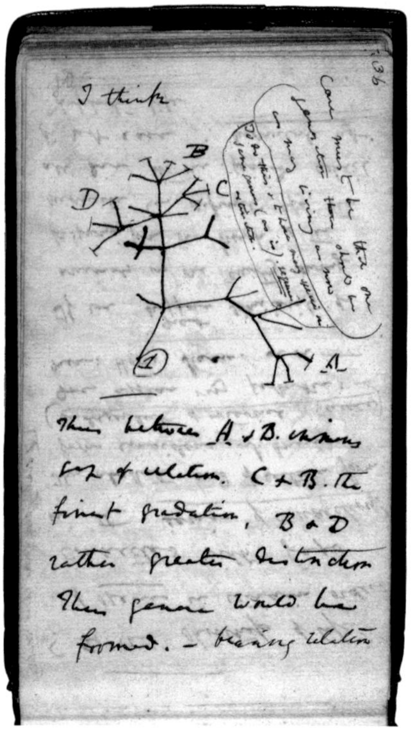

fig. 21

A diagram of an evolutionary tree, from Charles Darwin, *First Notebook on Transmutation of Species*, 1837

This simple sketch alludes to the branching of lineages, similar to the splitting of a tree's trunk, and marks the first-known representation of an evolutionary tree. Above the sketch, Darwin wrote, "I think."

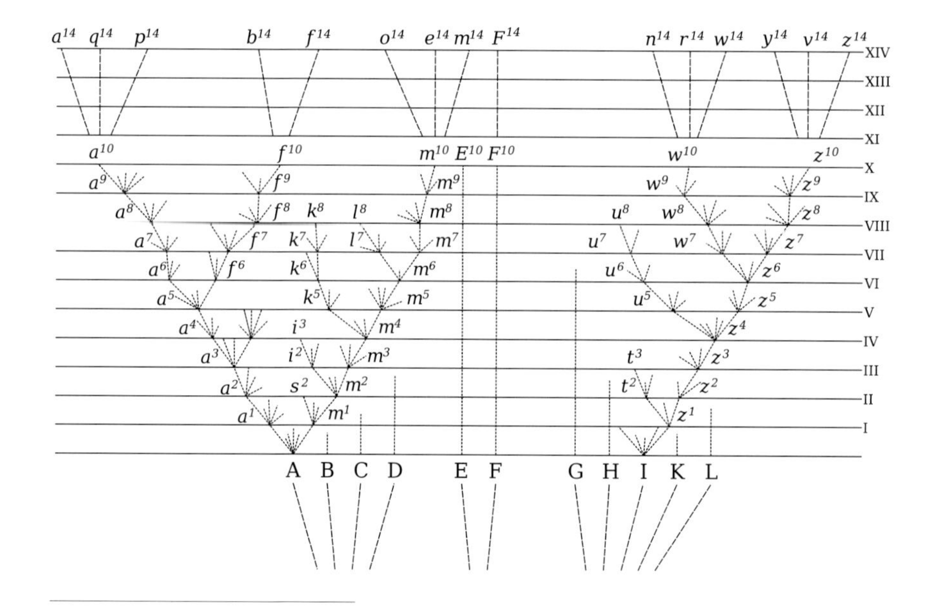

fig. 22

Tree of Life, from Charles Darwin, *The Origin of Species*, 1859

This illustration, the only one featured in the first edition of Darwin's masterpiece, popularized the notion of the evolutionary tree and later catalyzed the field of phylogenetics.

> The affinities of all the beings of the same class have sometimes been represented by a great tree. I believe this simile largely speaks the truth. The green and budding twigs may represent existing species; and those produced during former years may represent the long succession of extinct species. At each period of growth all the growing twigs have tried to branch out on all sides, and to overtop and kill the surrounding twigs and branches, in the same manner as species and groups of species have at all times overmastered other species in the great battle for life.[27]

Darwin was well aware of the importance of the tree schema and considered it at the core of his thinking. In a letter to his publisher John Murray, sent on May 31, 1859, a few months before the publication of his scientific landmark, Darwin writes: "Enclosed is the Diagram which I wish engraved on Copper on *folding* out Plate to face latter part of volume.—It is an odd looking affair, but is *indispensable* to show the nature of the very complex affinities of past & present animals."[28] This indispensable illustration is, according to Darwin, not a secondary element to his narrative but a crucial symbol of his idea. fig. 22

The notion of a tree as a classification system was nothing new and neither was its applicability to the organization of species. We had seen it before in the Great Chain of Being, and much of the work developed by Aristotle, Leibniz, and Linnaeus. fig. 23 However, Darwin introduced a critical unsettling element to the equation: time. Darwin's evolutionary tree was no longer a static immutable image of the present but a shifting dynamic model, encompassing years and years of change and adaptation. This genealogical variable, however, brought a big dilemma to biological sciences, as Sigrid Weigel points out in her brilliant "Genealogy: On the Iconography and Rhetorics of an Epistemological Topos." Biologists had to not only strive for the constancy required in a classification system agreed by all but also embrace variation and modification as part of evolutionary change. This was understandably a cumbersome endeavor.

The continuous challenge of combining uniformity and variety in an integrated model has recently suffered a major quiver, with scientists having to rethink their approach to phylogenetics. The first significant change is technical and relates to new methods of analysis. While most early evolutionary trees were based on morphological attributes (appearance and physical traits), the more recent trees are constructed using molecular data (based on genetic and molecular sequencing). By providing an alternative way of looking at the shared relationships between species, this approach has introduced uncertainty to established labeling conventions. The second, and the most current, upheaval runs much deeper and has caused a major shift in the way we conceive the classification of biodiversity.

In June 2005 a group of researchers from the Computational Genomics Group at the European Bioinformatics Institute (EBI) published an influential paper entitled "The Net of Life: Reconstructing the Microbial Phylogenetic Network." In this study, the EBI team, lead by Christos Ouzounis, set up a new vision for evolution classification based on networks, rather than trees, in which "the genomic history of most microbial species is a mosaic, with a significant amount of horizontal gene transfer present."[29]

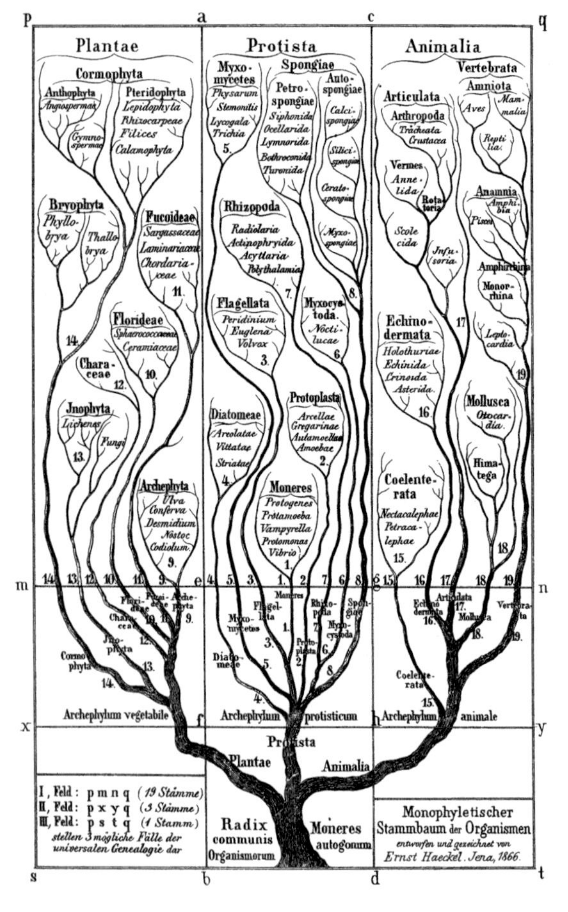

fig. 23

A diagram of the tree of life. From Ernst Haeckel, *Generelle morphologie der organismen* (General morphology of organisms), 1866.

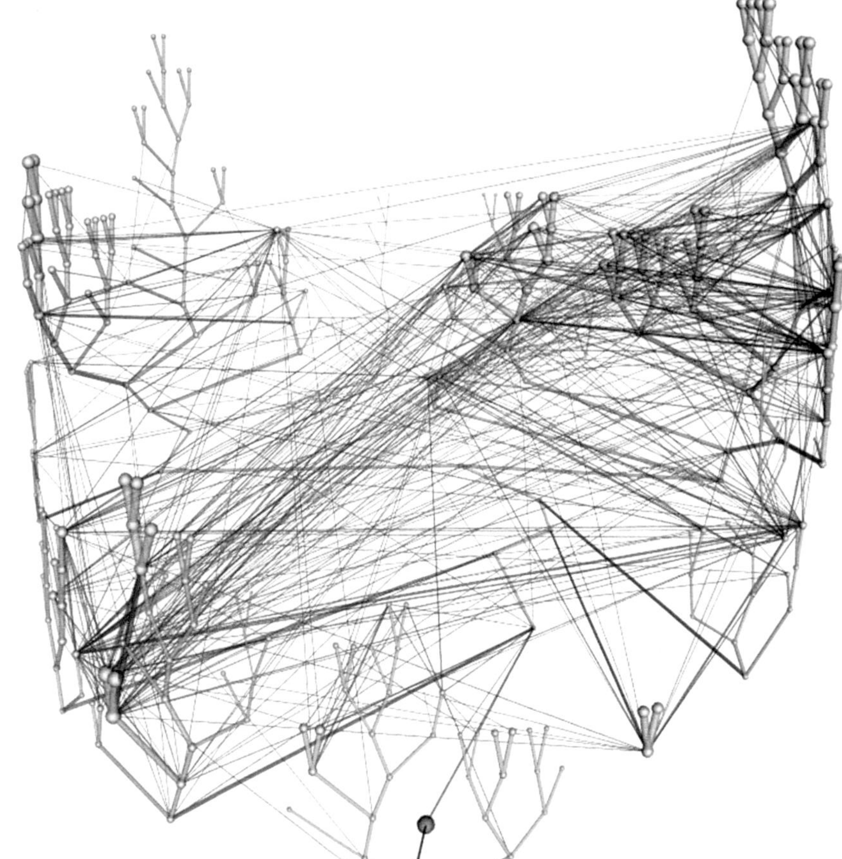

fig. 24

A three-dimensional representation of the net of life—an alternative version to the common tree of life. Red lines, depicting horizontal gene transfer, tie individual bacteria and archaea, which all originate from a common root depicted in orange. From V. Kunin, L. Goldovsky, N. Darzentas, and C. A. Ouzounis, "The Net of Life: Reconstructing the Microbial Phylogenetic Network," *Genome Research* 15, no. 7 (July 2005): 954–59.

Horizontal gene transfer (HGT) is a recurrent process in nature and occurs when a living being incorporates genetic material from a different organism without being its offspring. As a prevailing form of genetic transfer in single-celled organisms, such as bacteria, HGT is the subject of much debate and study. Since roughly 90 percent of the cells in the human body are nonhuman organisms—essentially bacteria—the impact of these studies in future assessments of evolutionary processes is monumental.

Phylogenetics, therefore, is in the midst of a reconstruction phase in which there is a vertical disintegration of the tree of life. According to biologist Johann Peter Gogarten, "the original metaphor of a tree no longer fits the data from recent genome research."[30] He also suggests that biologists "use the metaphor of a mosaic to describe the different histories combined in individual genomes and use the metaphor of a net to visualize the rich exchange and cooperative effects of HGT among microbes."[31]

In the study conducted by EBI, an arresting image was produced showing a hybrid between the original tree of life and the new HGT mesh. In this model the various horizontal links, or "vines," cross the tree in a form of rhizomatic contamination. fig. 24 Based on these remarkable advancements, an international set of rules for phylogenetic nomenclature, called the PhyloCode, is currently being devised by the International Society for Phylogenetic Nomenclature. This considerable effort might well be the basis for a new shift in species classification, soon replacing the old tree metaphor with a novel network representation: the net of life.

Network Thinking

In the various cases explored throughout this chapter, we saw how previous conceptions based on hierarchical and centralized tree organizations are giving way to new ideas that are able to address the inherent complexities of modern society. Cities, the brain, the World Wide Web, social groups, knowledge classification, and the genetic association between species all refer to complex systems defined by a large number of interconnected elements, normally taking the shape of a network. This ubiquitous topology, prevalent in a wide range of domains, is at the forefront of a new scientific awareness of complexity, epitomizing the third stage of science described by Weaver. Networks are not just an omnipresent structure but also a symbol of autonomy, flexibility, collaboration, diversity, and multiplicity. As nonhierarchical models, networks are embedded with processes of democratization that stimulate individuality and our appetites for learning, evolving, and communicating. They are, in essence, the fabric of life.

However, even though a significant transition from trees to networks has occurred in a variety of fields, the two models are not necessarily conflicting. "There are knots of arborescence in rhizomes, and rhizomatic offshoots in roots," proclaim Deleuze and Guattari with respect to this occasional overlap.[32] In some of the aforementioned cases, network thinking denotes an alternative and possibly complementary view of the analyzed system; in others, it embodies a drastic departure from the existing modus operandi.

In order to tackle problems of an increasingly complex and interconnected nature, we need to consider new methods of analysis, modeling, and simulation. More importantly, we need to consider an alternative way of

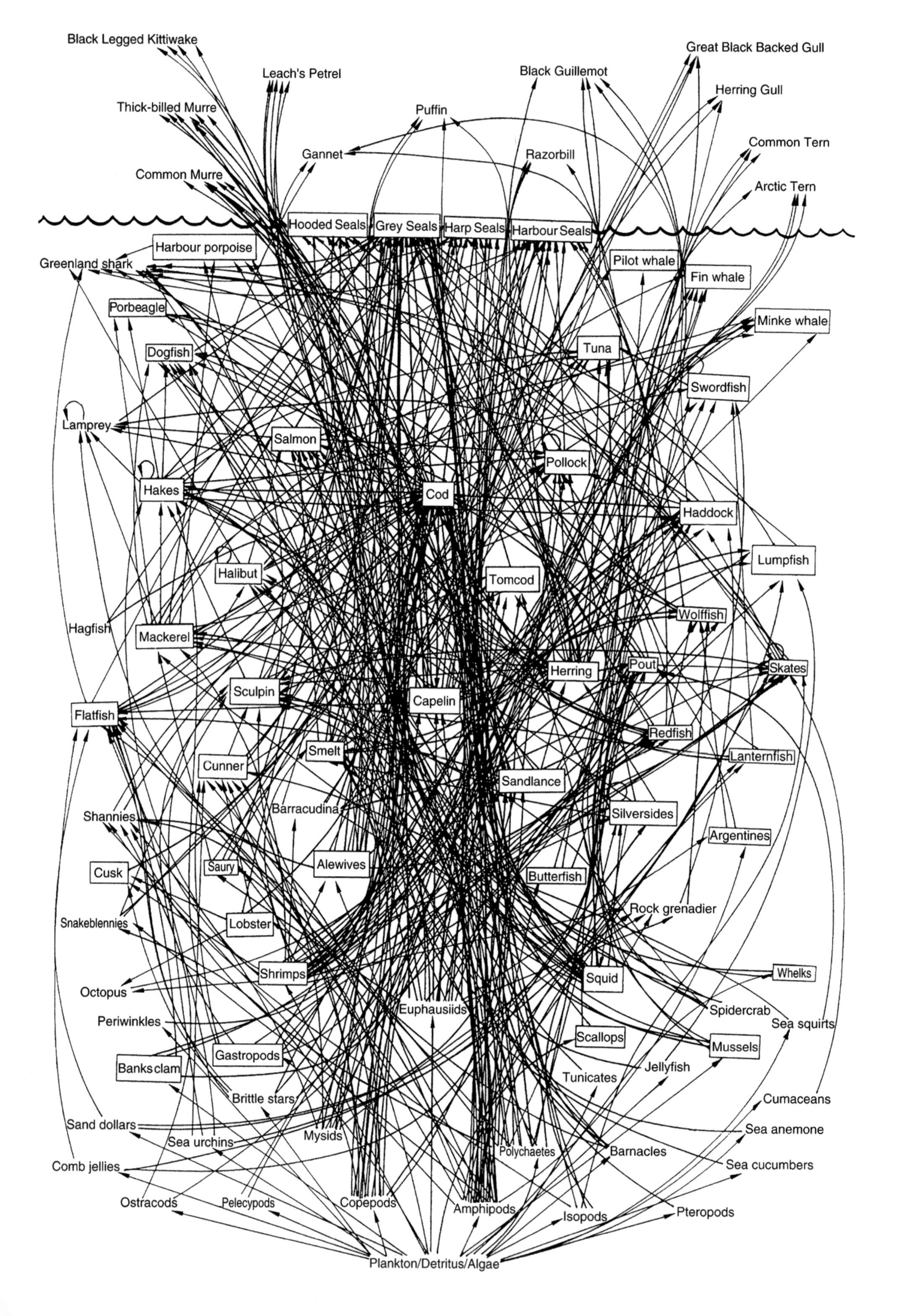

Black Legged Kittiwake
Leach's Petrel
Thick-billed Murre
Common Murre
Gannet
Puffin
Black Guillemot
Razorbill
Great Black Backed Gull
Herring Gull
Common Tern
Arctic Tern
Hooded Seals
Grey Seals
Harp Seals
Harbour Seals
Harbour porpoise
Greenland shark
Pilot whale
Fin whale
Minke whale
Porbeagle
Dogfish
Tuna
Swordfish
Lamprey
Salmon
Pollock
Hakes
Cod
Haddock
Lumpfish
Halibut
Tomcod
Hagfish
Mackerel
Wolffish
Herring
Pout
Skates
Sculpin
Capelin
Flatfish
Redfish
Smelt
Lanternfish
Cunner
Sandlance
Barracudina
Silversides
Shannies
Argentines
Cusk
Saury
Alewives
Butterfish
Rock grenadier
Snakeblennies
Lobster
Shrimps
Squid
Whelks
Octopus
Euphausiids
Spidercrab
Periwinkles
Sea squirts
Scallops
Gastropods
Mussels
Banksclam
Jellyfish
Tunicates
Brittle stars
Cumaceans
Sand dollars
Sea anemone
Mysids
Sea urchins
Polychaetes
Barnacles
Comb jellies
Sea cucumbers
Ostracods
Pelecypods
Copepods
Amphipods
Isopods
Pteropods
Plankton/Detritus/Algae

thinking. We act and live in networks, so it makes sense that we start thinking in networks. fig. 25 By truly embracing network thinking we can not only dissect a variety of interdependent natural systems, including our own brain, but also apply the same knowledge in the development of future endeavors. In reference to the challenge of social networks, leading social scientist Jacob Moreno stated back in 1933 that "until we have at least determined the nature of these fundamental structures which form the networks, we are working blindly in a hit-or-miss effort to solve problems."[33] The global effort of constructing a general theory of complexity is tremendous and may lead to major improvements in health, stability, and security of most systems around us. As physicist and complex-network expert Albert-László Barabási declares in *Linked* (2003): "Once we stumble across the right vision of complexity, it will take little to bring it to fruition. When that will happen is one of the mysteries that keeps many of us going."[34]

Notes

1 Lecointre and Le Guyader, *The Tree of Life*, 17.
2 Ibid., 21.
3 Ibid.
4 Ibid.
5 Alexander, "A City is Not a Tree."
6 Ibid.
7 Ibid.
8 Ibid.
9 Jacobs, *The Death and Life of Great American Cities*, 3.
10 Ibid., 433.
11 Greenfield, *The Human Brain*, 53.
12 Lehrer, "Out of the Blue."
13 Fildes, "Artificial brain '10 years away.'"
14 Lehrer, "Out of the Blue."
15 Ibid.
16 Artz and Kamalipour, *The Globalization of Corporate Media Hegemony*, 118.
17 Berners-Lee, "Tim Berners-Lee."
18 Brafman, *The Starfish and the Spider*, 18.
19 Tapscott and Williams, *Wikinomics*, 1.
20 Lovink, *The Principle of Notworking*, 11.
21 Crosby, *The Measure of Reality*, 63.
22 Ibid.
23 Ibid.
24 Ibid.
25 Morville, *Ambient Findability*, 139.
26 Lecointre and Le Guyader, *The Tree of Life*, 5.
27 Darwin, *The Origin of Species*, 171.
28 Darwin Correspondence Project Database.
29 Ibid.
30 Gogarten, "Horizontal Gene Transfer."
31 Ibid.
32 Deleuze and Guattari, *A Thousand Plateaus*, 20.
33 *New York Times*, "Emotions Mapped by New Geography."
34 Barabási, *Linked*, 238.

fig. 25

A partial food web, depicting predator-prey relationships between species at the Scotian Shelf in the Northwest Atlantic off of the east coast of Canada. Species names enclosed in rectangles are of those exploited by humans, with cod at the heart of the convoluted network. Despite its remarkable intricacy, this food web is incomplete, because the feeding habits of all participants have not been fully described. Furthermore, not all species—including some of the marine mammals—spend the entire year in the area.

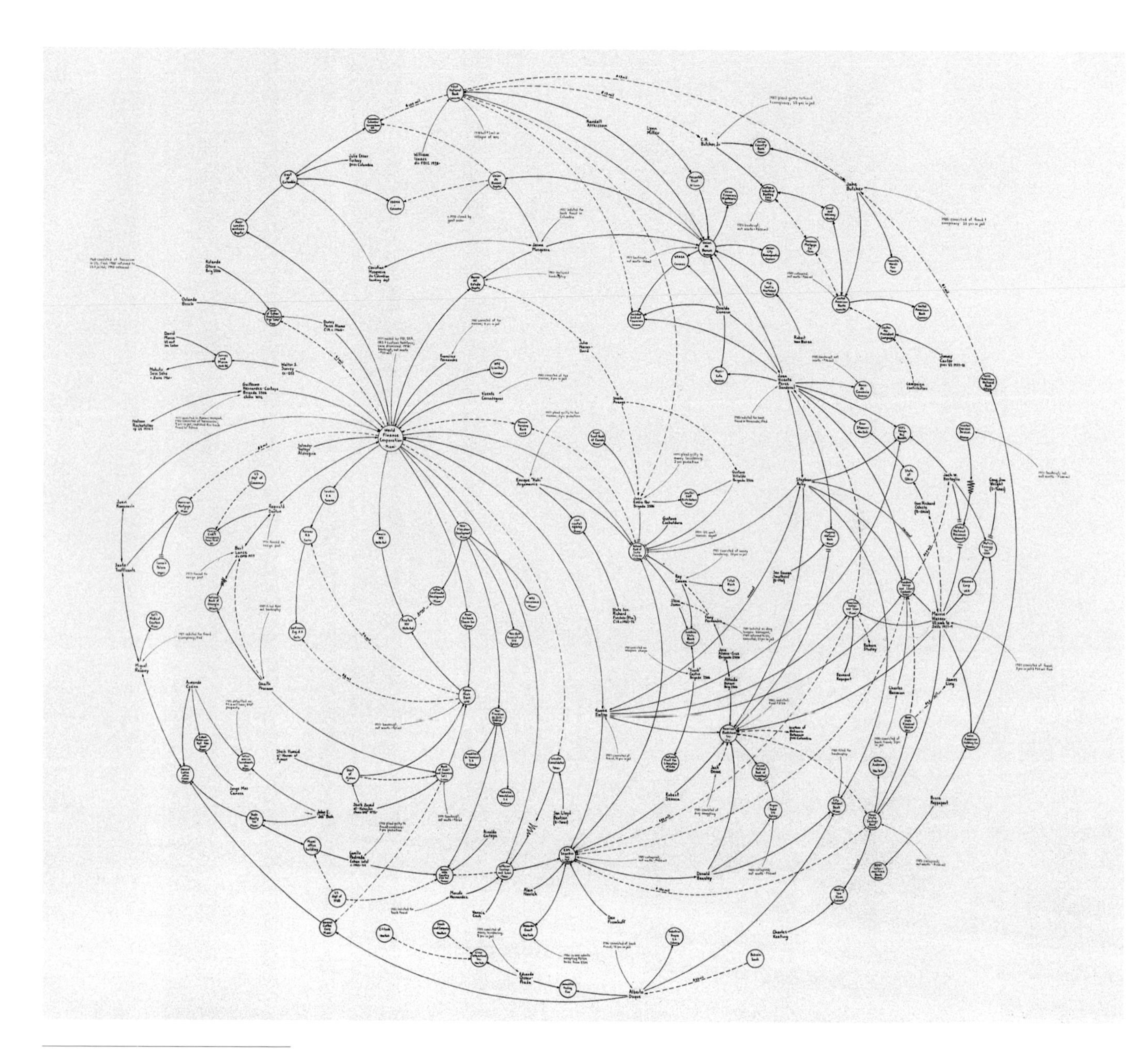

Mark Lombardi, *World Finance Corporation and Associates, ca. 1970–84: Miami, Ajman, and Bogota-Caracas (Brigada 2506: Cuban Anti-Castro Bay of Pigs Veteran) (7th version)*, 1999

03 | Decoding Networks

If we ever get to the point of charting a whole city or a whole nation, we would have an intricate maze of psychological reactions which would present a picture of a vast solar system of intangible structures, powerfully influencing conduct, as gravitation does bodies in space.

—Jacob Moreno

The graphic is no longer only the "representation" of a final simplification, it is a point of departure for the discovery of these simplifications and the means for their justification. The graphic has become, by its manageability, an instrument for information processing.

—Jacques Bertin

Networks are everywhere. It is a structural and organizational model that pervades almost every subject, from genes to power systems, from social communities to transportation routes. This ubiquitous topology is the object of study in network science, a new thriving discipline aiming to uncover the inherent principles and behaviors that regulate a variety of natural and artificial systems, normally characterized by a multitude of interconnecting elements. As an important driving force for understanding the complex connectedness of modern society, network science has innumerous applications in fields such as physics, economics, biology, computer science, sociology, ecology, and epidemiology. Although the discipline's considerable expansion occurred only fairly recently, its roots go back to the first half of the eighteenth century.

The two epigraphs to this chapter are drawn from *New York Times*, "Emotions Mapped by New Geography"; and Bertin, *Semiology of Graphics*, 4.

The Birth of Network Science

Although human beings have previously envisioned models of networklike structures, the first documented mathematical analysis of the process occurred in 1736 by Leonhard Euler (1707–1783). Euler was a prolific mathematician and a key contributor to the fields of calculus, optics, fluid dynamics, astronomy, and geometry. But it was in the explanation of a witty mathematical problem that Euler became forever associated with network science.

Founded in 1255, the city of Königsberg—in modern-day Kaliningrad, Russia—sat on the banks of the Pregel River. Within the river were two large islands, which were connected to each other and the adjacent riverbanks by seven bridges. fig. 1 A popular pastime of Königsberg's citizens in the eighteenth century was to find a route where one could cross all seven bridges without crossing the same one twice. Euler was amused by this dilemma and was determined to solve it:

> Concerning these bridges, it was asked whether anyone could arrange a route in such a way that he would cross each bridge once and only once. I was told that some people asserted that this was impossible, while others were in doubt; but nobody would actually assert that it could be done. From this, I have formulated the general problem: whatever be the arrangement and division of the river into branches, and however many bridges there be, can one find out whether or not it is possible to cross each bridge exactly once?[1]

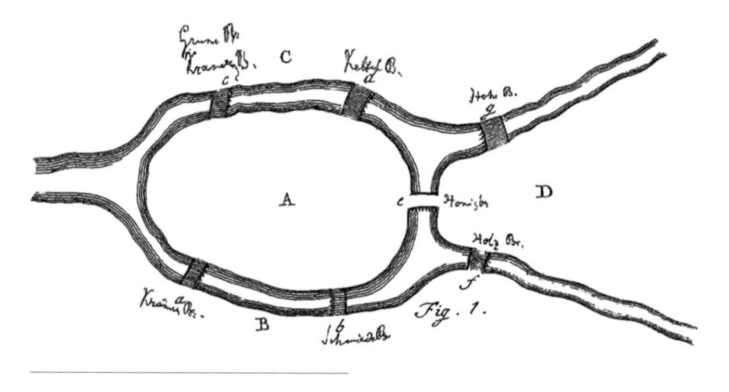

fig. 1

The original sketch of the seven bridges of Königsberg, from Leonhard Euler, *Solutio problematis ad geometriam situs pertinentis* (Solution of a problem relating to the geometry of position), 1736

Euler's answer to the problem was shown in a paper published in 1736 entitled *Solutio problematis ad geometriam situs pertinentis* (Solution of a problem relating to the geometry of position), in which he rigorously proves by means of a pioneering analytical method that such a path does not exist. In an extended English translation of the paper in *Graph Theory 1736–1936* (1976), by Norman L. Biggs, E. Keith Lloyd, and Robin J. Wilson, Euler starts his argument by shunning the conventional procedure of "making an exhaustive list of all possible routes, and then finding whether or not any route satisfies the conditions of the problem."[2] As he explains, due to the large number of possible paths, this solution would be too exhausting and probably impossible to execute in a scenario with more than seven bridges. Euler rejects this approach and suggests a new, groundbreaking one:

> My whole method relies on the particularly convenient way in which the crossing of a bridge can be represented. For this I use the capital letters A, B, C, D, for each of the land areas separated by the river. If a traveler goes from A to B over bridge a or b, I write this as AB—where the first letter refers to the area the traveler is leaving, and the second refers to the area he arrives at after crossing the bridge.

Thus, if the traveler leaves B and crosses into D over bridge f, this crossing is represented by BD, and the two crossings AB and BD combined I shall denote by the three letters ABD, where the middle letter B refers to both the area which is entered in the first crossing and to the one which is left in the second crossing.[3]

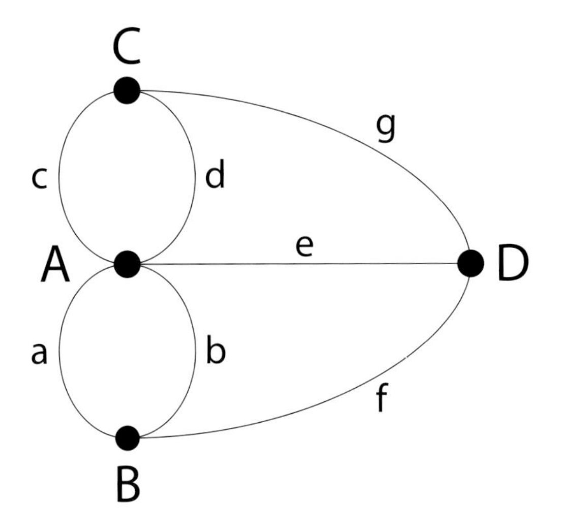

Euler essentially reformulated the problem in abstract terms, isolating the seven bridges as a series of edges (links) connecting the different landmasses represented by vertices (nodes). fig. 2 Even though Euler did not use any of these modern terms, he did however deconstruct the problem in such a way that suggests a type of simplified scheme, commonly called a graph in mathematics. Euler's ability to look at the problem from a topological perspective—by conceiving the bridges challenge as a graph—laid the foundation for graph theory (the study of graphs in mathematics and computer science) and, consequently, network science.

In the evolution of graph theory, there has been a great deal of work developed after Euler's foundational approach, particularly by mathematical giants such as Augustin Louis Cauchy, William Rowan Hamilton, Arthur Cayley, Gustav Robert Kirchhoff, and George Pólya.[4] But most of these efforts remained within the confines of mathematics. Strangely enough, even the burst of originality in mid-nineteenth-century cartography—with the mapping of statistical data, originally pioneered by William Playfair's seminal work *The Commercial and Political Atlas* (1786)—remained immune to the appeal of the network diagram. Network science would have to wait a couple of centuries for another pioneer in network representation to provide a significant breakthrough.

Psychological Geography

Psychologist Jacob Moreno was born in 1889 in the city of Bucharest, Romania, and spent most of his youth and early career in Vienna, Austria, where he graduated with a medical degree in 1917. While studying at the University of Vienna, Moreno attended Sigmund Freud's (1856–1939) lectures and became an early challenger of his theories. While Freud favored meeting people individually in the artificial setting of his office, Moreno believed in the power of group settings for therapy, which could only be accurately conducted in their natural environment—the street, the park, the community. The latter also opposed the emphasis on the unconscious mind: "Moreno was more interested in the conscious process, the here and now, the creativity of the person, than the unconscious process, the past and the resistance of the 'patient,'" writes Moreno's biographer René Marineau.[5]

In order to further pursue his ongoing research on his theory of interpersonal relations, Moreno moved to New

fig. 2

Graph theory was born thanks to Leonhard Euler's deconstruction of the mathematical problem of the seven bridges in an abstract graph. In it, the land areas are represented by four nodes connected by seven links, which correspond to the bridges.

York City at the age of thirty-six. The following years saw Moreno become increasingly motivated by the prospect of visually representing social structures, and seven years after his arrival in the United States, at a convention of medical scholars, Moreno presented one of his most famous creations: the sociogram. Moreno's sociogram introduced a graphic representation of social ties between a group of boys and girls from one elementary school, marking the beginning of sociometry, which later came to be known as social network analysis—a field of sociology dealing with the mapping and measuring of relationships between people (e.g., kinship, friendship, common interests, financial exchange, sexual relationships). The idea of a measurable sociogram became a decisive turning point in the quantitative evaluation of an individual's role in a community, but it also demonstrated, for the very first time, the enticing power of network visualization.

Moreno's network depiction was so captivating that it was printed in a 1933 article, "Emotions Mapped by New Geography: Charts Seek to Portray the Psychological Currents of Human Relationships," in the *New York Times*. The novel practice merited the label "psychological geography" by the journalist, who was impressed with the diagram presented by Moreno: "A mere glance at the chart shows the strange human currents that flow in all directions from each individual in the group toward other individuals, from group to group, and from the entire group toward the individuals. Each group has its popular and unpopular members, and here and there an individual stands totally alone, isolated from the rest of the group."[6] The article was illustrated with one of Moreno's sociograms, which showed two independent groups, boys and girls, and links within each group and between the groups. fig. 3 "With these charts," explained Moreno, "we will have the opportunity to grasp the myriad networks of human relations and at the same time view any part or portion of the whole which we may desire to relate or distinguish."[7]

A year later Moreno expanded many of his initial ideas in what came to be known as the paramount work on sociometry, *Who Shall Survive? A New Approach to The Problem of Human Interrelations* (1934). The work contains some of the earliest graphical depictions of social networks and exposes Moreno's appreciation for the power of visualization. fig. 4 In his discourse, Moreno explains that the sociogram is not simply a method of presentation but a "method of exploration.... It is at present the only available scheme, which makes structural analysis of a community possible."[8]

Canadian psychologist Mary Northway dedicated almost three decades of research to the topic of sociometry, while working as an associate professor at the University of Toronto. Her book *A Primer of Sociometry* (1953) elaborates on Moreno's foundational work, advancing a sociometric test consisting of asking each person in a group whom they would choose to associate with for a particular activity or event. With the results of the inquiry, each person could be attributed a specific sociometric score based on their choices, depending on how many people they selected and how many people selected them.[9] These results would then be plotted in a sociogram—now called a social network diagram.

A few years before *A Primer of Sociometry*, Northway had already introduced a new visualization technique, called the target sociogram. fig. 5 fig. 6 This method is composed of four concentric circles, corresponding to different scoring outcomes in the aforementioned

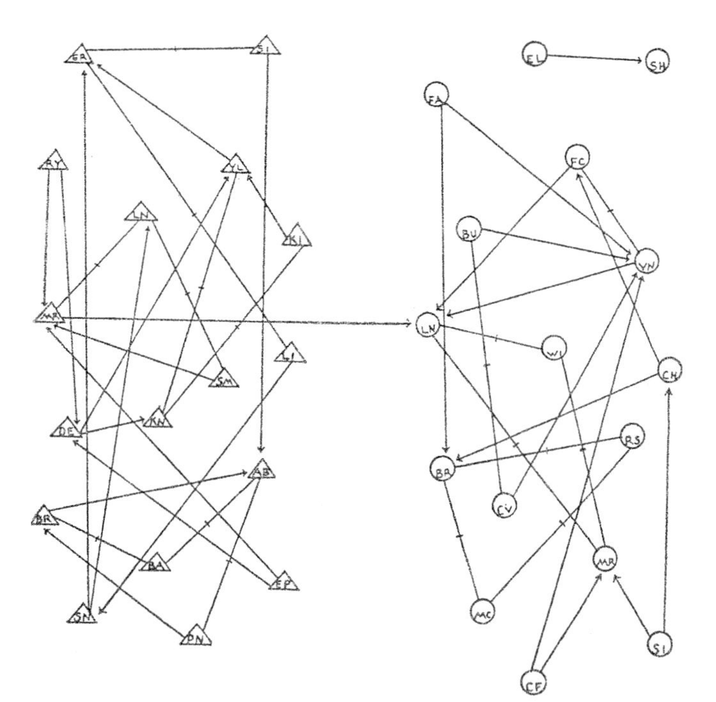

fig. 3

Jacob Moreno, one of the first sociograms, published in the *New York Times*, April 3, 1933, showing the relationships within a class of fourth graders. Boys (triangles) are on the left and girls (circles) on the right.

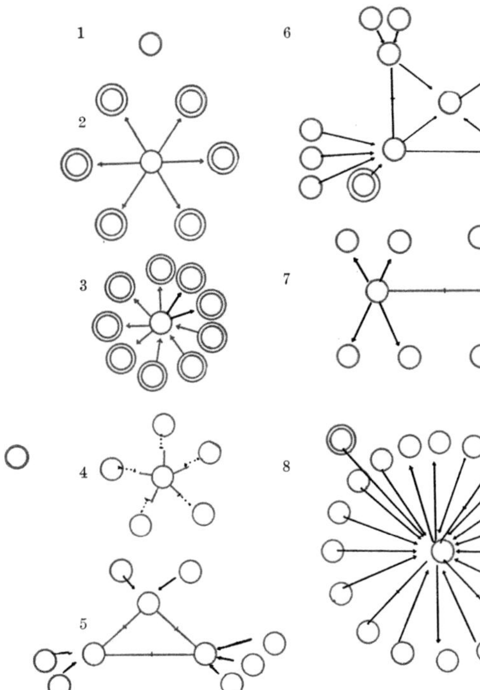

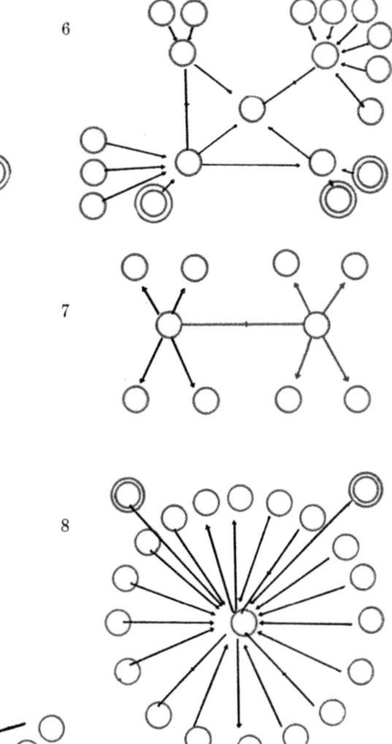

fig. 4

Typical Structures within Groups, from Moreno, *Who Shall Survive?*, 1934

This visual lexicon, from Moreno's seminal book, shows a set of possible sociogram constructs and scenarios, ranging from total isolation (1), to an individual being attracted to six others outside of her group (2), to a more complex structure defined by an individual rejecting six and also being rejected by fifteen individuals within and outside of her own group (8).

fig. 5

Target sociogram of a nursery school, from Mary Northway, *A Primer of Sociometry*, 1953

The target sociogram shows four concentric circles and four quadrants relating to four groups: senior boys, senior girls, junior boys, and junior girls.

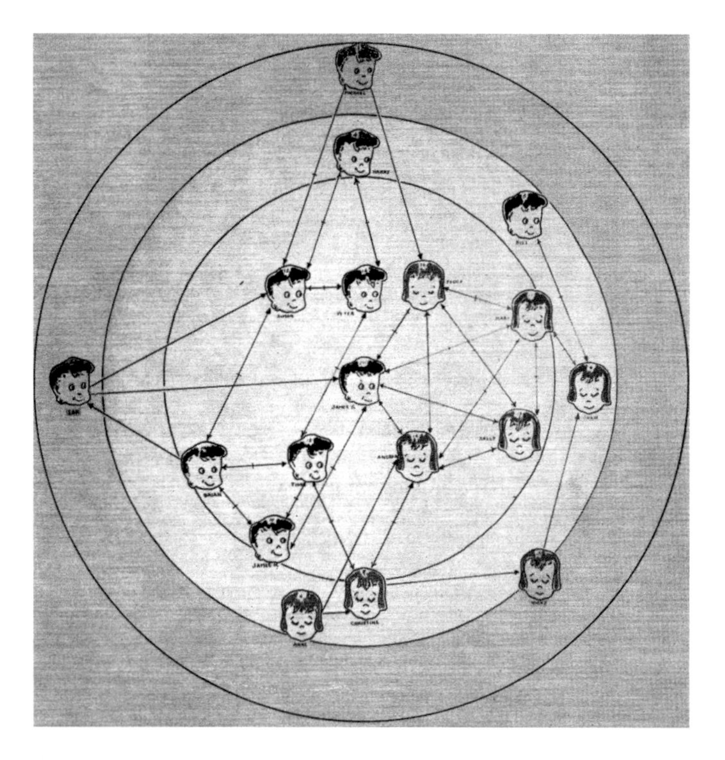

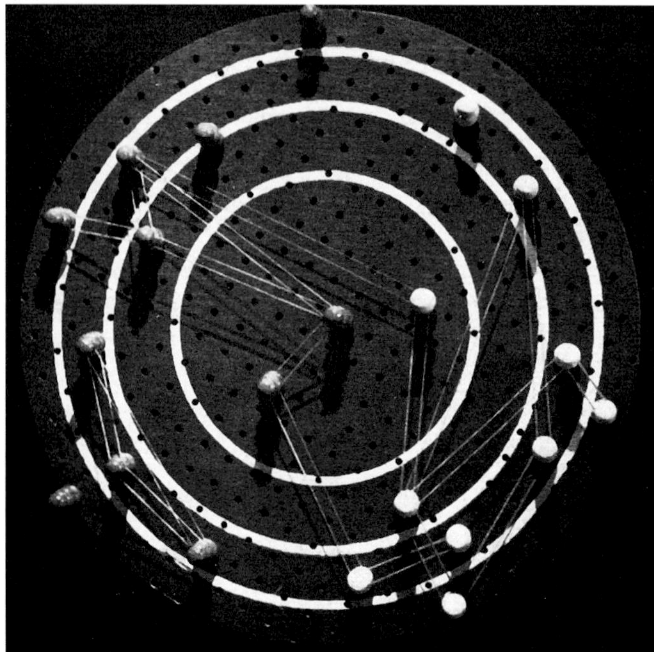

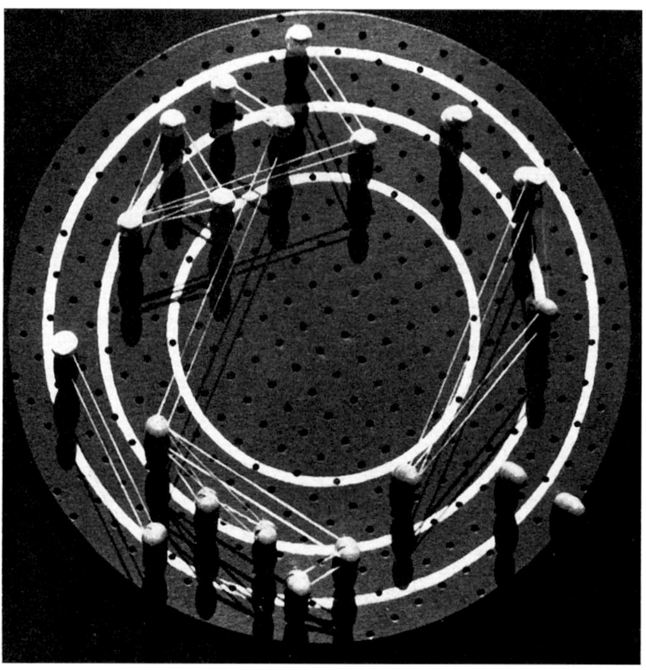

fig. 6 (top)

Target sociogram of a first-grade class, from Northway, *A Primer of Sociometry.* An original interpretation of a target sociogram by one of Northway's students, based on her teachings. Faces replaced the conventional triangles and circles, and the arrows between faces draw attention to the center, where key people are located.

fig. 7 (bottom row)

Target sociogram board, from Northway, *A Primer of Sociometry.* This apparatus—a set of movable physical pins connected by elastic bands—facilitated the quick prototyping of sociograms.

sociometric test: "Each circle may be used to represent the four quartiles or the four levels of probability, significantly above chance, above chance, below chance, and significantly below chance."[10] This way, the mere placement of nodes on a target sociogram—small triangles for males and small circles for females—could easily portray the likelihood of being selected among the different members of a community while still maintaining a high level of graphic clarity. Inspired by these developments, Dorothy McKenzie, the supervisor of the nursery school at the Institute of Child Study, a laboratory school and research institute at the University of Toronto where Northway taught, designed and constructed a target sociogram board. fig. 7 Modeled on a child's pegboard, this physical device allowed the easy mock-up of a target sociogram by means of movable pegs (standing for nodes) and rubber bands (linkages), in what can be considered a forerunner of modern-day computerized, interactive depictions.

The Cartography of Networks

Since these developments pioneered by Moreno and Northway, many other researchers have dedicated their time and energy to the depiction of network diagrams, increasingly through the use of computer software algorithms. Today network representation is commonly pursued under two main areas: graph drawing (under graph theory) and network visualization (under information visualization). In both disciplines *graph* is the preferred term to describe the pictorial depiction of a network through a set of vertices (nodes) connected by edges (links). But while graph drawing, as the name implies, deals primarily with the mathematical drawing of graphs, network visualization extends beyond the mere geometric construct, employing elementary design principles aimed at an efficient and comprehensible representation of the targeted system.

Networks have multiple interpretations and definitions, usually depending on the particular discipline responsible for studying the network. There are also numerous insights that can be extracted from these structures: What are the nodes doing? How are they interacting? How many connections do they have? What are they sharing? This series of queries can lead to the identification of a taxonomy, or topological truth, of the analyzed network. In this pursuit, network visualization can be a remarkable discovery tool, able to translate structural complexity into perceptible visual insights aimed at a clearer understanding. It is through its pictorial representation and interactive analysis that modern network visualization gives life to many structures hidden from human perception, providing us with an original "map" of the territory. Even though social networks (relationships of friendship, kinship, collaboration, common interest) have the longest history of quantitative study and analysis, network visualization explores numerous phenomena, particularly in technological networks (the World Wide Web, train systems, air routes, power grids), knowledge networks (classification systems, information exchange, semantic relationships between concepts), and biological networks (protein-interaction networks, genetic-regulatory networks, neural networks).

A highly influential tradition for network visualization, besides the intellectual legacy of graph theory and the recent advancement of computer graphics, is cartography. From the outstanding contribution of Ptolemy's *Geographia* (Geography) (ca. 150 AD), almost two millennia ago, and the notable mapmakers of the Age of Exploration—which took place during the fifteenth, sixteenth, and seventeenth centuries—to the explosion of statistical mapping, or "thematic mapping," in the mid-nineteenth century, the ancient heritage of cartography provides a rich setting for the present development of network visualization. The bond between both areas may even be strengthened when historians examine the current efforts many decades from now. After all, this burst of innovation, as network visualization embraces a multitude of attempts at decoding complex systems, resembles a new golden age of cartography led by invigorating aspirations for knowledge. Even though we feel the need to label our contemporary endeavors in network visualization as a unique and original practice, cartography might simply incorporate them as an evolutionary step in its long practice.

Cartography has commonly been used as a vehicle for the depiction of various abstract concepts and imaginary places. Nonetheless, its roots are in the representation of physical features of the natural environment: coastlines, mountains, rivers, cities, and roads. Cartography is an illustration of the tangible world—an abstraction of the thing itself—which ties back to philosopher Alfred Korzybski's well-known expression that "the map is not the territory."[11] Korzybski's assertion triggers an age-old concern that equally applies to network visualization, warning against the disproportionate belief in the trustworthiness of certain maps. Any system can be depicted and interpreted in multiple ways, and a specific map delivers only one of many possible views. But network visualization is also the cartography of the indiscernible, depicting intangible structures that are invisible and undetected by the human eye, from eccentric visualizations of the World Wide Web to representations of the brain's neural network. In some cases, the maps of these hidden structures are the only visual reference we have, constituting its own alternative territory.

There is a lot network visualization can learn from cartography, particularly as an exemplary case of harmoniously combining science, aesthetics, and technique. A brief overview of the grammar of maps highlights the indubitable relationship between the two disciplines, as most maps, similar to network representations, employ three basic types of graphical markers: areas, line features, and point features.[12] Not only are their ingredients similar, but also many of their aspirations. Mapmaking and the "charting" of networks are fundamentally bounded by similar goals of simplifying, clarifying, communicating, exploring, recording, and supporting.

So what are the specific purposes of network visualization? As a potential visual decoder of complexity, the practice is commonly driven by five key functions: document, clarify, reveal, expand, and abstract.

Document

Map a system that has never been depicted before. A result of our inherent human curiosity, this goal is tied to the most ancient cartographic ambition: to portray a new unfamiliar territory. The map of a particular system can stimulate interest and awareness of a subject matter while naturally opening the door for further discoveries and interpretations. A key drive for many projects is the prospect of documenting and recording the surveyed structure for posterior knowledge.

Clarify

Make the system more understandable, intelligible, and transparent. The central objective in this context is simplification—

to explain important aspects and clarify given areas of the system. By communicating in a simple, effective way, the network visualizations become powerful means for information processing and understanding.

Reveal

Find a hidden pattern in or explicit new insight into the system, or in other words, a polished gem of knowledge from a flat data set. The goal of revealing should concentrate on causality by leading the disclosure of unidentified relationships and correlations while also checking initial assumptions and central questions.

Expand

Serve as a vehicle for other uses and set the stage for further exploration. The subsequent expansion might relate to the portrayal of multidimensional behaviors, or the depicted structure might simply become a complementary part of a larger work. In this context, the network is seen as the means to an end, the underlying layer of additional visualizations able to integrate multivariate data sets. Nodes and edges become the terrain, in the same way many web-mapping services serve as the initial outline for further building and expansion.

Abstract

Explore the networked schema as a platform for abstract representation. Network visualization can be a vehicle for hypothetical and metaphorical expression, depicting a variety of intangible concepts that might not even rely on an existing data set.

Principles of Network Visualization

The pursuit of a rigorous classification of graphical methods is not a contemporary ambition. Almost one hundred years ago, Willard Brinton, in his outstanding book *Graphic Methods for Presenting Facts* (1914), pursued a much-needed taxonomy for a growing discipline:

> The rules of grammar for the English language are numerous as well as complex, and there are about as many exceptions as there are rules. Yet we all try to follow the rules in spite of their intricacies. The principles for a grammar of graphic presentation are so simple that a remarkably small number of rules would be sufficient to give a universal language. It is interesting to note, also, that there are possibilities of the graphic presentation becoming an international language, like music, which is now written by such standard methods that sheet music may be played in any country.[13]

At the end of the book, he identifies that "though graphic presentations are used to a very large extent today, there are at present no standard rules by which the person preparing a chart may know that he is following good practice. This is unfortunate because it permits every one making a chart to follow his own sweet will."[14] In response to this problem, he outlines twenty-five rules, simply described as "suggestions... until such time as definite rules have been agreed upon and sanctioned by authoritative bodies."[15] From broad guidelines suggesting a clear chart title where no misinterpretation is possible, to defining a specific line thickness for curve graphs, Brinton's list is insightful and extremely worth reading. He was well aware that it would take some time for the appropriate bodies to come together and consent to a system of graphical rules abided by all practitioners.

Some time has passed—almost a century to be precise—and we are still as far from that target as Brinton was in 1914. In fact, some might argue that it is not yet the time for a fixed set of norms. After all, one of the thrills of being involved in an emerging field is the sense that it is being defined with every execution. But as the past decade witnessed a meteoric rise in the visual representation of networks, it has made more pressing the need to reflect on what has been done and to propose ways to improve it.

Not all readers of this book will pursue their own network-visualization projects, but for those who do, the following list of eight principles is meant to encourage and support their endeavors. The first four are larger universal considerations that, due to their broad assessment, can be applied in a variety of graphical representations. The subsequent four encompass detailed principles, tackling explicit challenges in the depiction of networks.

1. Start with a Question

"He who is ashamed of asking is afraid of learning" is a famous Danish proverb. A great quality for anyone doing work in the realm of visualization is inquisitiveness. Every project should start with an inquiry that leads to further insights about the system and perhaps answer questions that were not originally asked. This investigation might arise from a personal quest or the specific needs of a client or audience, but there should always be a defined query to drive the work.

As Ben Fry states in his book *Visualizing Data* (2008), "the most important part of understanding data is identifying the question that you want to answer. Rather than thinking about the data that was collected, think about how it will be used and work backward to what was collected. You collect data because you want to know something about it. If you don't really know why you're collecting it, you're just hoarding it."[16] It is only from the problem domain that we can ascertain that a layout may be better suited and easier to understand than others. The initial question works as a yardstick of your efforts, constantly evaluating the effectiveness of your project as a measure for naturally filtering the essential from the superfluous.

The definition of a question is vital and ties back to the need for a clear purpose or goal in every execution. And sometimes, an initial question can lead to new compelling ones. This particularly explorative path is one of the most engaging and fascinating traits of visualization.

2. Look for Relevancy

After defining a central question, what normally follows is the quest for relevancy, which acts as a constant thread throughout the project. As Dan Sperber and Deirdre Wilson exposed in their influential *Relevance: Communication and Cognition* (1996), human cognition is relevance oriented: we pay attention to information that seems relevant to us. This drives a natural expectation of relevance in every act of communication. Something is said to be relevant if it serves as an effective means to a particular purpose or, more specifically, if it increases the likelihood of achieving an underlying goal. The measure of relevance is therefore primarily based on the intent of the project and the validation of the initial question that set it forward. A central responsibility of visualization is to fulfill this expectation in the most effortless manner possible.

In the context of visualization, relevancy comes into place when selecting two central elements: the supporting data set (content) and the subsequent visualization techniques (method). Choosing the most relevant data set is naturally dependent on the goal of the execution, but appropriateness does not necessarily translate into a direct correlation between data and purpose. Sometimes we need to look laterally for alternatives in order to find the content that can most appropriately answer our question. In the 2006 visualization *Tracing the Visitor's Eye*, researcher and computer scientist Fabien Girardin leveraged the rich image bank of Flickr to map the most popular paths made by tourists in Barcelona; in *Just Landed* (2009) (see page 150), digital artist and designer Jer Thorp looked into the content of Twitter messages, particularly those starting with "just landed in," to make a chart of popular travel destinations around the world. These examples illustrate the idiosyncratic nature of this process, recurrently involving a creative approach accentuated by lateral thinking.

The selection of the most suitable visualization method for the project is largely determined by the central question. However, this particular quest is equally dependent on the end users, their immediate context and expressed needs. Acknowledging the different contexts of use—when, where, and how the final execution will be used—is crucial in the pursuit of relevancy.

If the relevancy ratio is high, it can increase the possibility of comprehension, assimilation, and decision-making, becoming a fundamental step in the transition from information into knowledge. The greater the processing effort, the lower the relevance. As linguistics professor Elly Ifantidou explains, "A speaker aiming at optimal relevance should try to formulate her utterance in such a way as to spare the hearer gratuitous processing effort, so that the first acceptable interpretation to occur to the hearer is the one she intended to convey."[17]

3. Enable Multivariate Analysis

In many cases the depiction of a network is seen as a binary system, where connections are simply turned on and off like transistors in a computer. But the ties among elements in a network are immensely rich and detailed, and the inclusion of additional information can be fundamental in the unveiling of many of these nuances. By embracing complementary data sets—able to provide additional information on the nature of nodes and respective ties—the system can easily expose causality in patterns and relationships, contributing decisively to the holistic understanding of the depicted topology.

Let's suppose you are creating a visualization of a network of rivers, where nodes represent key locations—e.g., neighborhoods, cities, districts, regions—crossed by the different streams. Now imagine the unique richness of one single stream and the number of its oscillating variables: color and temperature of water, pollution levels, tides, current speed, and width of stream bed, among many other facets. Such a multivariate approach would be critical in the interpretation of particular behaviors and potential accidents, by determining, for instance, the causes for water contamination, a sudden blockage, or overflow. But such an affluent contemplation is not unique to this scenario; it applies broadly to most types of networks. Try to think of how many elements you could consider when mapping your own network of friends and the immense qualitative richness that underlies each individual relationship, and you might have a better grasp of the importance of integrating multivariate data sets.

Multivariate analysis is a prerequisite in a variety of ongoing scientific endeavors that involve the interweaving of a vast assortment of elements, and, as Bruce Mau states in *Massive Change* (2004): "When everything is connected to everything else, for better or worse, everything matters."[18] In the end, we are multivariate beings involved in multivariate actions inhabiting a multivariate world. "Nearly all the interesting worlds (physical, biological, imaginary, human) we seek to understand are inevitably multivariate in nature," elucidates statistician Edward Tufte.[19] As a longtime advocate of this principle, Tufte explains in detail why this consideration is vital: "The analysis of cause and effect, initially bivariate, quickly becomes multivariate through such necessary elaborations as the conditions under which the causal relations holds, interaction effects, multiple causes, multiple effects, causal sequences, sources of bias, spurious correlations, sources of measurement error, competing variables, and whether the alleged cause is merely a proxy or a market variable."[20]

4. Embrace Time

Time is one of the hardest variables to map in any complex system. It is also one of the richest. If we consider a social network, we can quickly realize that a snapshot in time can only tell us a bit of information about that community. Alternatively, if time were to be properly measured and mapped, it would provide us with a comprehensive understanding of the social group's changing dynamics. Out of the existing panoply of social-network-analysis tools available, very few offer the ability to explore the network over time, investigate how it expands or shrinks, how relationships evolve, and how certain nodes become more or less prominent. This, of course, should change.

Networks are evolving systems, constantly mutating and adapting. As physicists Mark Newman, Albert-László Barabási, and Duncan J. Watts explain, "Many networks are the product of dynamical processes that add or remove vertices or edges....The ties people make affect the form of the network, and the form of the network affects the ties people make. Social network structure therefore evolves in a historically dependent manner, in which the role of the participants and the patterns of behavior they follow cannot be ignored."[21] In some cases, the changes do not take weeks or months, but minutes or hours. And it is not only the network that adapts; whatever is being exchanged within the system also fluctuates over time (e.g., information, energy, water, a virus).

If we consider the vast hidden networks that sustain our biosphere, we can truly understand how critical the dimension of time really is. After all, it is the particularly dynamic nature of interconnecting ecosystems around the world that poses one of the most difficult challenges to our enduring effort to understand the intricacies of our planet. Even something seemingly as stable as the human brain is continuously adding or removing synapses—the connections between neurons—in a process associated with cognitive learning. Not to mention the internet, with its constant flux of information and vast landscape of servers, frequently adds or disconnects machines from the network. And time analysis does not only cover historical evolution; it equally applies to real-time dynamics and oscillations. fig. 8, fig. 9

fig. 8 (opposite)

Skye Bender-deMoll and Dan McFarland, *Social Network Image Animator (SoNIA)*, 2004.

A series of images from a movie made using SoNIA showing classroom interactions between teacher and students using 2.5-minute time slices.

The data set consists of repeated observations in more than one hundred fifty high school classrooms during the 1996–97 school year.

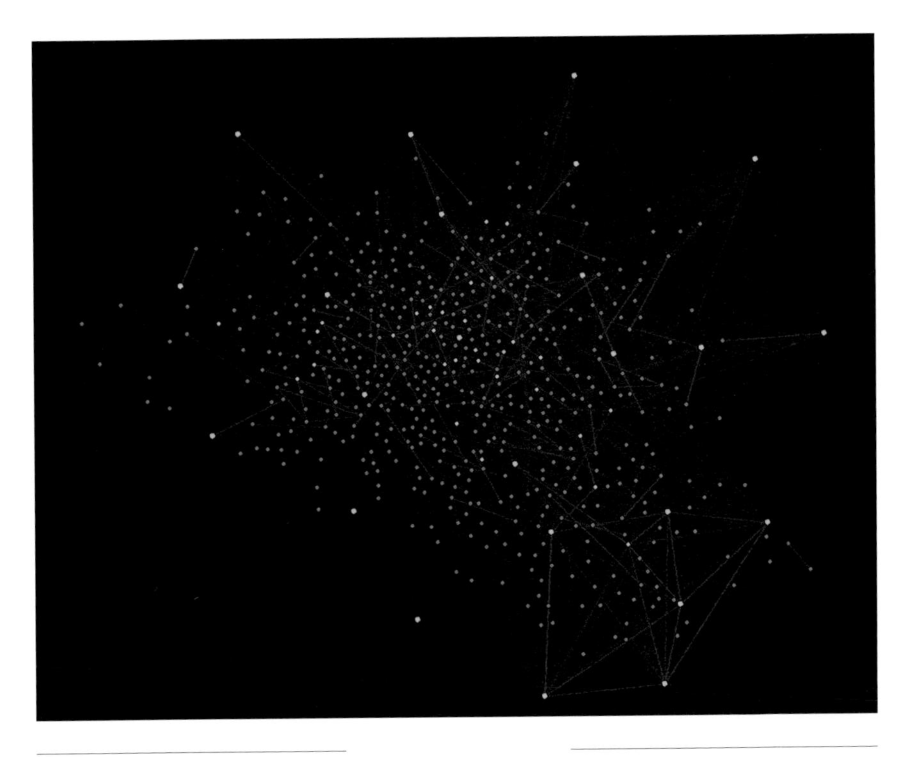

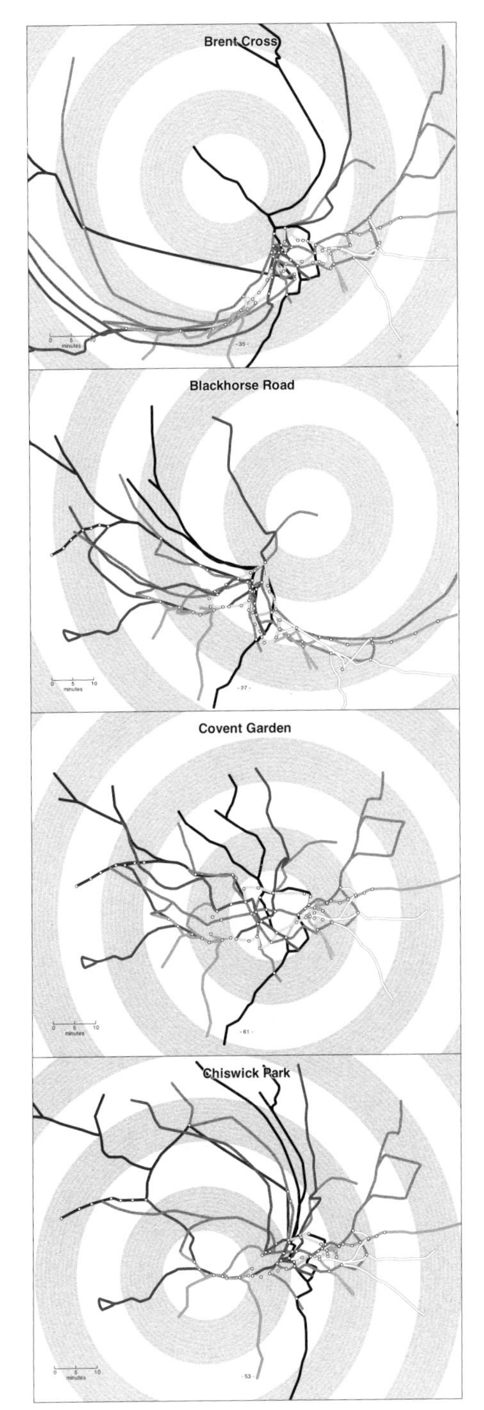

fig. 9

W. Bradford Paley and Jeff Han, *Trace Encounters*, 2004

A real-time visualization debuted on September 3, 2004, in Linz, Austria, at the one-day Ars Electronica Festival. It mapped live social interaction among the event's participants by means of limited-range infrared stickpins embedded on the participant badges.

fig. 10 (right)

Tom Carden, *Travel Time Tube Map*, 2005

Four frames of an innovative interactive map of the London Underground system. Once a station is selected, the entire map adjusts to show the time it takes from the chosen station to any other in the system. Time is represented by concentric circles demarcating ten-minute intervals.

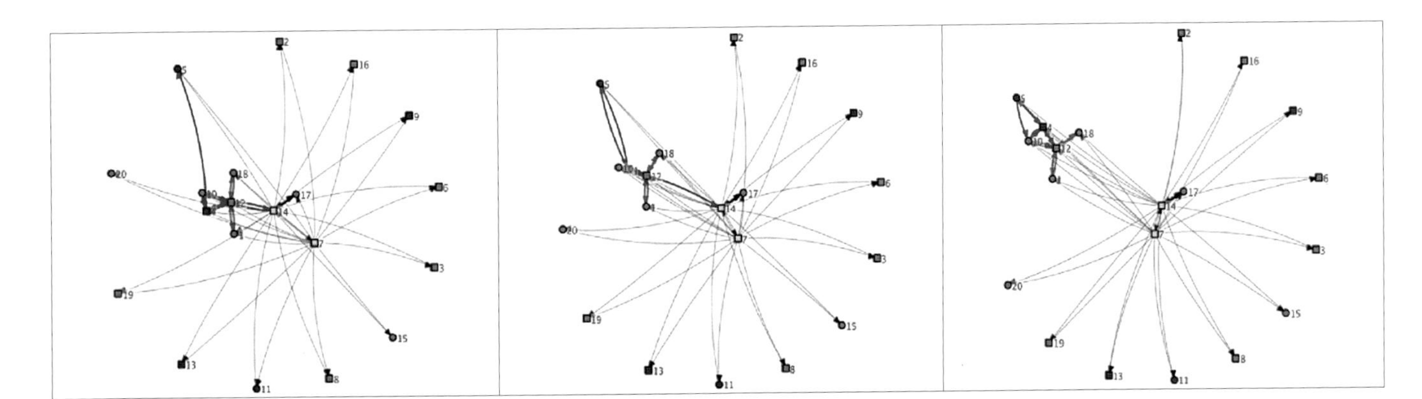

But mapping time in any network, as computer scientist Chaomei Chen recognizes, "is one of the toughest challenges for research in information technology....[It is] technically challenging as well as conceptually complex."[22] Due to the extremely demanding nature of charting the passage of time within a network, most scientists and designers feel apprehensive about incorporating this dimension in many of their executions, which in part explains the lack of projects in this realm. There is no doubt that when we embrace time, the difficulty of the task at hand increases tenfold, but if visualization is to become a fundamental tool in network discovery, it needs to make this substantial jump. Most networked systems are affected by the natural progression of time, and their depiction is never complete unless this critical dimension becomes part of the equation. fig. 10

5. Enrich Your Vocabulary

Whenever considering the representation of a network, there are two vital elements to consider: nodes (vertices) and links (edges). While the recipe is simple enough, these two essential ingredients are rarely used to their fullest potential. Usually represented by mere circles or squares and indistinguishable connecting lines, nodes and links are too often mistreated. A consideration of a full spectrum of visual properties—color, shape, size, orientation, texture, value, and position, as outlined in Jacques Bertin's list of seven graphical attributes from his seminal work *Semiology of Graphics* (1984)—can and should be used comprehensively, always reinforced by a specific semantics able to tie the different data attributes to corresponding visual elements.

Richer nodes

Nodes are the atomic units of a graph, the objects within the system. Instead of being depicted as empty squares or circles, they can be made more intelligible with an appropriate use of color and graphical features. They can also become responsive and provide important contextual information through the use of interactive features. Most graphic variables (size, color, shape, position) can express the type and prominence of a node, as well as its natural affordances: Is the node interactive? Does it have hidden links? Does it contain additional details? By embracing interactivity, there is also a spectrum of pertinent features to explore. Nodes can expand or shrink, show or hide relevant information, and ultimately morph according to the user's criterion and input. fig. 11, fig. 12

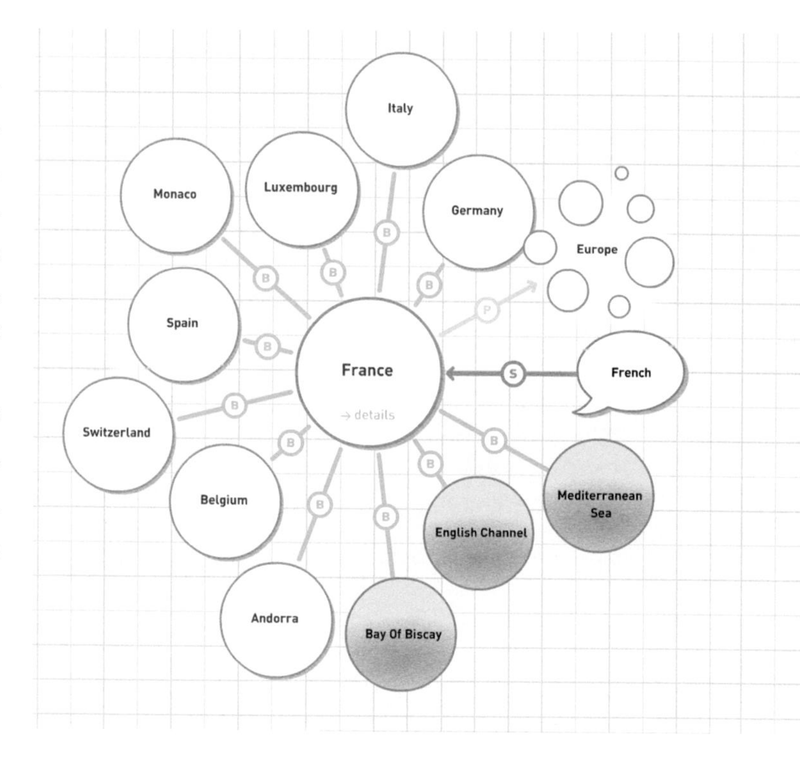

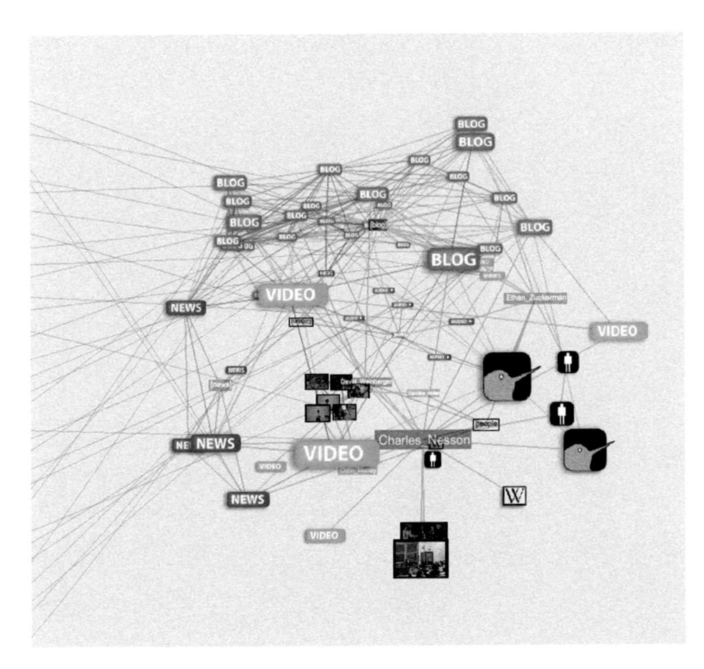

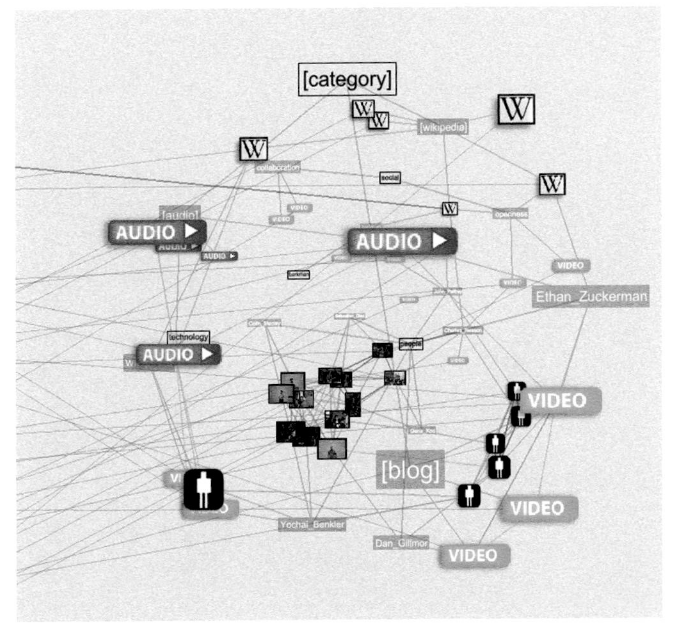

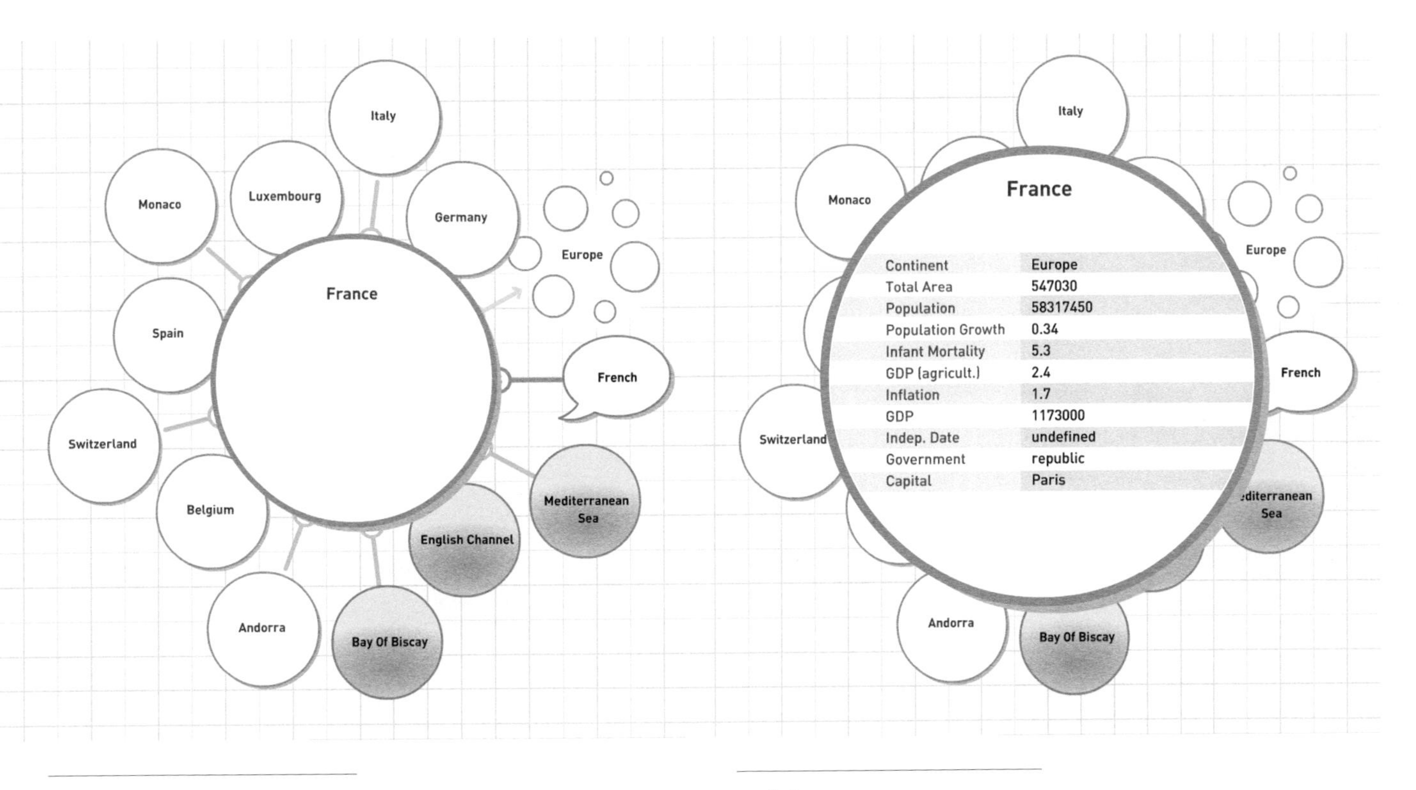

fig. 11 (top row)

Bestiario, *B10*, 2008

Two frames of a dynamic visualization created for the tenth anniversary of Harvard University's Berkman Center for Internet & Society. It showcases various ideas and media gleaned from conferences organized by the Berkman Center between 2007 and 2008. The graphical richness of the nodes is evident in this execution, making it easy to identify a variety of categories, like blog posts, video files, audio files, news articles, Wikipedia articles, tweets, images, and people.

fig. 12 (bottom row)

Moritz Stefaner, *CIA World Factbook Visualization*, 2004

An interactive map of geographic boundaries and linguistic ties (B, has a border with; P, is part of; S, is spoken in) between countries found in the CIA world factbook database. Any country name, upon selection, immediately expands to reveal detailed information about it.

Expressive edges

Edges are connectors in a graph and are a vital element in any network representation—without them nodes would simply be hollow elements in space. But edges can express much more than a single connection between entities. For every relationship between nodes, there are innumerous layers of quantitative and qualitative information pertaining to the nature of the connection (e.g., geographical or emotional proximity, frequency of communication, duration of friendship).

Cartography is a great source for inspiration when examining the portrayal of edges. In a conventional country map, a number of associations are evident: two major cities can be connected by a variety of line segments—primary and secondary roads, train tracks, rivers, and alternative paths—all depicted in a unique, discernable way. fig. 13 A similar process could be applied for network visualization. The following factors should be considered in visualizing edges: *length* to suggest a gradation of values, such as physical proximity, degree of relationship, strength, similarity, or relatedness; *width* to express density or intensity of flow, or an alternative gradation of values; *color* to differentiate or highlight particular groups, categories, and clusters, or alternatively, singular connections; *shape* to communicate the type of relationship (e.g., family, friends, coworkers). fig. 14, fig. 15, fig. 16

Clear visual language

One of the caveats behind the implementation of diverse graphical attributes is to beware of creating a visual language that might not be immediately recognized by everyone. We can flatten out the required learning curve by simply embracing a widespread cartographic technique: the legend. A map legend is simple, yet vital, allowing for a quick interpretation of the various graphic components. This prevalent mapping practice should be widely adopted by network visualization, making the vocabulary intelligible and facilitating an immediate understanding of the final piece.

6. Expose Grouping

The ability to showcase variation in a depicted system is a central attribute of network visualization. This can be achieved not only by enriching the visual vocabulary but also by exploring the potentialities of spatial arrangement. Spatial relationships are as important as explicit visual ties and are a critical element in exposing contrast and similarity. The concept of grouping is particularly significant in this context, allowing for the improved apprehension of clusters, islands, prominent patterns, and the general distribution of nodes and links. The idea of grouping is simply to combine several units of information into related chunks in order to reinforce relationships, reduce complexity, and improve cognition.

Grouping can be pursued with a variety of criteria in mind, ultimately depending on the central goal of the execution. But in most cases, elements can be grouped in five distinct ways: alphabetically, by time, by location, by a particular continuum (or scale), and by a specified category (e.g., images, videos, text). This procedure, first proposed by Richard Saul Wurman in *Information Anxiety* (2000), is known as the five hat racks, and it delivers an effective way to organize most types of information.

Another remarkable source of knowledge on the notion of grouping comes from Gestalt psychology. Emerging in the early twentieth century, by the hands of prominent psychologists Max Wertheimer, Kurt Koffka, and

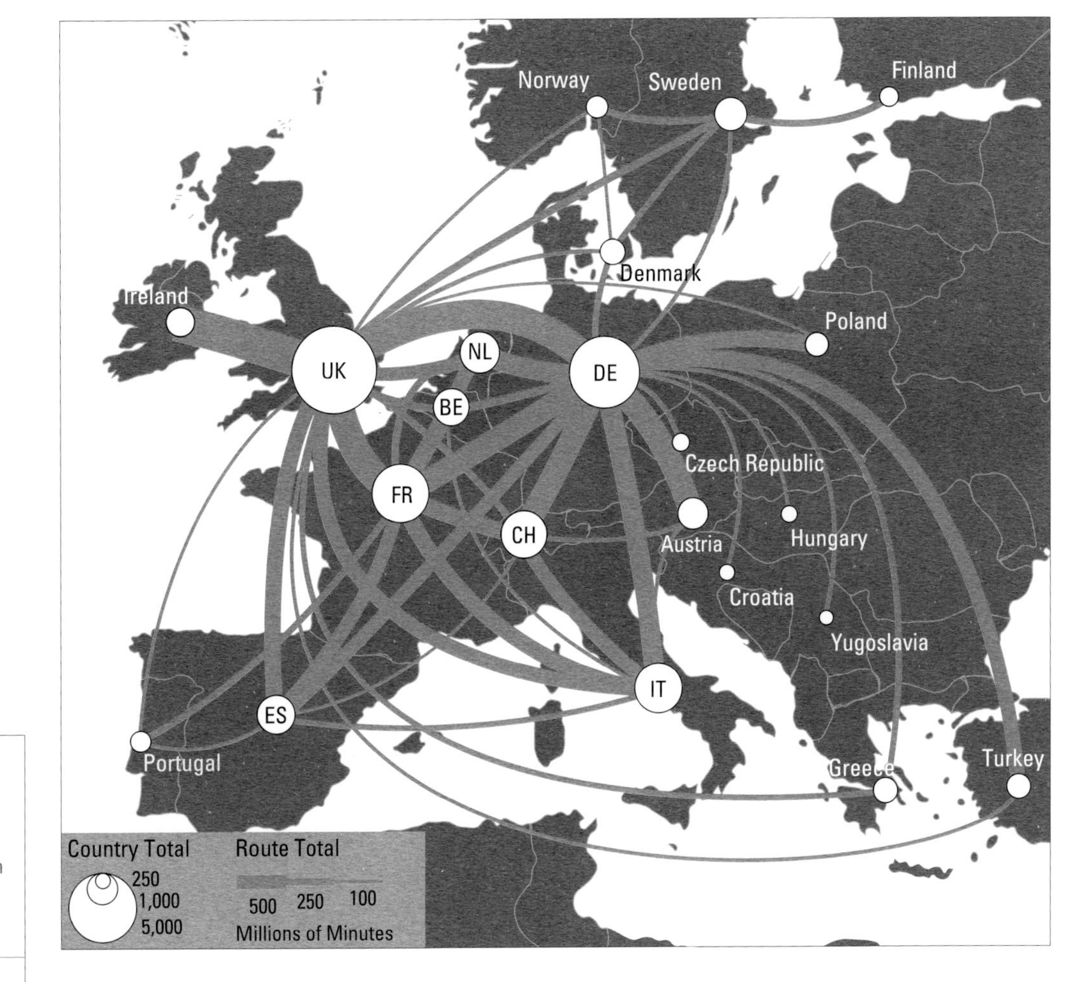

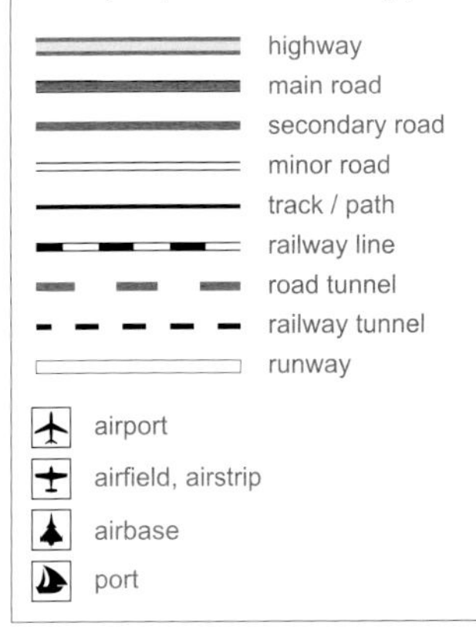

fig. 13 (left)

Eric Gaba and Bamse, *Maps Template*, 2009

A legend of the WikiProject Maps initiative, which provides advice and templates for the creation of geographic and topographic maps on Wikipedia Commons, an online repository of free-use images, sound, and other media files. This excerpt of the legend shows the graphical diversity usually applied to different types of points, e.g., capital, city, village, as well as line features, e.g., highway, secondary road, railway line, in many cartographic projects.

fig. 14 (right)

TeleGeography, *Traffic Flow Map*, 2000

A map of the communication-traffic flow between European countries. The width of the orange lines is proportional to the annual volume of traffic between countries, measured as one unit equaling one hundred million minutes of voice telecommunication. Circular symbols, located on the capital city, encode the country's total annual outgoing traffic to all other countries.

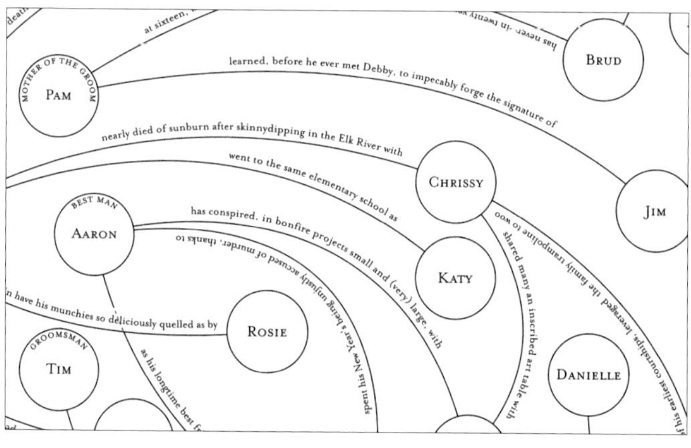

fig. 15 (top)

Andrew Coulter Enright and Heather Samples, *Dramatis Personae*, 2007

A map of relationships between guests at Andrew Coulter Enright and Heather Samples's wedding. With the goal of helping start conversations, the couple produced a chart of the tightly knit group of family and friends at the wedding, connecting guests based on favorite shared stories, which they had collected and solicited from their parents. By tying people with stories, this original treatment of edges brings an appealing qualitative and personal layer to the execution.

fig. 16 (bottom)

Detail of *Dramatis Personae*

Wolfgang Kohler, Gestalt psychology was a serious attempt at comprehending our perception of visual patterns. Some of the results from their studies are still of immense importance to most forms of visual communication. Of particular relevance are the devised rules of perceptual organization, also known as Gestalt laws of grouping, which, by explaining how people perceive a well-organized pattern, can easily translate into concise design principles. Three of the Gestalt laws—similarity, proximity, and common fate—are particularly important rules in exposing groups in network visualization.

Law of similarity (graphical treatment)

The law of similarity asserts that elements that are similar—either in terms of color, shape, or size—are perceived to be more related than elements that are dissimilar. This Gestalt principle highlights the need for a differentiated graphical vocabulary in the depiction of nodes, as a critical measure for spotting similarities and differences and in order to apprehend the overall distribution within the system.

Law of proximity (spatial arrangement)

The law of proximity states that elements that are close together are perceived as being more related than elements that are farther apart. This organizing principle proves that relatedness is not only expressed by graphical properties but also by spatial proximity. The mere placement of homologous nodes closer to each other suggests inherent relationships not solely manifested by edges (links).

Law of common fate (motion)

The law of common fate proclaims that elements that move simultaneously in the same direction and at the same speed are perceived as being more related than elements that are stationary or that move in different directions. This notion is particularly pertinent when trying to highlight contrast through animation (e.g., depicting the changing dynamics of a network over time).

7. Maximize Scaling

One of the biggest misconceptions in network visualization is the notion that a representation that works at one scale will also work at a larger or smaller scale. This scaling fallacy is a fundamental cause for many misguided projects. Not only do networks showcase different patterns and behaviors at different scales, but also the user's needs vary depending on his or her particular position with respect to the network. When representing a network, it is important to consider three fundamental views in line with a specific method of analysis: macro view, relationship view, and micro view. fig. 17

Macro view (pattern)

A well-executed macro view does not have to provide a detailed understanding of individual links, and less so of individual nodes. It should provide a bird's-eye view into the network and highlight certain clusters, as well as isolated groups, within its structure. As the common entry point to a particular representation, it needs to facilitate an understanding of the network's topology, the structure of the group as a whole, but not necessarily of its constituent parts. In most cases, the use of color (within nodes or edges) and relevant positioning (grouping) is enough to provide meaningful insight into the network's broad organization.

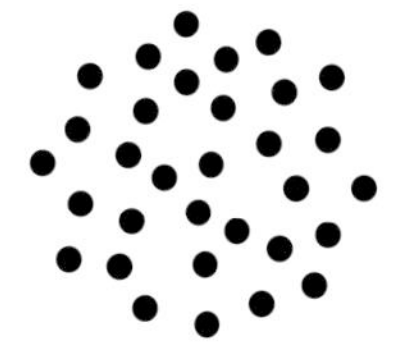

Macro Analysis
(pattern)

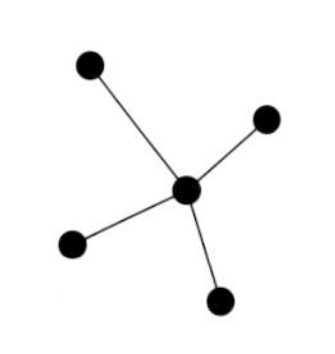

Relationship Analysis
(connectivity)

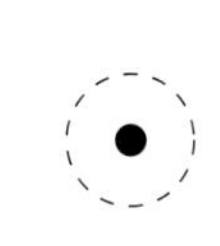

Micro Analysis
(entities)

fig. 17

The three critical views of network visualization

Relationship view (connectivity)

The relationship view is concerned with an effective analysis of the types of relationships among the mapped entities (nodes). It not only indicates the existence of connections but also offers further revelation, such as proximity between the nodes, and type and intensity of association. This is a fundamental view of network visualization and normally requires analysis from different perspectives or points of view in order to obtain a solid grasp of the different topologies. While the main concern in the macro view is synthesis (to grasp the whole with one look), the relationship view is about analytics (to efficiently dismantle the system and discover the interconnections between the parts).

Micro view (individual nodes)

The last layer of insight provided by an efficient network visualization relates to the disclosure of an individual node's qualitative attributes. A micro view into the network should be comprehensive and explicit, providing detailed information, facts, and characteristics on a single-node entity. This qualitative exposure helps clarify the reasons behind the overall connectivity pattern, from an isolated node to one highly connected to a large number of other nodes.

A successful network visualization does not have to possess all three views, but the more questions it is able to answer, the more successful it will be. If a representation focuses exclusively on a macro view, it can still provide relevant insights into a network's topology, but by leaving out the other two views, it is neglecting a set of possible answers. The three views also shape a natural continuum of processing, where the increasing detail of information, from macro to micro, can be a critical measure for reasoning.

8. Manage Intricacy

Even though the three main views for network visualization appear to be autonomous, it is imperative that users are able to navigate between them effortlessly, in a seamless wayfinding approach. Ben Schneiderman's renowned visual-information-seeking mantra is a great place to start. Proposed in his seminal paper "The Eyes Have It: A Task by Data Type Taxonomy for Information Visualizations" (1996), the mantra reads as follows: "Overview first, zoom and filter, then details on demand."[23] This apparently obvious rule, instinctually practiced by different design practitioners, even those unfamiliar with Schneiderman's mantra, is an excellent strategy for network visualization. As computer scientist Riccardo Mazza explains, "It is necessary to provide a global overview of the entire collection of data so that users gain an understanding of the entire data set, then users may filter the data to focus on a specific part of particular interest."[24]

Underlying Schneiderman's maxim is the notion of progressive disclosure—a widespread interaction-design technique aiming at simplification that allows additional content and options to be revealed gradually, as needed, to the user. This technique is particularly relevant if we consider Hick's Law, put forth by psychologist William Edmund Hick, which states that the time required to make a decision increases as the number of variables increases. Alluding to the risk of displaying a full, convoluted network at once in one single view, Hick's Law is an important point of awareness of the perceptual limits of network visualization. Even though other methods can and should be devised, there are three important concepts that can help minimize intricacy and unify the three views of network visualization.

Adaptive zooming

This widely used modern cartographic technique—strongly tied with the notion of progressive disclosure—enables the system to render a different set of visual elements depending on the present zooming view. An interactive web map, for example, typically starts with regional view; as you zoom in, more detailed elements progressively appear: primary roads, secondary roads, road names, and points of interest. A similar method could be employed in the depiction of networks, by focusing on a gradation from macro view to micro view, showing the most prominent nodes first, and then slowly disclosing additional graphical and textual elements: major hubs and primary links, labels, secondary nodes and links, tertiary nodes, and so on. fig. 18

Overview and detail

A common interaction-design technique, seen in a variety of contemporary mapping tools, overview and detail usually comprises a primary viewing area (detail) that allows for different levels of zoom, accompanied by a smaller macro view (overview), which permits users to see where they are in the general context. This is particularly relevant in reassuring users they are free to navigate the system without getting lost. fig. 19

Focus and context

This widely used information-visualization concept is one of the field's strongest contributions and its most studied technique. It simultaneously provides a detailed view (focus) and a macro view (context) within a single configuration. Popularized by the widespread fish-eye view, this method merges both views in the same space without the need to

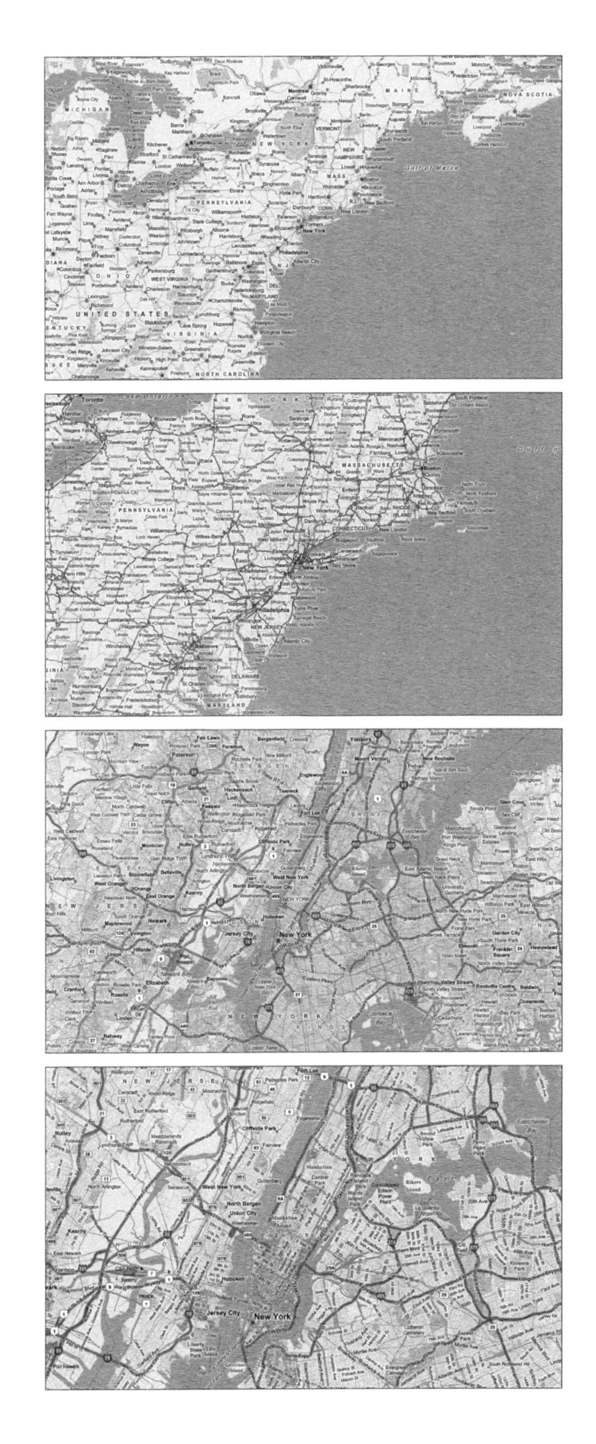

fig. 18

Bing Maps, *Adaptive Zooming*, 2010

Microsoft Bing Maps, like many other web-mapping services, makes use of adaptive zooming. The displayed information changes at different zooming levels, showing progressively more detailed facts—from names of cities to names of roads and neighborhoods—as users zoom in.

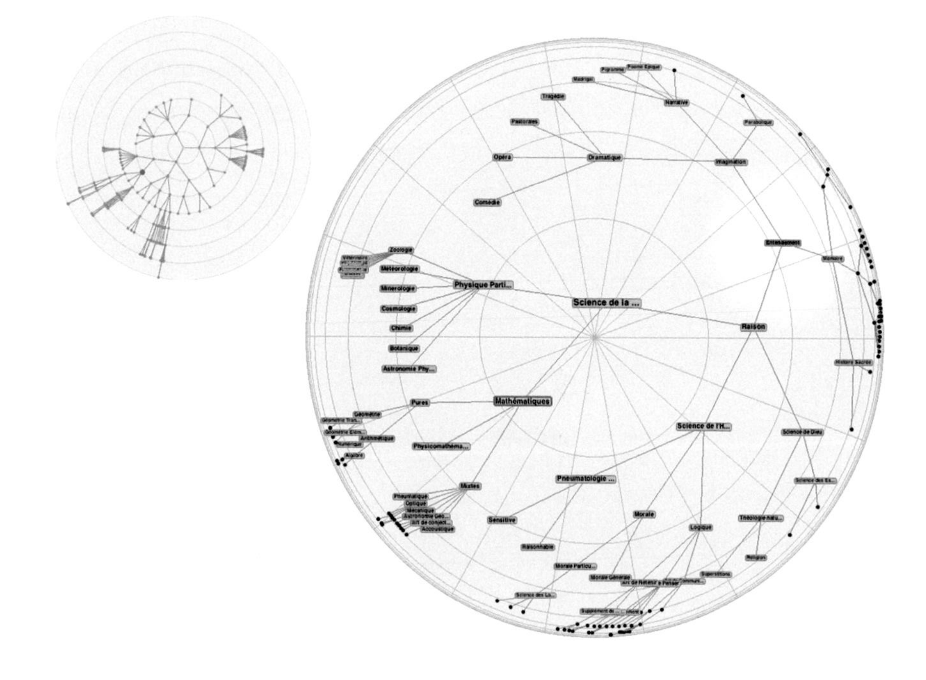

fig. 19

Christophe Tricot, *EyeTree*, 2006

An interactive fish-eye view of Diderot and D'Alembert's *Figurative System of Human Knowledge*. This project makes great use of both overview and detail as well as focus and context methods.

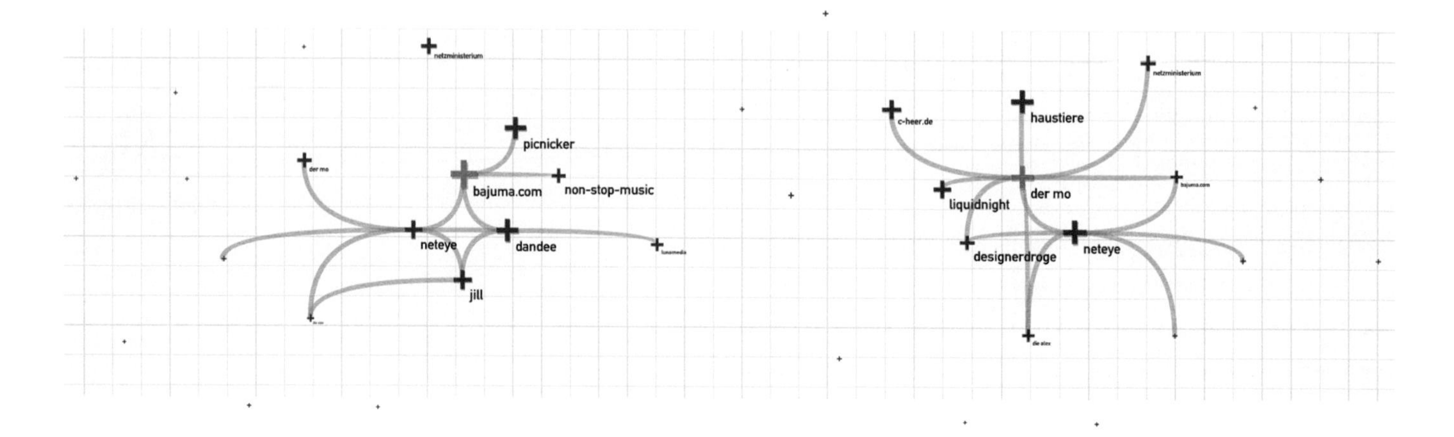

fig. 20

Moritz Stefaner, *Organic Link Network*, 2006

An example of the focus and context technique, in which the activated nodes (crosses) are given greater prominence by enlarging in size, while the remaining ones are left to the periphery at a reduced scale

segregate them. This is usually achieved by enlarging the detailed view—the user's focus of attention—while leaving the other nodes and edges to the periphery. fig. 20

On the Principles

The main premise of the previous list of principles is that networks are very difficult to visualize, but we do not need to make them more complex in the process of trying. Graphs are, as of today, the most suitable method for the depiction of networks due to their intrinsic organization based on nodes and links, but they are far from perfect. Many of the current limitations—such as resolution and screen size—can quickly lead to cluttered and indecipherable displays. This drawback, paired with a long-existing emphasis on process rather than outcome, explains the challenging state of affairs in network visualization. Even though the field has traditionally placed a strong emphasis on mathematics and the generation of computer algorithms, these are merely a means to an end. The end should always be a useful depiction able to fulfill its most fundamental promise of communicating relevant information.

If sometimes an intricate outcome can reveal a considerable appeal, it might equally raise critical issues of clarity and function. The adoption of interactive techniques solves some but not all of the problems. In order for the general usability of network visualization to improve, we need to embrace the existing body of knowledge from graphic design, cartography, and visual perception, including notions of color theory, composition, typography, layout, and spatial arrangement. The aim is not to merely create an algorithm capable of sustaining copious amounts of nodes and links, but also to select the most appropriate scheme based on well-founded design principles and appropriate interactive methods.

The eight principles exposed in this chapter are not meant to be restrictive but generative. They are meant to inspire and influence current practitioners and to be the basis for further study and exploration, superseding many of the field's limitations. There is still a great amount of work to be done in network visualization, but we can collectively improve it, step by step, node by node.

Notes

1 Biggs and Wilson, *Graph Theory 1736–1936*, 3.
2 Ibid., 3–4.
3 Ibid., 4.
4 Barabási, *Linked*, 12.
5 Marineau, *Jacob Levy Moreno 1889–1974*, 31.
6 *New York Times*, "Emotions Mapped by New Geography."
7 Ibid.
8 Moreno, *Who Shall Survive?*, 95–96.
9 Ibid., 22.
10 Ibid.
11 Korzybski, "A Non-Aristotelian System."
12 Ware, *Information Visualization*, 215.
13 Brinton, *Graphic Methods for Presenting Facts*, 3.
14 Ibid., 361.
15 Ibid.
16 Fry, *Visualizing Data*, 4.
17 Ifantidou, *Evidentials and Relevance*, 64.
18 Mau and Institute Without Boundaries, *Massive Change*, 129.
19 Tufte, *Beautiful Evidence*, 129.
20 Ibid.
21 Newman and Watts, *The Structure and Dynamics of Networks*, 7.
22 Chen, *Information Visualization*, 69.
23 Ware, *Information Visualization*, 317.
24 Mazza, *Introduction to Information Visualization*, 106.

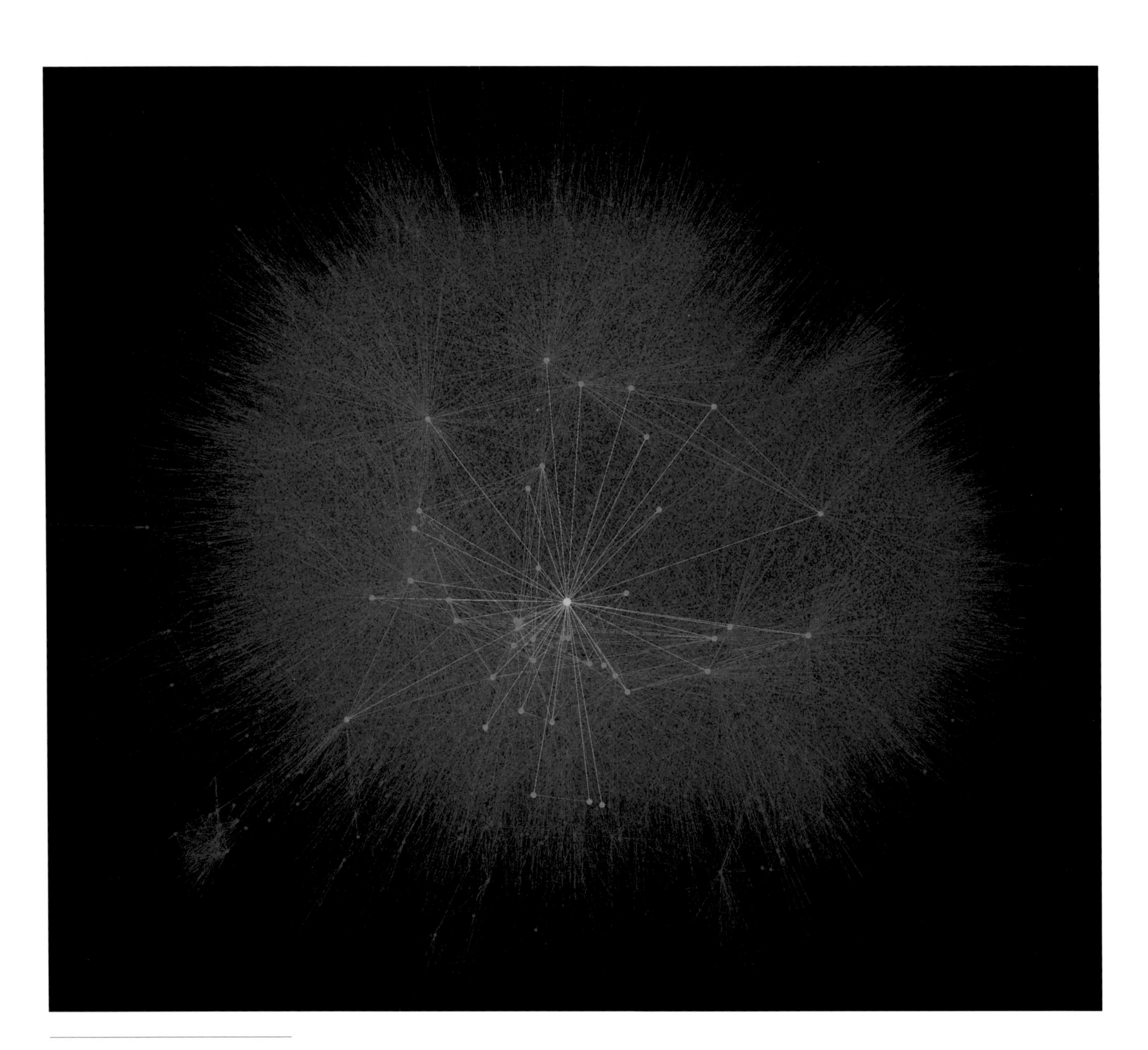

Jeffrey Heer, *Personal Friendster Network*, 2004

A map of Heer's three-level Friendster network—friend, friend-of-a-friend, and friend-of-a-friend-of-a-friend social structure—consisting of 47,471 people connected by 432,430 friendship ties. The data was collected between October 2003 and February 2004.

04 | Infinite Interconnectedness

When everything is connected to everything else, for better or worse, everything matters.
—Bruce Mau

The dynamic of our society, and particularly our new economy, will increasingly obey the logic of networks. Understanding how networks work will be the key to understanding how the economy works.
—Kevin Kelly

Over recent years network visualization has shed light on an incredible array of subject areas, and by doing so has drawn the attention back onto itself. Driven by a surge in computing power and storage, increasingly open and accessible data sets, a large adoption by mainstream media and online-social-network services, and most importantly, our never-ending eagerness for measurement and quantification, visualization is currently at a tipping point. This drastic growth symbolizes a new age of exploration, with the charting of innumerous undiscovered territories. While the vast majority of network visualizations are illustrated by the common graph—expressed by an arrangement of nodes and links—the range of depicted subjects is astonishing. People, companies, websites, emails, IP addresses, routers, species, genes, proteins, neurons, scientific papers, books, or words—bonded by a multiplicity of lines—expressing anything from social ties to bibliographic citations, communication flows to hyperlinks. Never before have we felt so strongly the sense of living in a highly interrelated and interdependent world. The following examples showcase the network as the ubiquitous model in the new age of infinite interconnectedness.

The two epigraphs to this chapter are drawn from Mau and Institute Without Boundaries, *Massive Change*, 129; and Kelly, *New Rules for the New Economy*, 9–10.

Blogosphere

Blogging presents one of the most interesting social phenomena of our time. This change in the flow of online information is radically changing the way we look at news providers and large media conglomerates. It also provides a remarkable laboratory to investigate how information spreads across online social communities. Most visualization projects under this theme map different aspects of the blogosphere, from charting the link exchange between political blogs to the dynamic blogspace of an entire country.

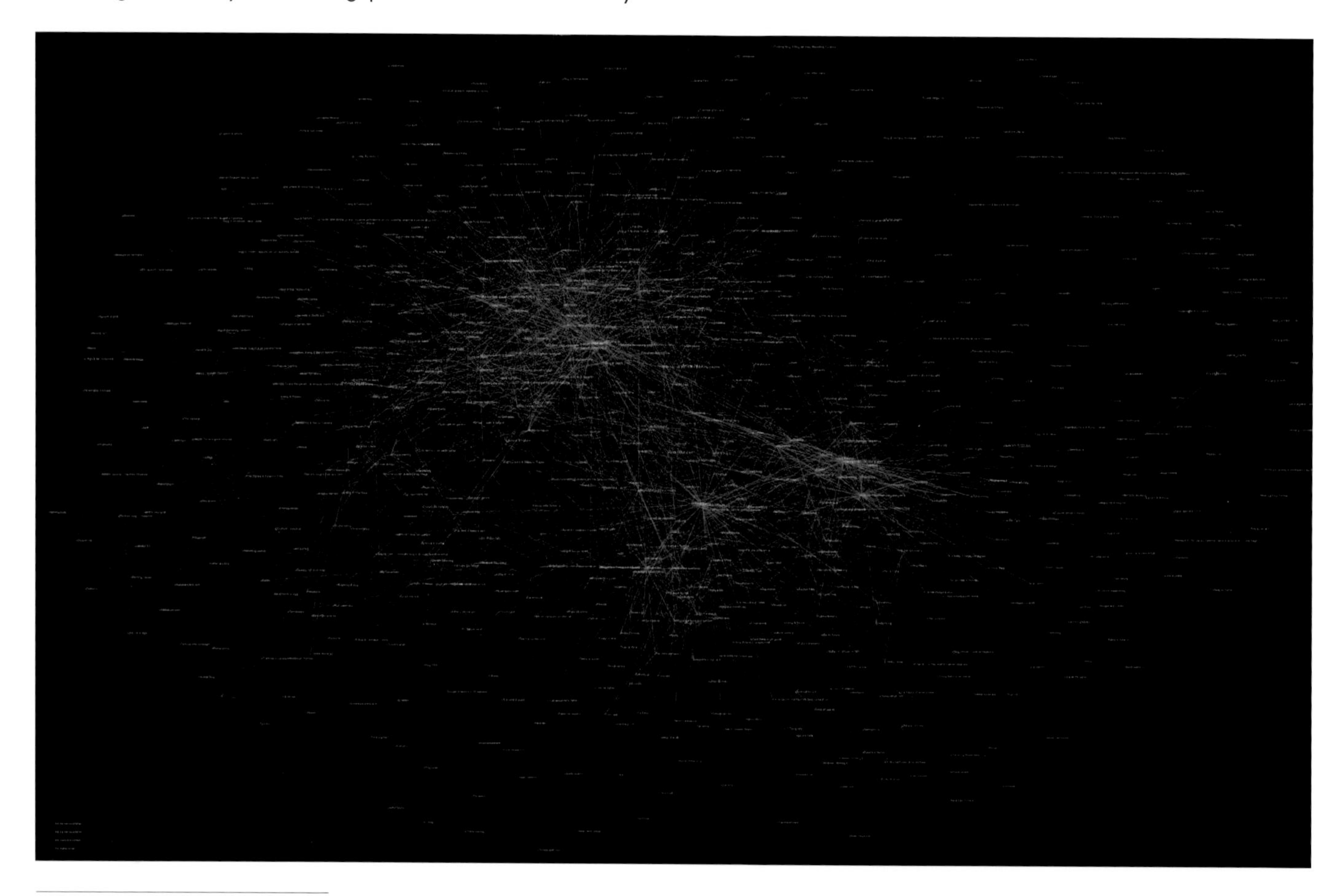

Ludovico Magnocavallo
Italian Blogosphere
2006

A chart of the five hundred most influential Italian blogs and their shared links, for a total of 2,500 blogs (above); detail (right)

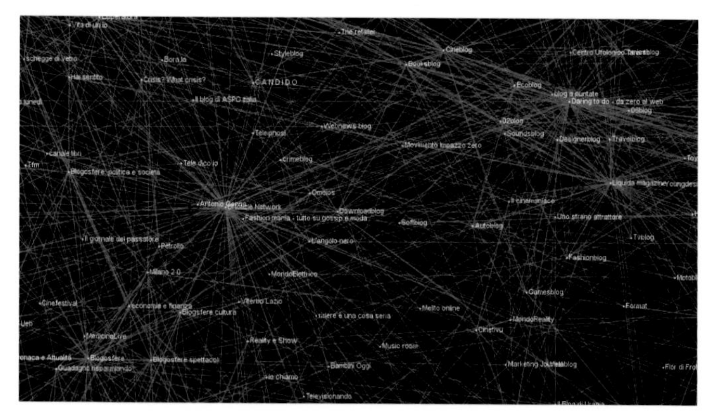

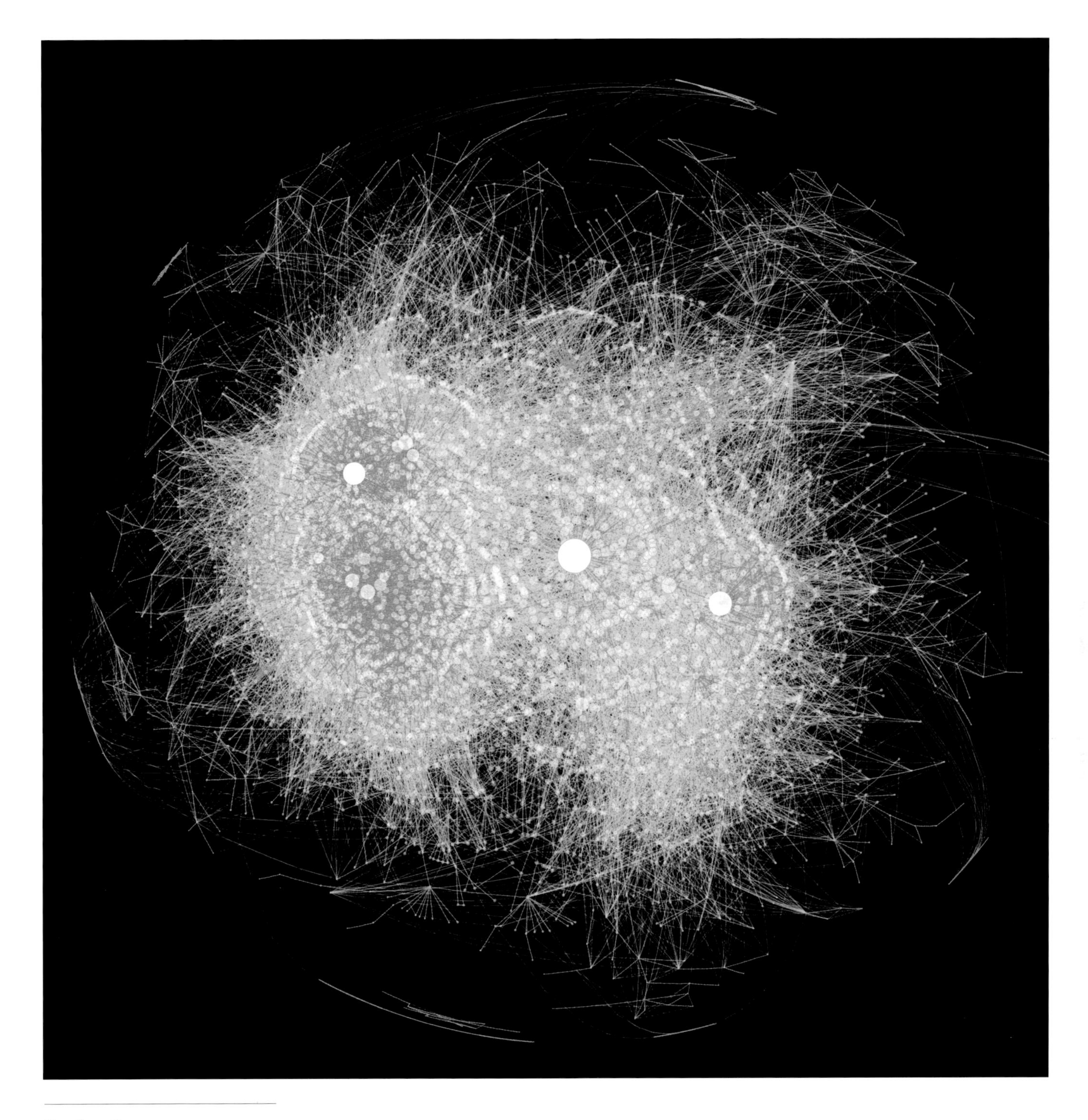

Matthew Hurst

The Hyperbolic Blogosphere

2007

This intricate map plots the most active and interconnected parts of the blogosphere from link data collected over six weeks.

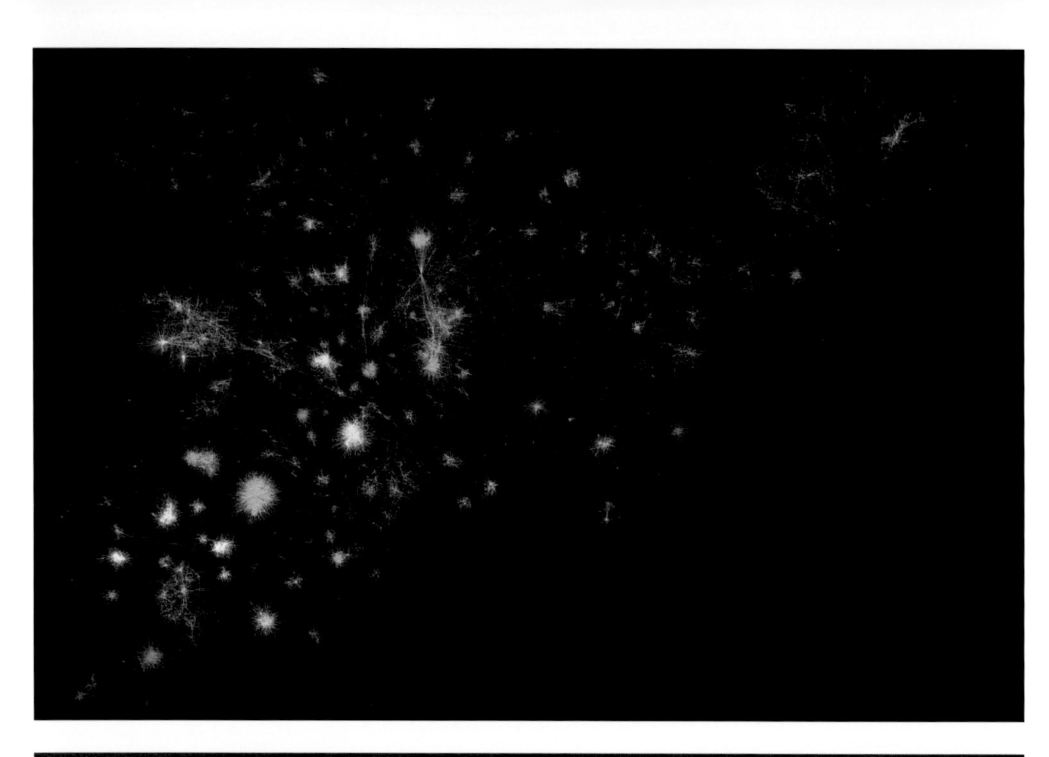

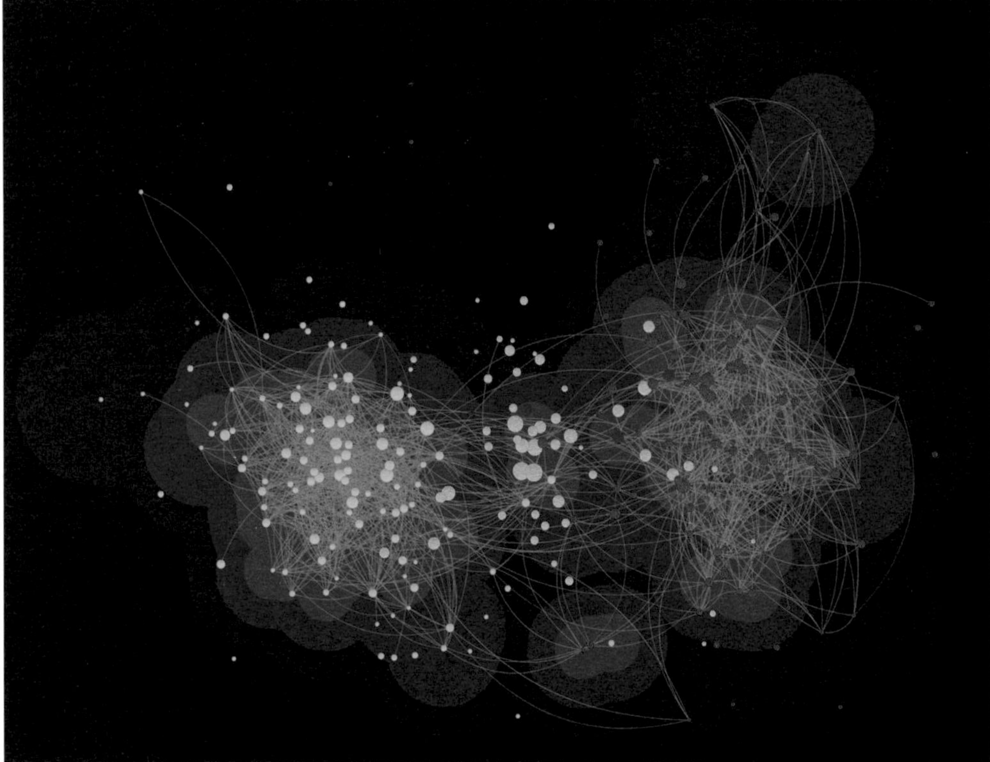

(top)

Makoto Uchida and Susumu Shirayama

Japanese Blogosphere

2006

This large survey of the Japanese blogosphere highlights the herd behavior of blogs. Each community (depicted by a different color) represents a group of blogs that discuss the same or similar topics. From Uchida and Shirayama, "Formation of patterns from complex networks," *Journal of Visualization* 10, no. 3 (August 2007): 253–56.

(bottom)

Antonin Rohmer

PoliticoSphere.net

2009

A map of the 2008 U.S. political blogosphere, showcasing the links between the 292 influential sites and opinion hubs that contributed to the online debate about the presidential race. The graph shows four communities depicted by different node colors. Three are based on their political orientation—progressive, independent, and conservative—and the smallest group represents mass-media websites.

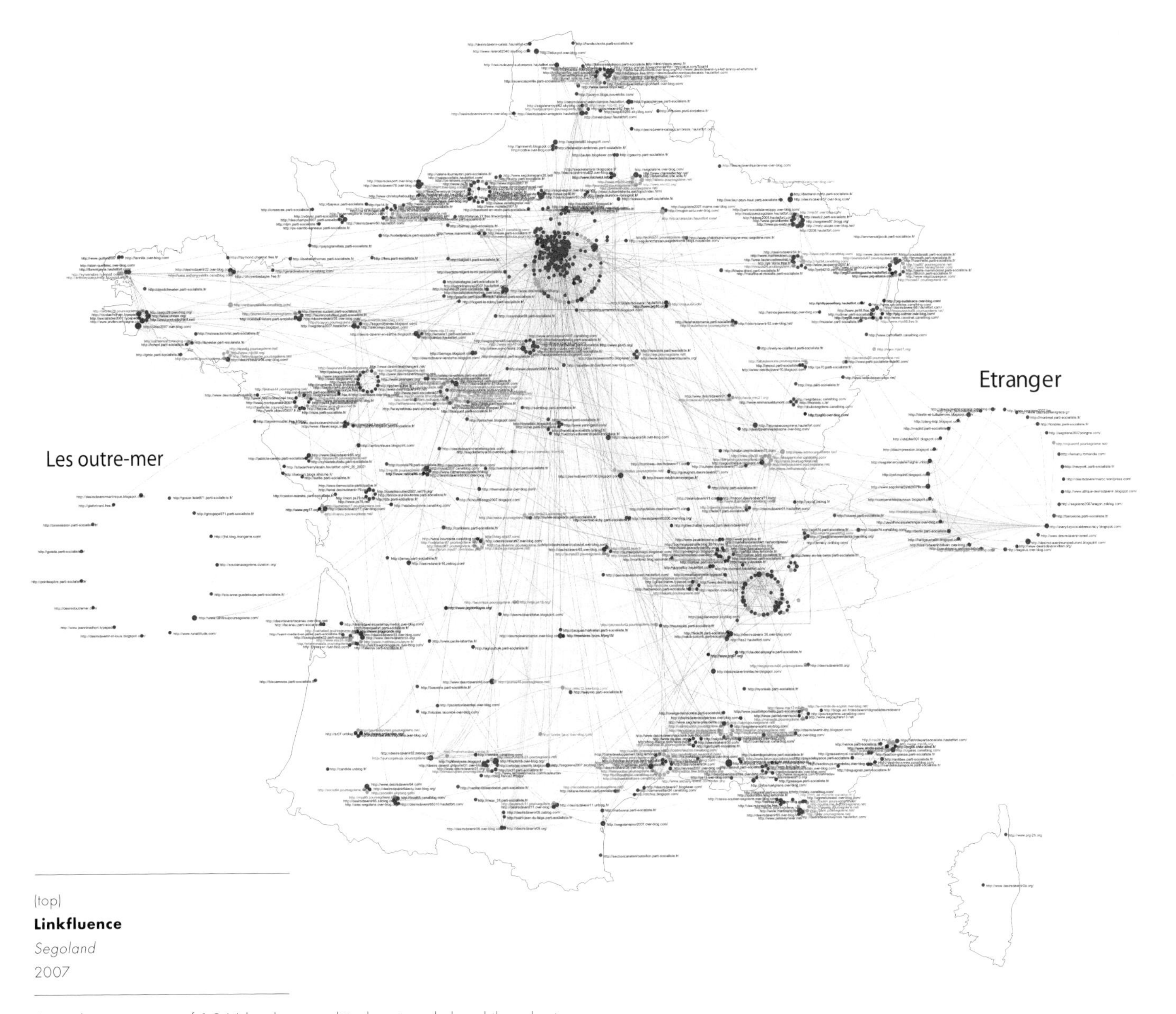

(top)

Linkfluence

Segoland

2007

A visual representation of 1,044 local blogs and websites that supported the 2007 French presidential candidate Segolene Royal. The depicted blogs, tied by shared links, are grouped by political party and placed throughout a map of the French territory according to their geographic coordinates.

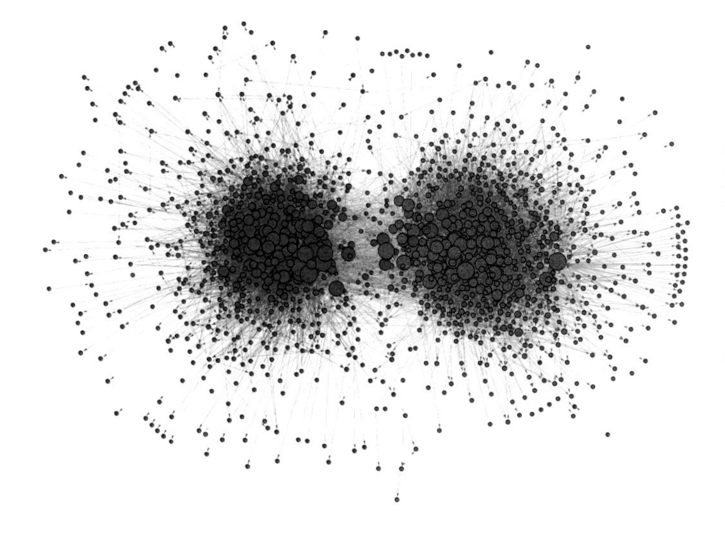

(left)

Lada Adamic and Natalie Glance

The 2004 U.S. Election and Political Blogosphere

2005

A map of the American political blogosphere. The authors analyzed the posts of forty A-list political blogs—of both liberal (blue) and conservative (red) parties—over the period of two months preceding the 2004 U.S. presidential election to calculate the frequency of reference of the name of a candidate by another candidate and to quantify the overlap in the topics discussed on the blogs, both within and across parties.

Citations

Bibliographic citation is a common practice in academic publications and an important measure of popularity and credibility among scholarly circles. It also makes for an accurate means to ascertain relationships of similarity between subjects. If two works are cited by a third, a connection can be inferred between the first two, even if they do not cite each other. This approach can be administered to a large body of books and research papers, creating vast matrices of association and highlighting proximity across domains.

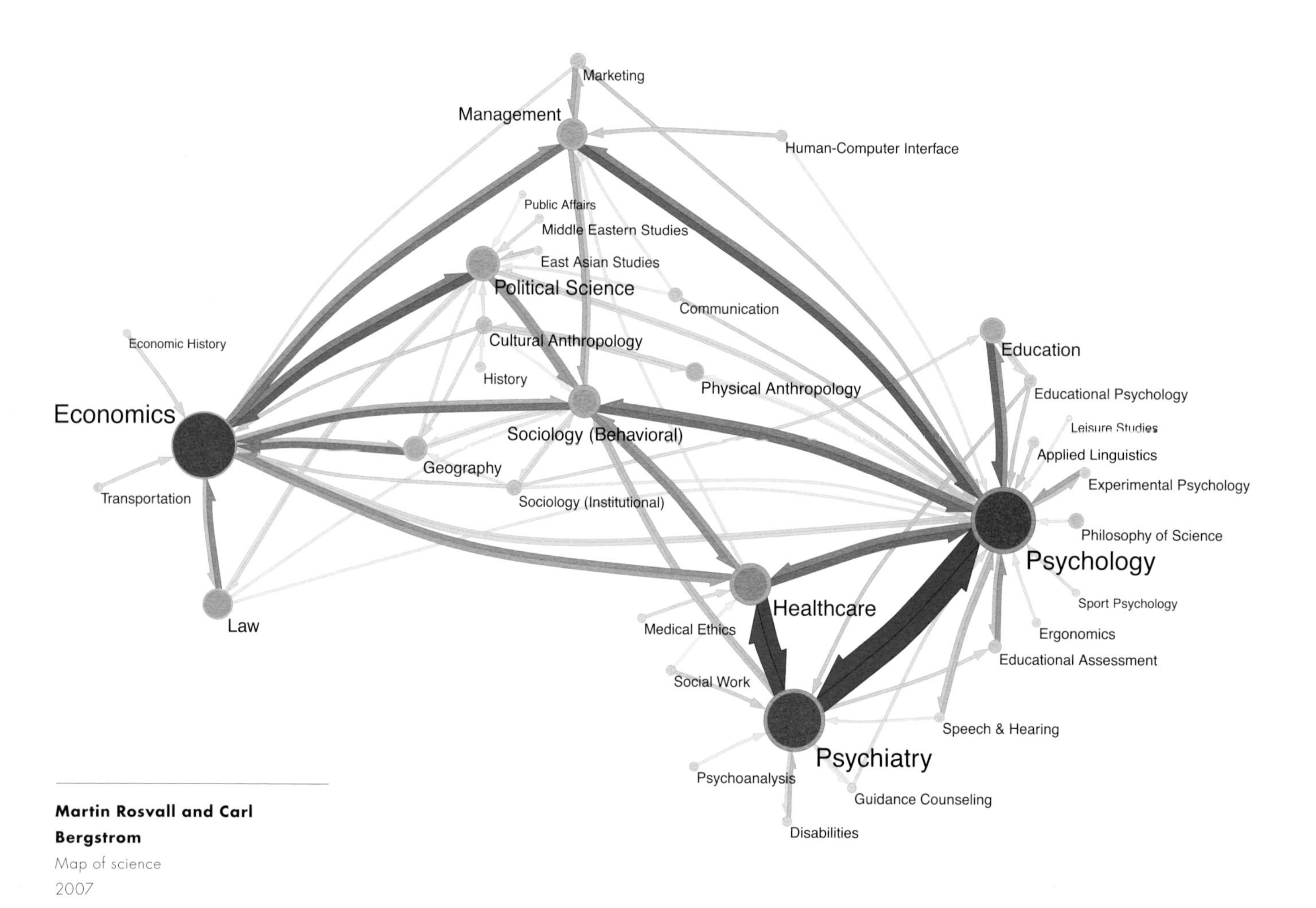

Martin Rosvall and Carl Bergstrom
Map of science
2007

A map of 6,434,916 citations in 6,128 scientific journals based on data from the Thomson Reuters's 2004 *Journal Citation Reports*. From Rosvall and Bergstrom, "Maps of Random Walks on Complex Networks Reveal Community Structure," *Proceedings of the National Academy of Sciences (PNAS) USA* 105, no. 4 (January 29, 2008): 1118–23.

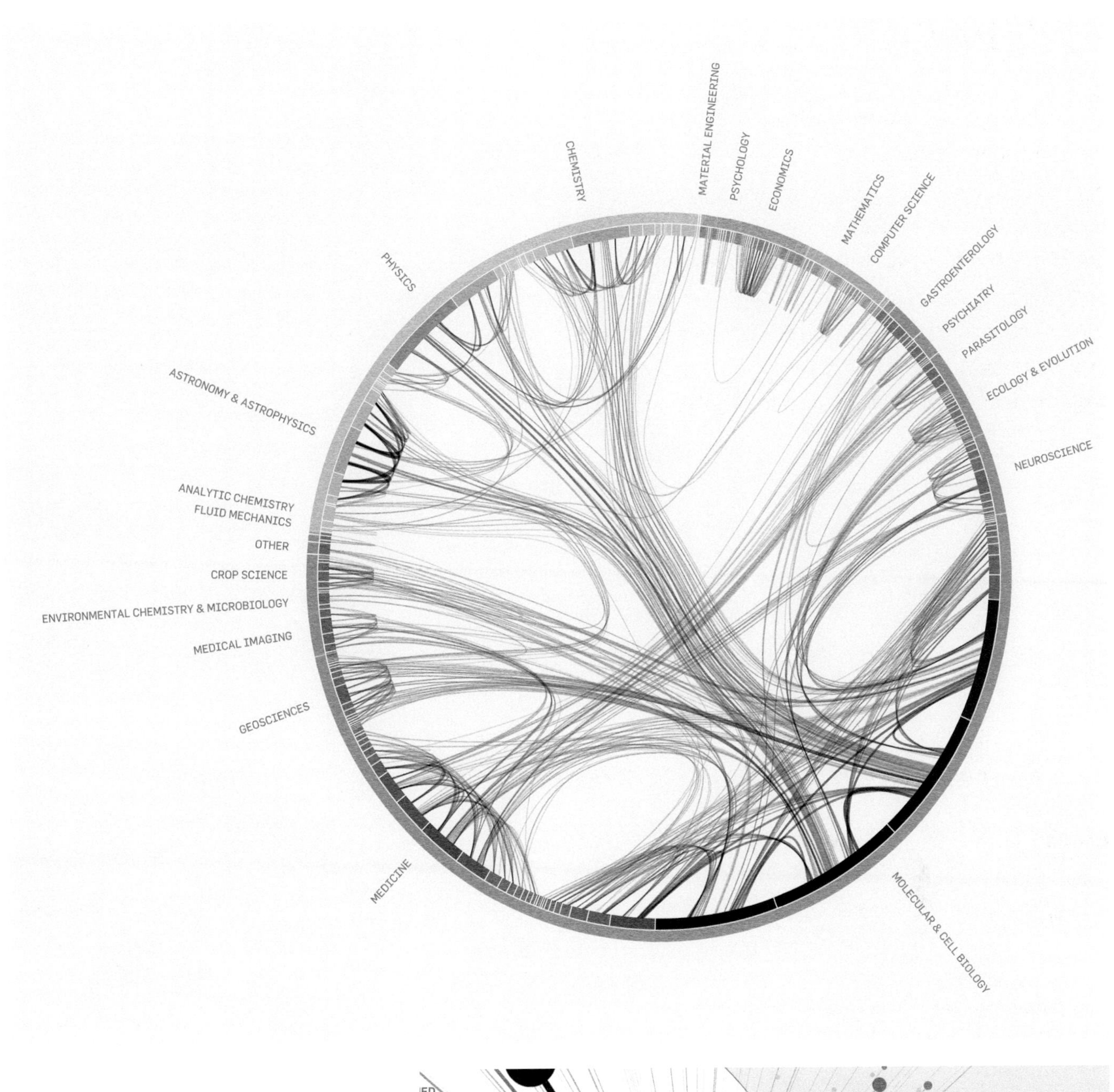

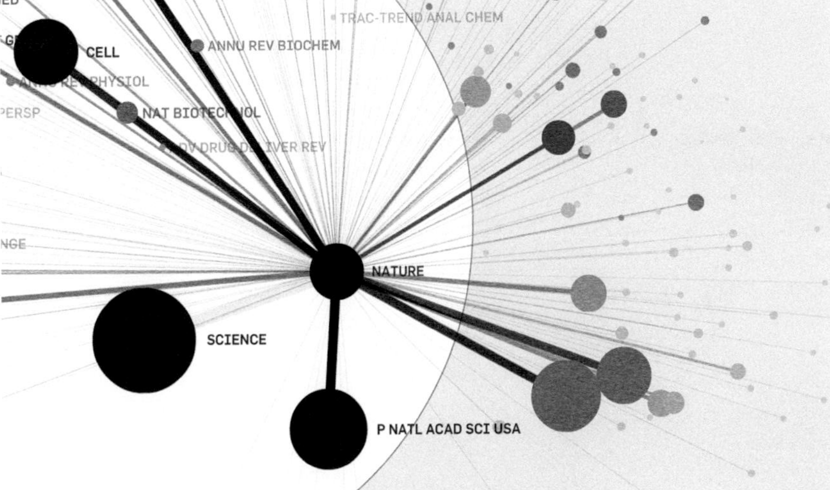

Eigenfactor.org and Moritz Stefaner

Visualizing Information Flow in Science

2009

A citation network of a subset of Thomson Reuters's *Journal Citation Reports* between 1997 and 2005 (above); detail (right)

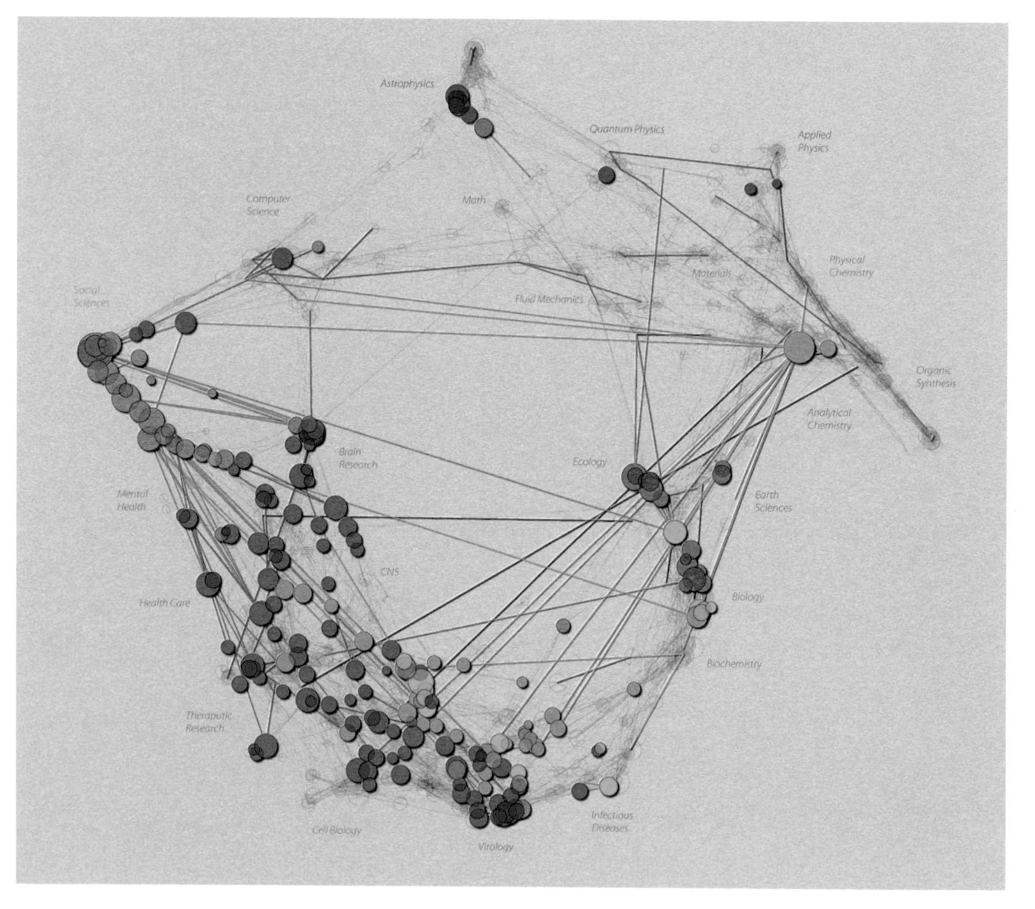

W. Bradford Paley, Dick Klavans, and Kevin Boyack
The Strengths of Nations
2006

A visualization of the most active scientific areas in the United States, based on references (how often articles were cited by authors of other articles) in roughly eight hundred thousand scientific papers. Scientific publications (colored nodes) are divided into twenty-three major scientific topics, such as astrophysics, mathematics, and biochemistry.

W. Bradford Paley, Dick Klavans, and Kevin Boyack
The Strengths of Nations
2006

Maps of the most prolific scientific areas (according to the number of published articles) in six countries: United Kingdom, France, China, Australia, Germany, Taiwan. Nodes and edges are highlighted if a nation publishes significantly more in that area.

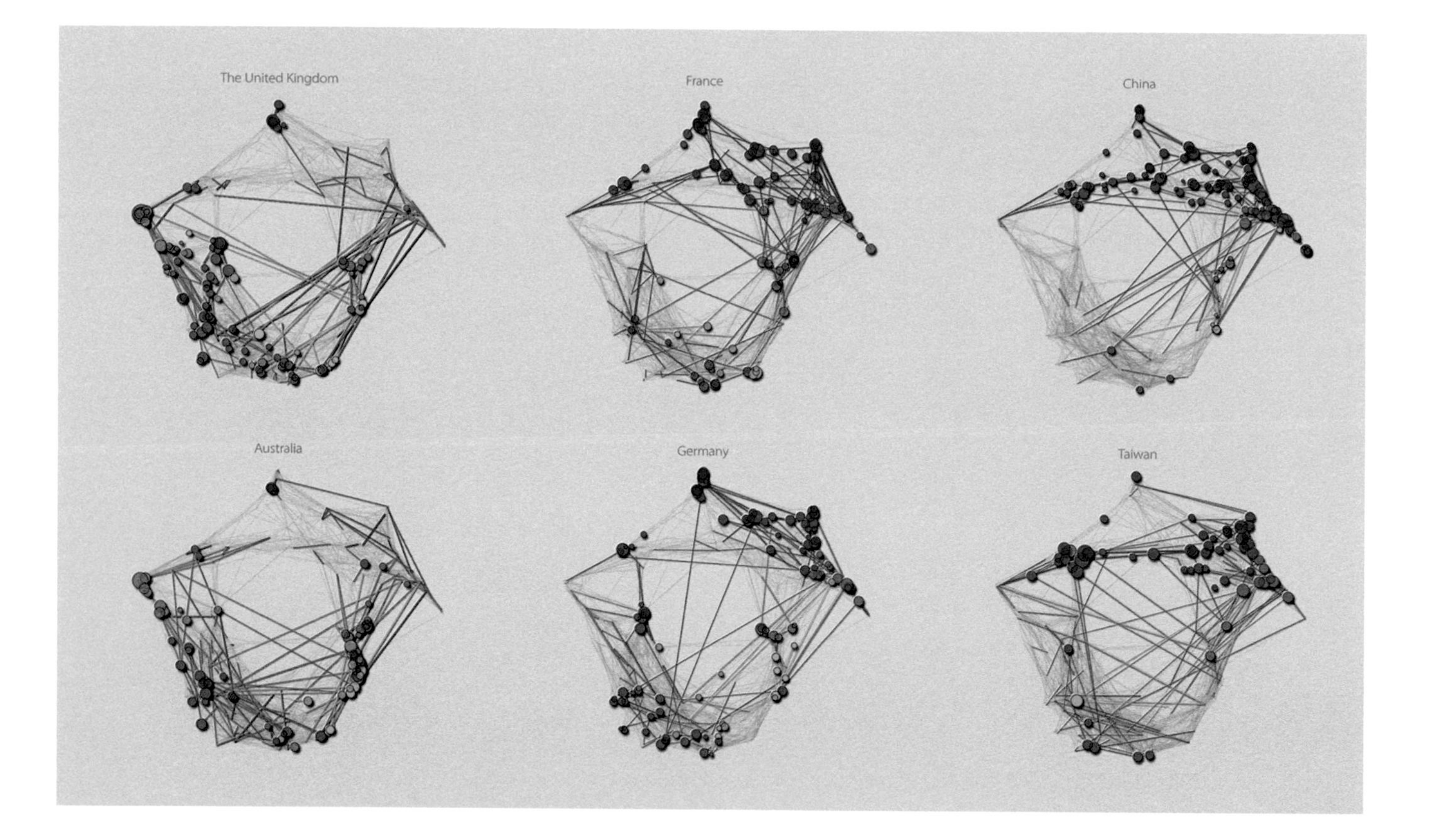

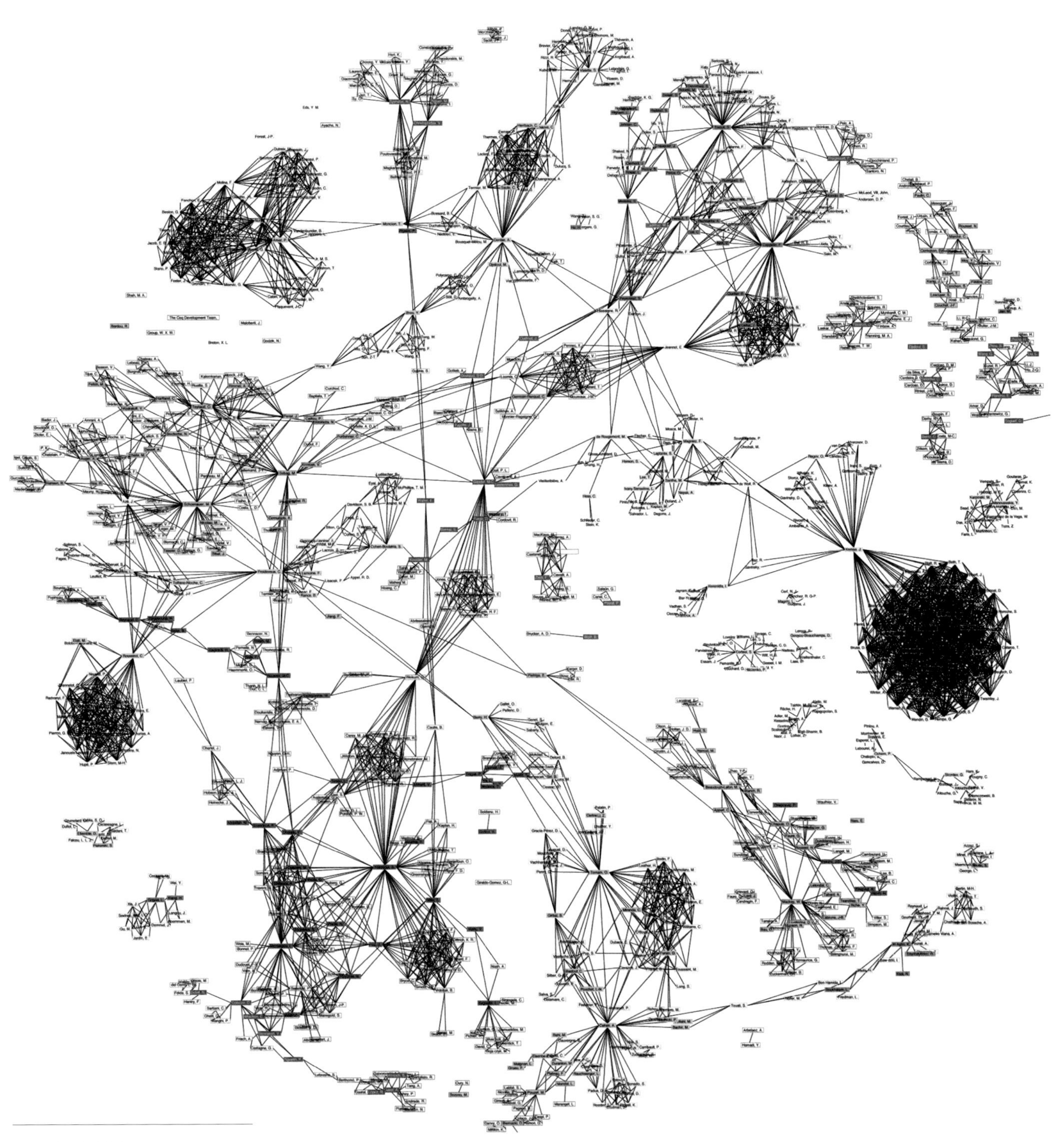

Jean-Daniel Fekete

LRI Co-authorship Network

2007

A graph showing the coauthorship of papers written by members of the Laboratoire de Recherche en Informatique (LRI), a computer science laboratory at the University Paris-Sud, France

Del.icio.us

Founded in 2003, Del.icio.us was a pioneer social-bookmarking system and one of the precursors of folksonomy. Like many subsequent bookmarking systems, Del.icio.us makes the recording of new information extremely easy. However, as the collection of bookmarks grows over time, it is easy to get lost in the pile of tags. Many visualization authors have tried to come up with alternative ways of visualizing their personal tagging systems, either to improve the retrieval process or simply to unearth their own indexing behaviors. Del.icio.us ultimately embodies an extraordinary epistemological laboratory where one can draw interesting insights on how humans collectively classify information.

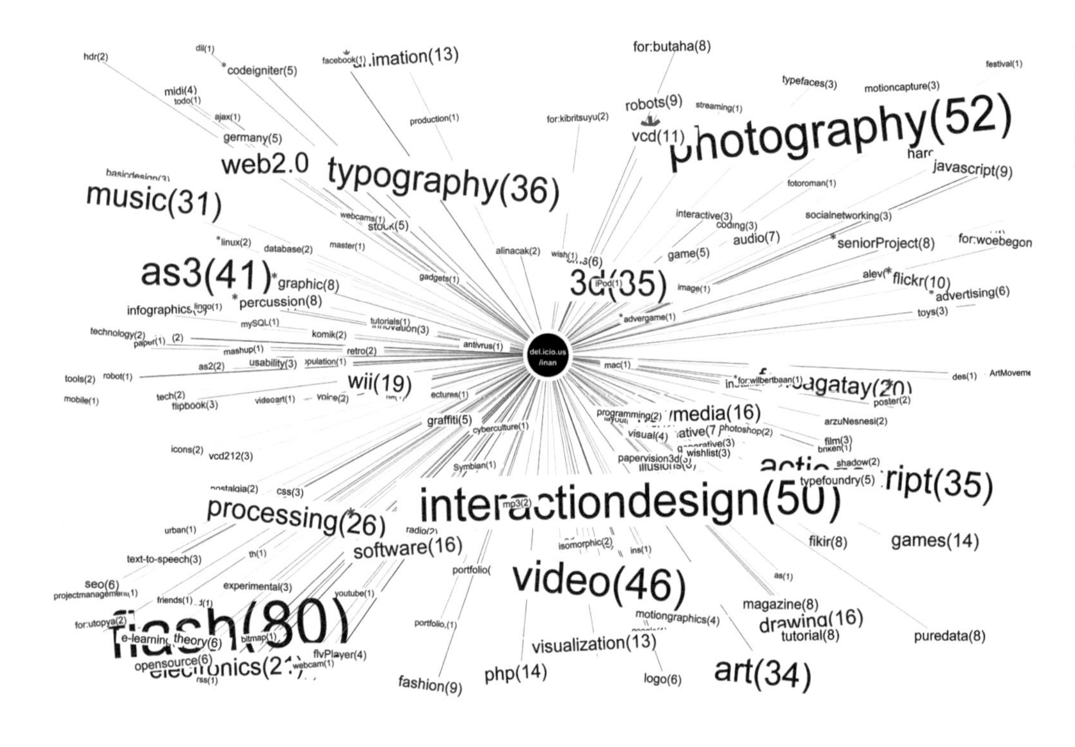

Inan Olcer
Delicious Tag Cloud
2006

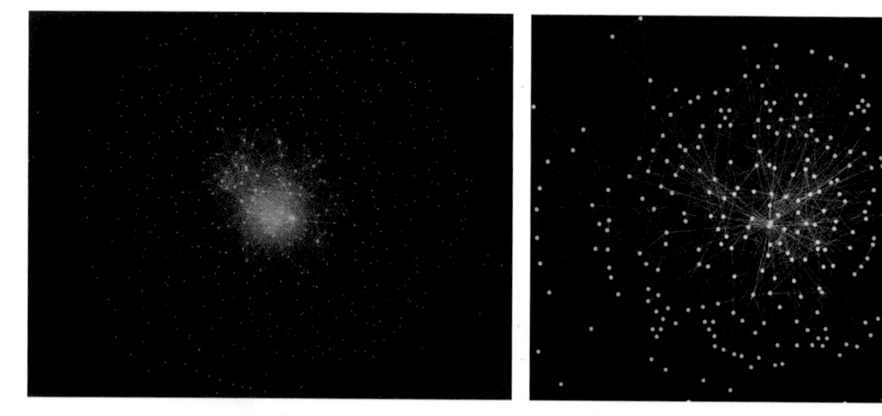

(left and right)
Kunal Anand
Looks del.icio.us.
2008

A graph of tag relationships in an individual's Del.icio.us account

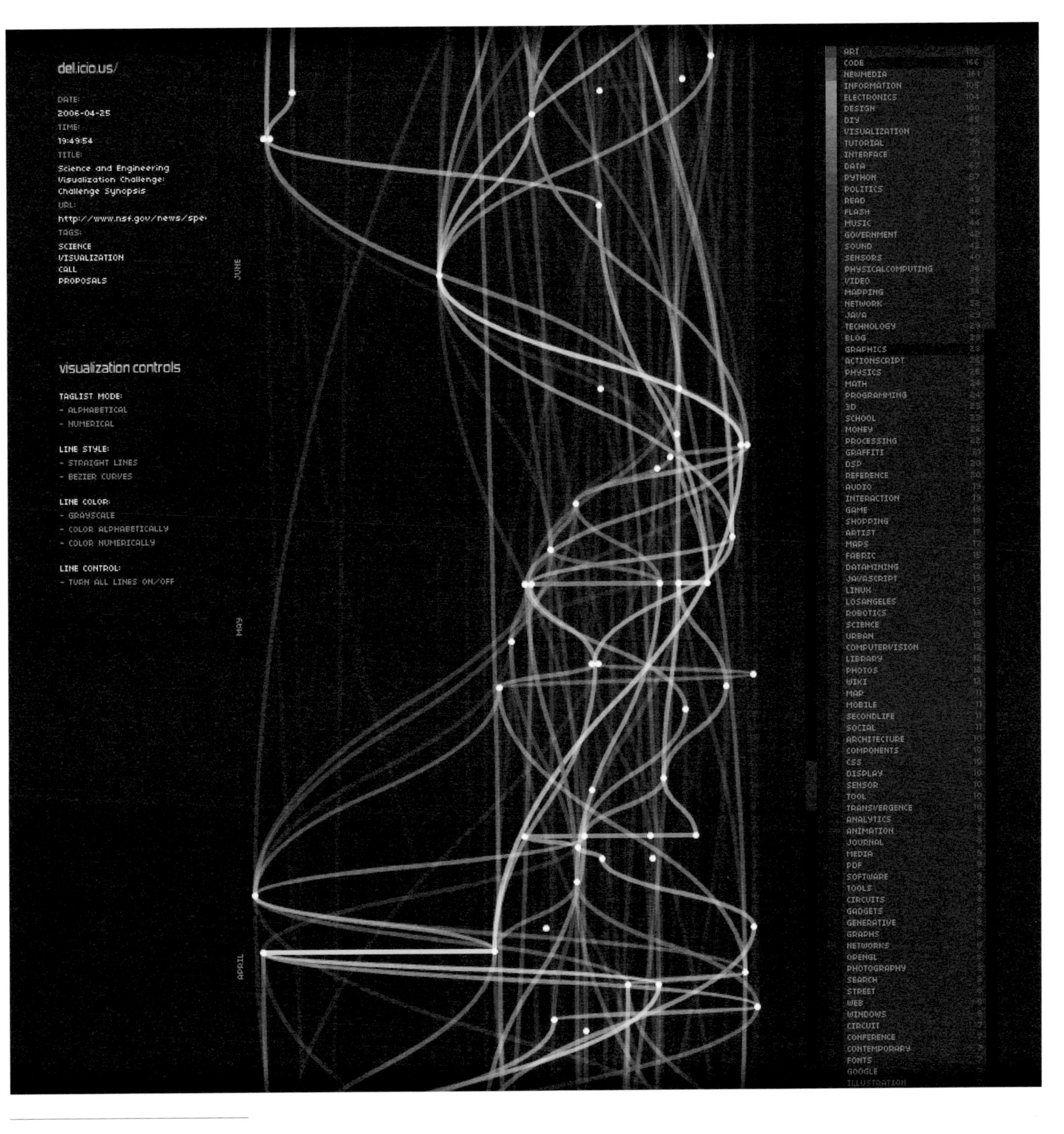

Aaron Siegel
Research Chronology 2
2006

A chronological visualization of an individual's bookmarking activity on Del.icio.us. It maps research activity along a time line, tying bookmarks together with colored tag lines. Each line represents a tag, which changes from yellow to red, depending on the number of bookmarks using that tag.

Ian Timourian
del.icio.us.discover
2006

A comparative analysis of tagging behavior between different Del.icio.us users

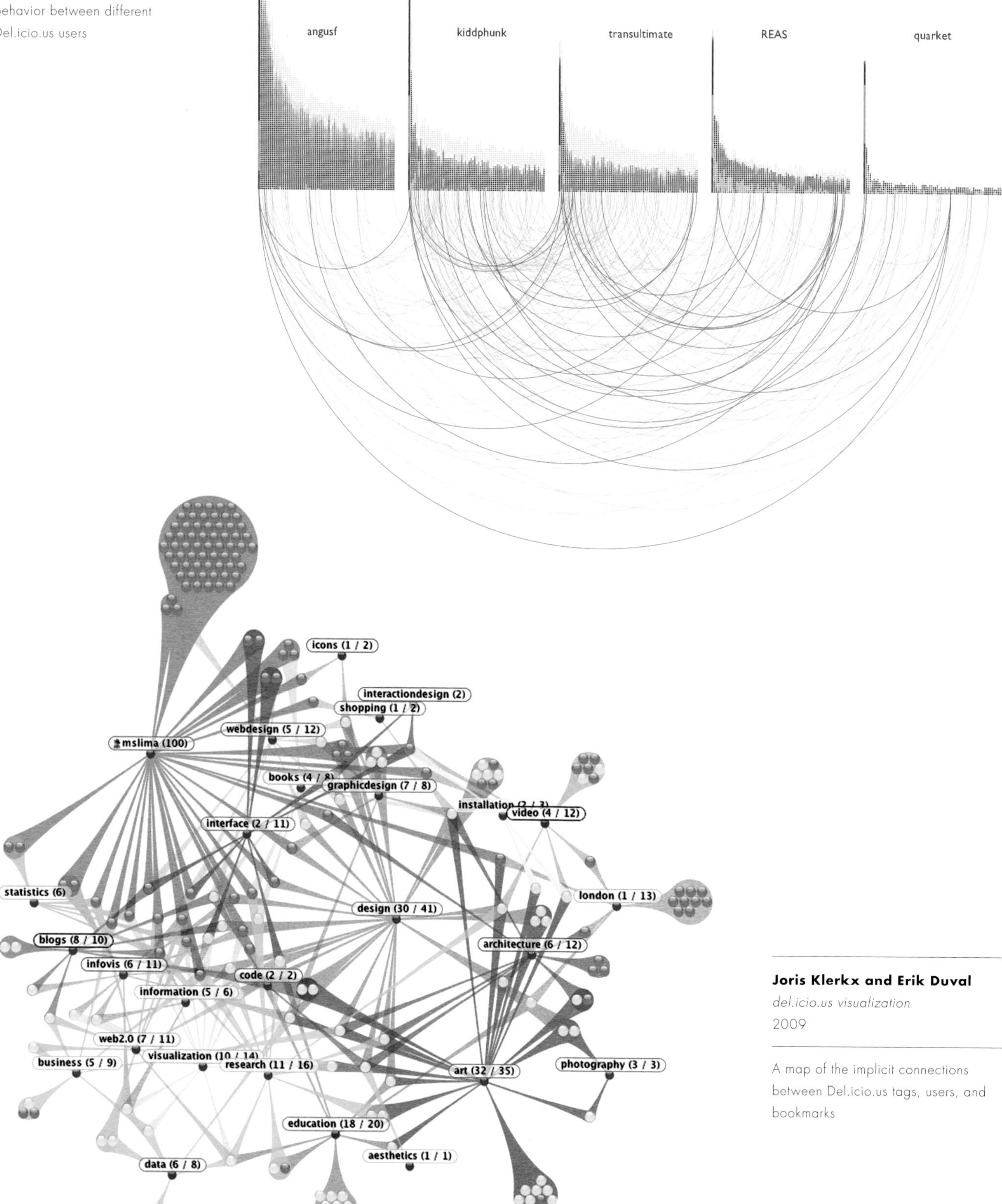

Joris Klerkx and Erik Duval
del.icio.us visualization
2009

A map of the implicit connections between Del.icio.us tags, users, and bookmarks

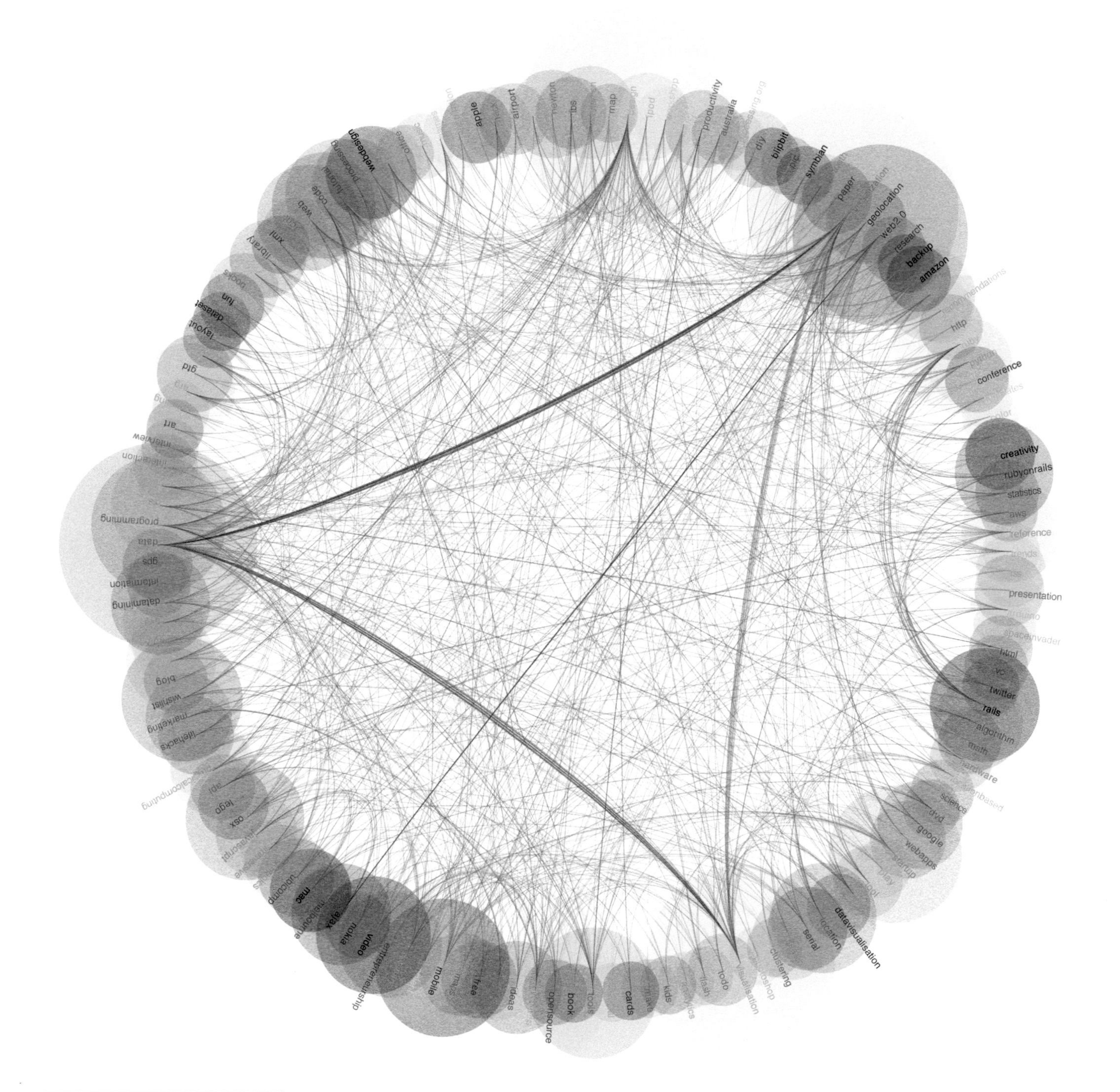

Flink Labs

Delicious Circle

2009

A visualization of an individual's Del.icio.us tags and the connections that emerge between them

Donations

An extraordinary outcome of the growing number of freely accessible data has been the outbreak of public visualization projects scrutinizing various business and political practices. Many of these initiatives convey the flow of private and public donations to political candidates but also explore lesser-known associations, such as the funding of political and environmental organizations by gas and oil companies.

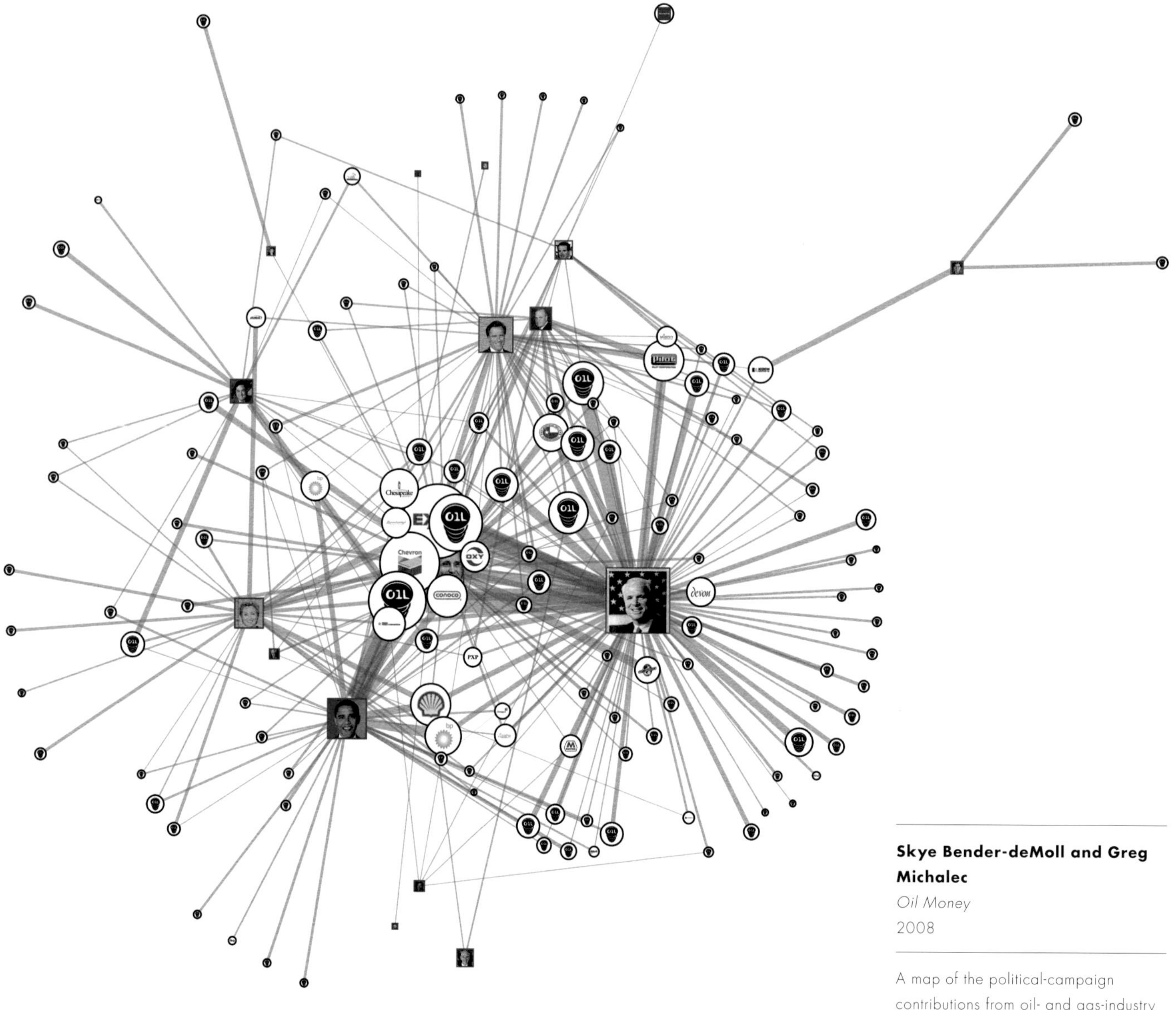

Skye Bender-deMoll and Greg Michalec
Oil Money
2008

A map of the political-campaign contributions from oil- and gas-industry companies to presidential candidates in 2008

Wesley Grubbs and Nick Yahnke

2008 Presidential Candidate Donations

2008

A map of all donations made between January 2007 and July 2008 to John McCain (in red) and Barack Obama (in blue). The area of the two inner semicircles represents the total amount of donations for the candidates, and the outer segments, fanning from the center, illustrate variations in the amounts donated. On either side of the inner semicircles, the top fan represents any donation between $1 and $100; the middle narrow fan, $101–$500; the bottom fan, $501–$1,000; and finally, all amounts over $1,000 are depicted by two lightweight fans on the very bottom of the semicircles.

Wesley Grubbs and Nick Yahnke
2008 Presidential Candidate Donations: Job Titles of Donors
2008

A map of all donations made between January 2007 and July 2008 to Barack Obama, sorted by the 250 most common job titles held by the donors

Josh On and Amy Balkin
Exxon Secrets
2005

A visualization platform developed for Greenpeace, exposing Exxon-Mobil's funding of climate-change-skeptic think tanks, conservative institutions, and other organizations

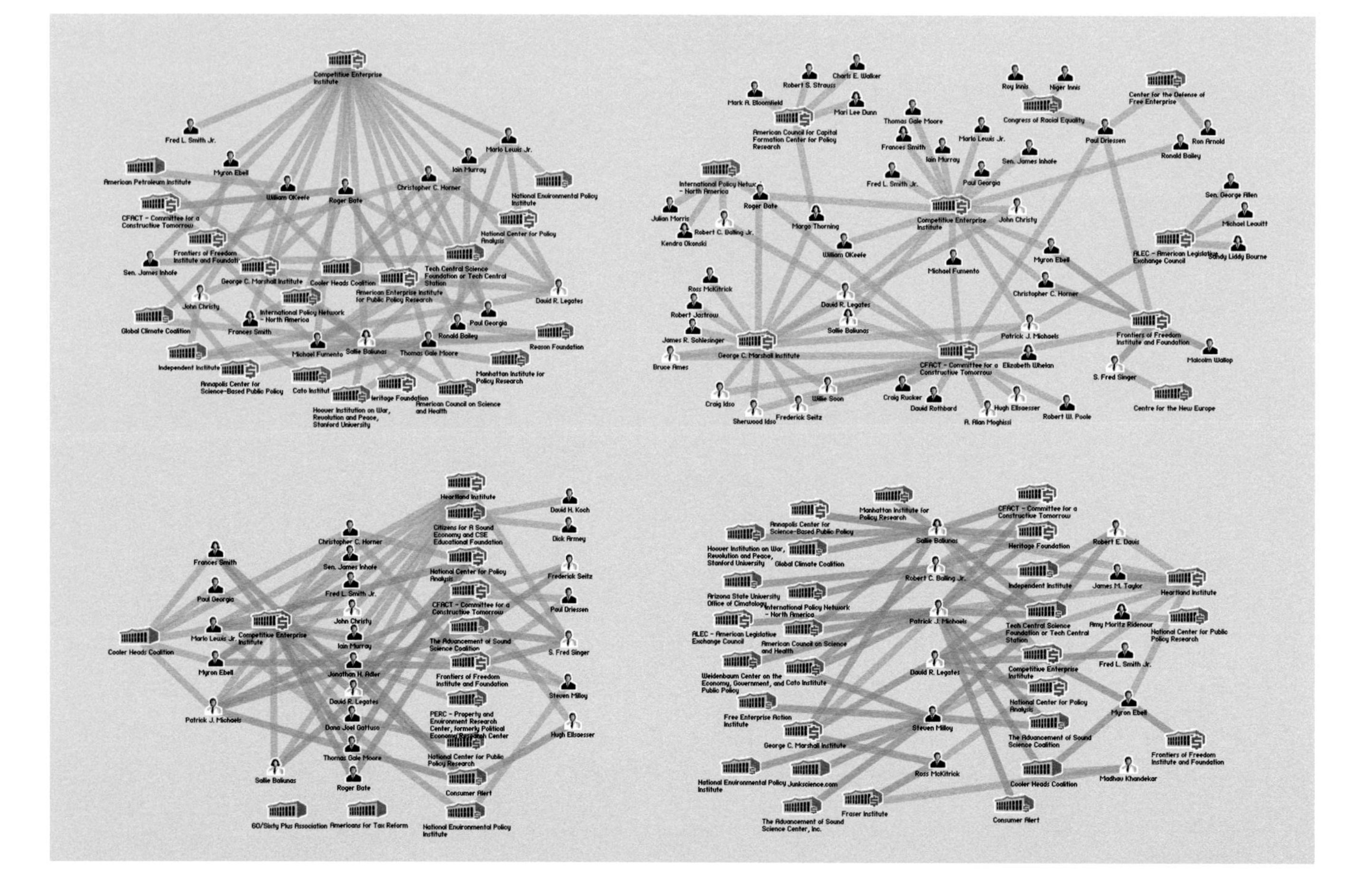

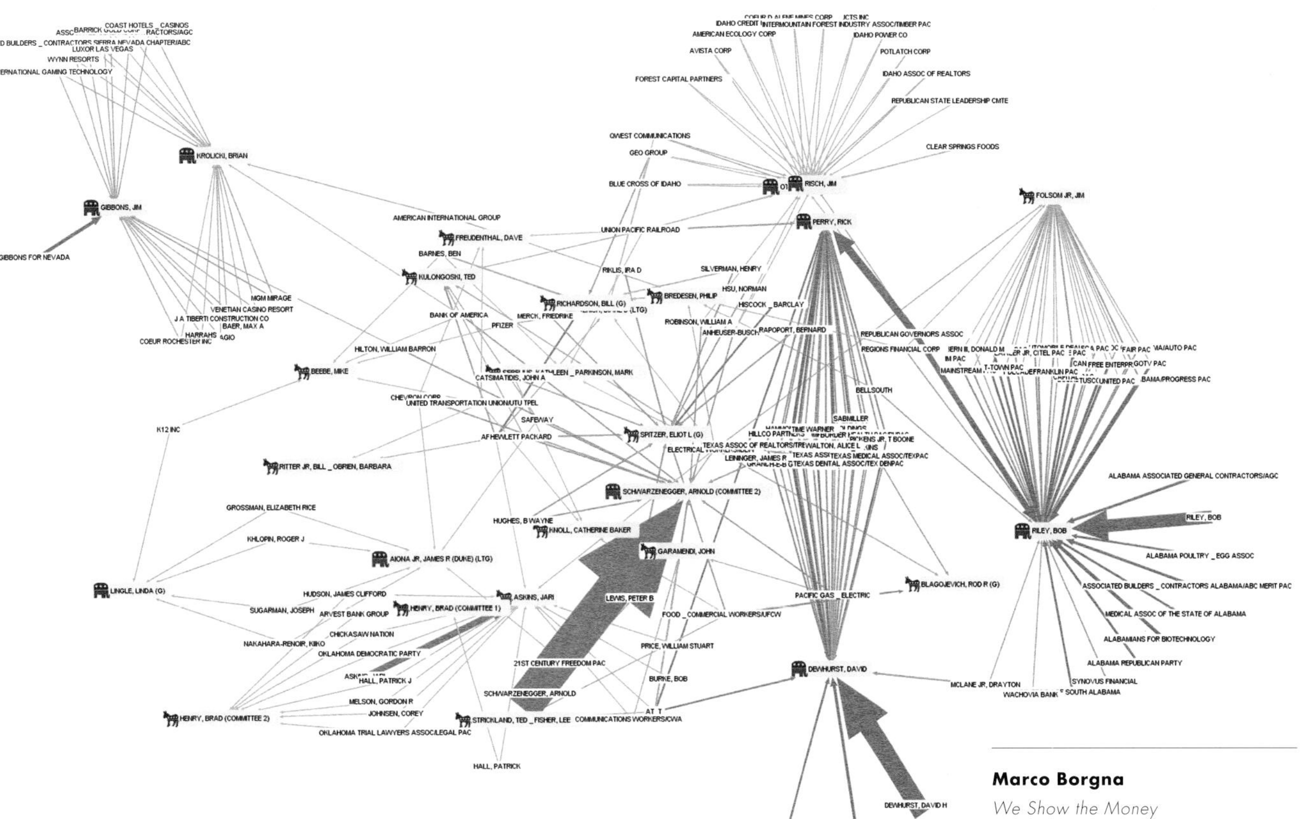

Marco Borgna

We Show the Money

2007

A visualization that reveals the connection between U.S. governors elected in 2006 and the top one hundred donors. It shows only those who donated to more than one candidate or made at least one single donation greater than $100,000.

Jute Networks

PublicMaps

2008

A sociogram of individual donations made in Asheville, North Carolina, to the 2008 U.S. presidential candidates

Email

Email is one of the most important communication channels in modern society. It is estimated that two million emails are sent every second across the world, and for many of us this number translates into copious amounts of messages that we send and receive every single day. In order to better understand the habits and social behavioral patterns of our in-boxes, many authors depict these rich containers in various ways, usually by looking at the complex social structures created within email lists of companies, schools, and institutions, with the Enron email data set being a popular example.

Matthias Dittrich
5 Years Designerlist
2008

Visualizations showing all of the emails exchanged in two twenty-four-hour time periods at the University of Applied Sciences Potsdam, Germany

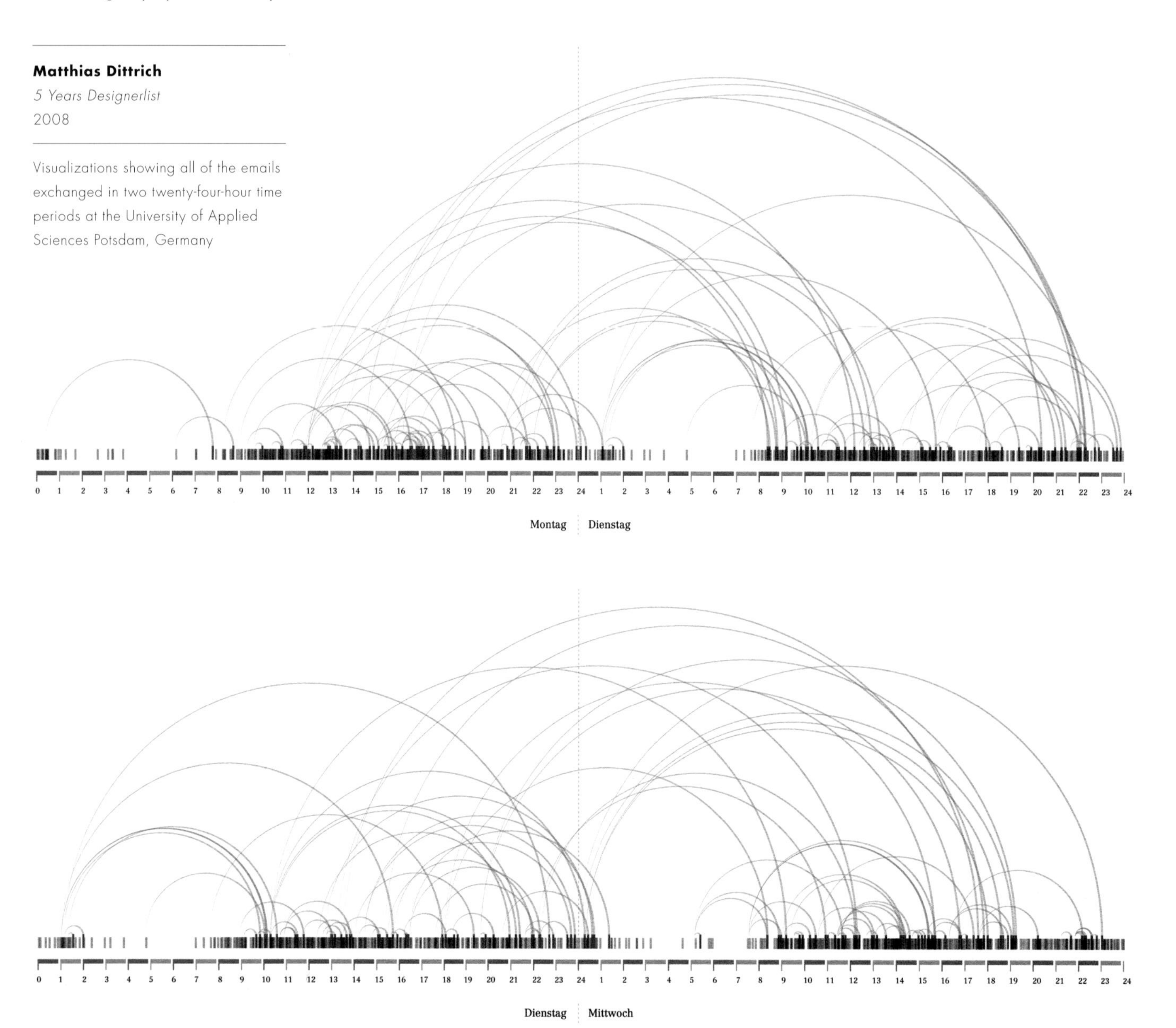

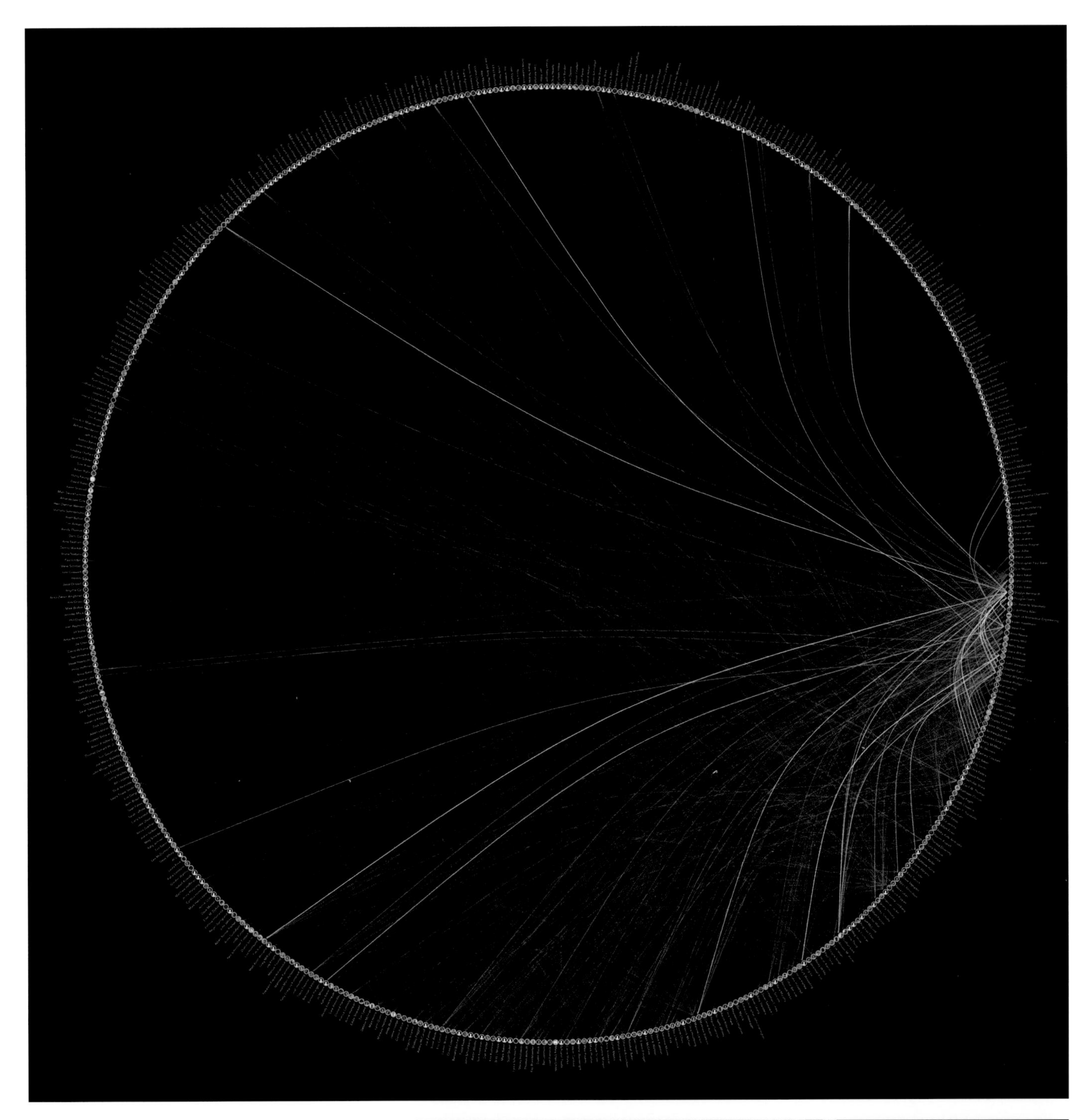

Christopher Paul Baker
Email Map
2007

A rendering of the relationships between Baker and individuals in his address book generated by examining the *to*, *from*, and *cc* fields of every email in his in-box archive (above); detail (right)

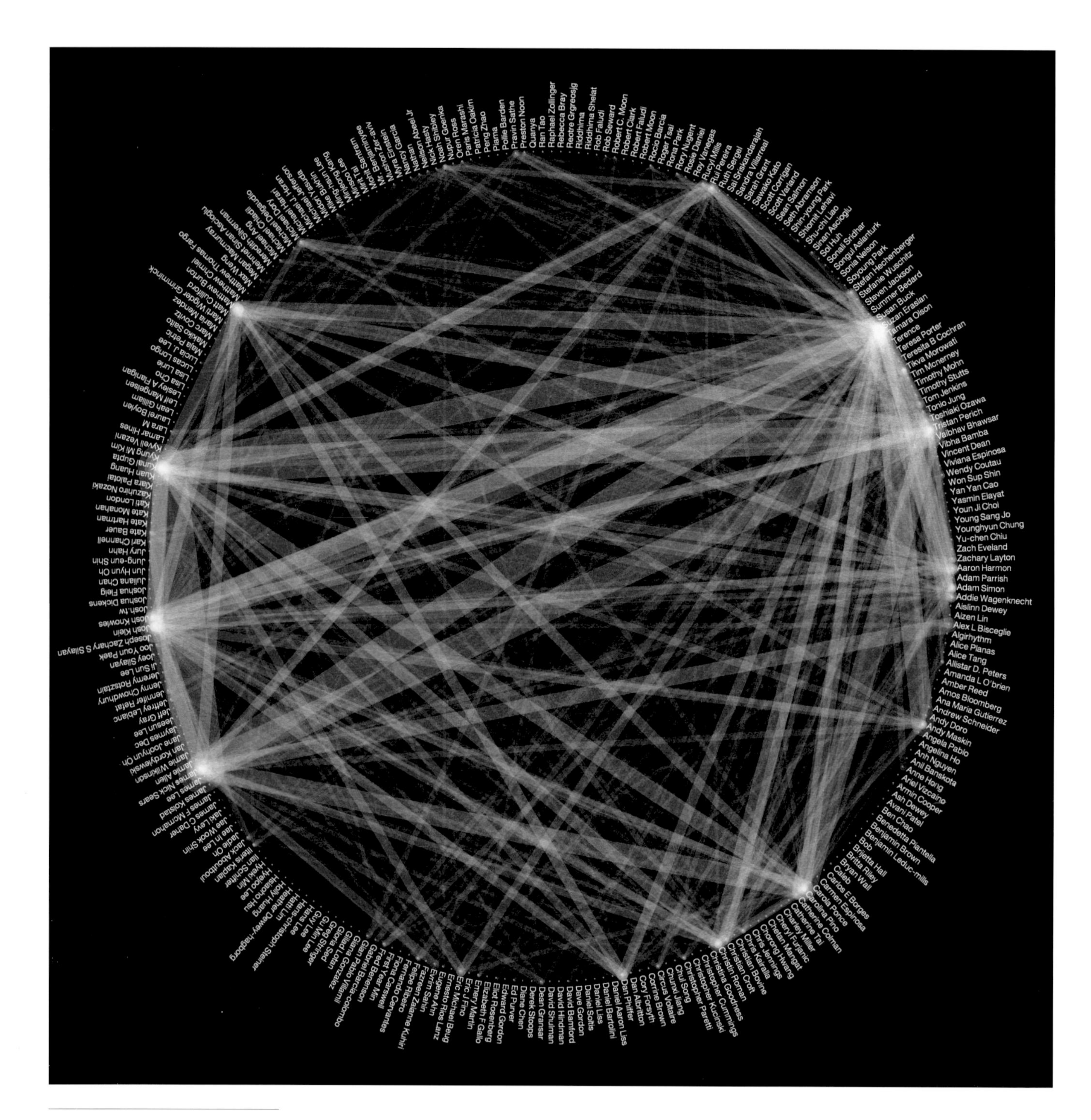

Josh Knowles
ITP Student List Conversations
2007

A visualization of the email conversations that occurred within the last four months of 2006 by the students of the Interactive Telecommunications Program (ITP) at the Tisch School of the Arts, New York University. The amount of conversation between two people was determined by the number of emails they each sent to the same discussion threads; the stronger the connection, the heavier the line between them.

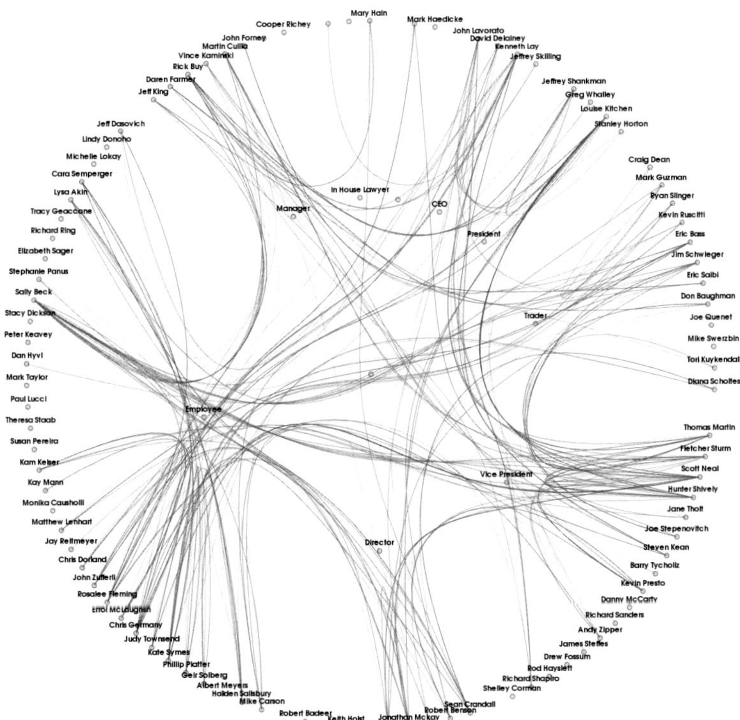

Kitware Inc.

Enron Communication Graph

2008

A map of the links between all the employees' in-boxes—the emails sent internally to each other—that were examined as a part of the Enron-crash investigation

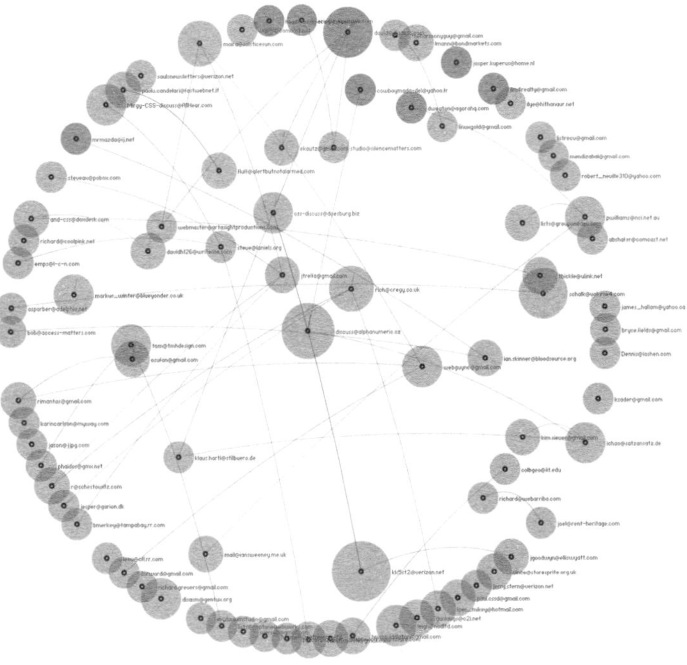

Marcos Weskamp

Social Circles

2003

A chart of social interactions (based on exchanged emails) within a mailing list. Nodes represent members of the mailing list, while each node's size illustrates the frequency of posts by that member. Links between members depict communication in reply to a thread.

Internet

The internet is an intriguing domain for many people. With its vast network of servers and routers, linked by copper wires and fiber-optic cables, this hidden landscape that spans the globe represents a noteworthy target in the new age of technological discovery. As with many newly exposed territories, its visual depiction is the very first step in awareness and understanding of its inherent structure. The question of what the internet looks like has compelled many authors to create striking visualizations of different facets of the system.

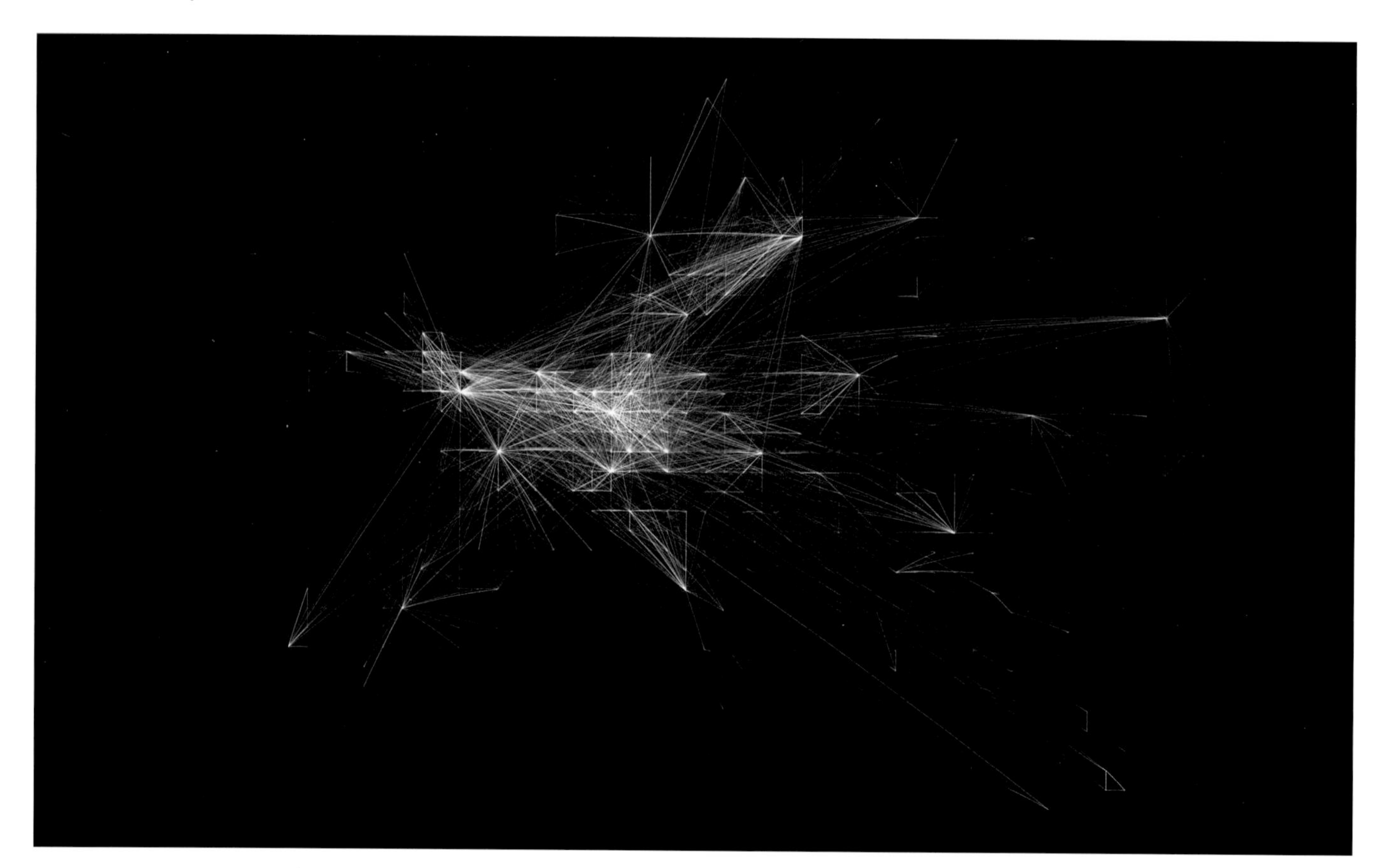

Chris Harrison
Internet Map
2007

A map of internet connectivity in Europe, based on router configuration. The opacity of the edge line reflects the number of connections between the two points. Data was collected from the Dimes Project, a scientific-research project aimed at studying the structure and topology of the internet.

Barrett Lyon

Opte Project

2003

A complete internet map from November 23, 2003, displaying over five million links across millions of IP addresses in several regions of the world: Asia Pacific (red), Europe/Middle East/Central Asia/Africa (green), North America (blue), Latin America/Caribbean (yellow), private networks (cyan), unknown (white).

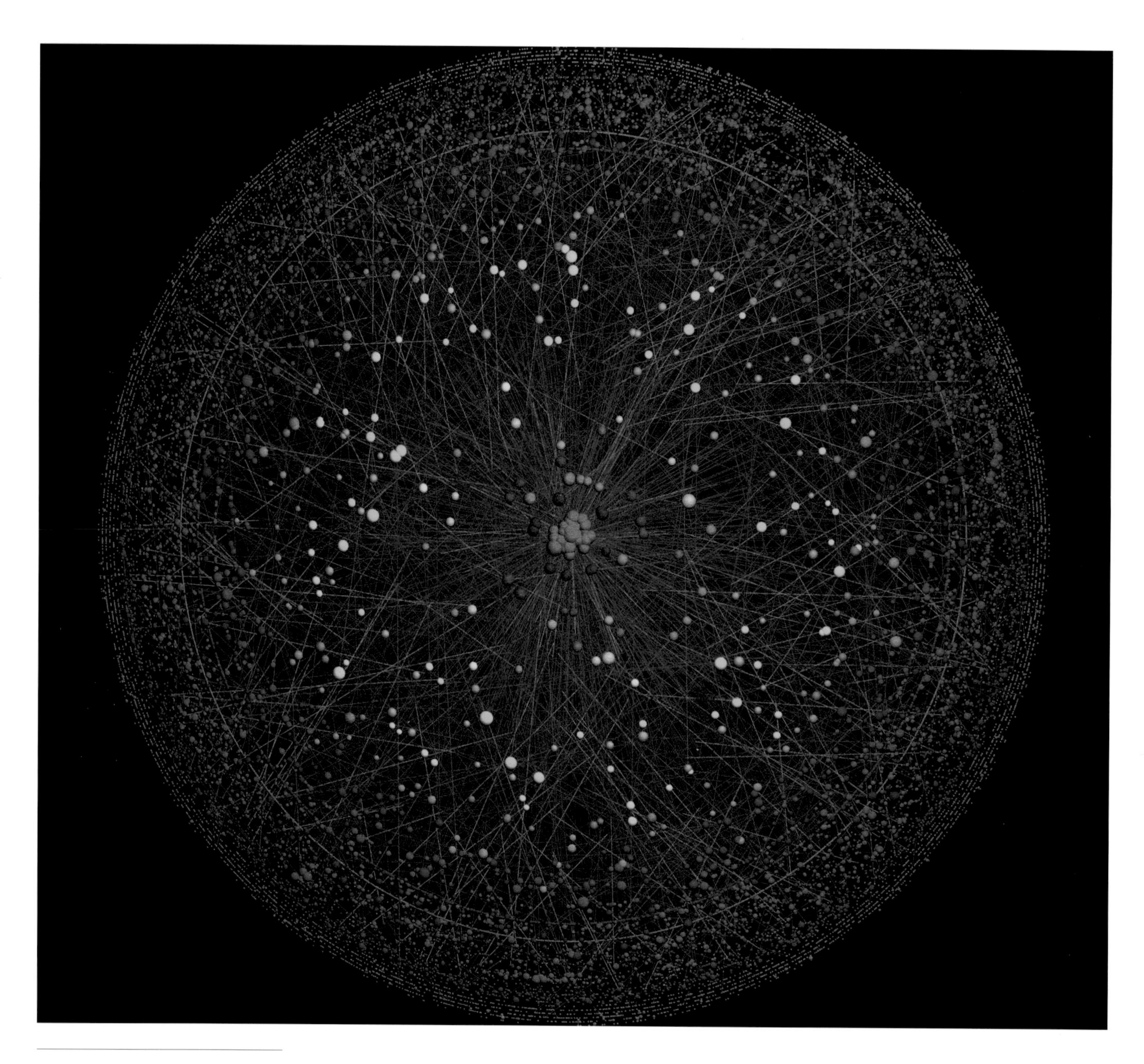

J. I. Alvarez-Hamelin, M. Beiró, L. Dall'Asta, A. Barrat, and A. Vespignani
Internet Autonomous Systems
2007

A map of interconnectivity between autonomous systems (a collection of IP prefixes controlled by one or more network operators) based on data obtained from Oregon Route Views Project, a tool for internet operators to obtain real-time information about the global routing system that makes up the internet. Red indicates the most connected nodes; violet, the least connected ones.

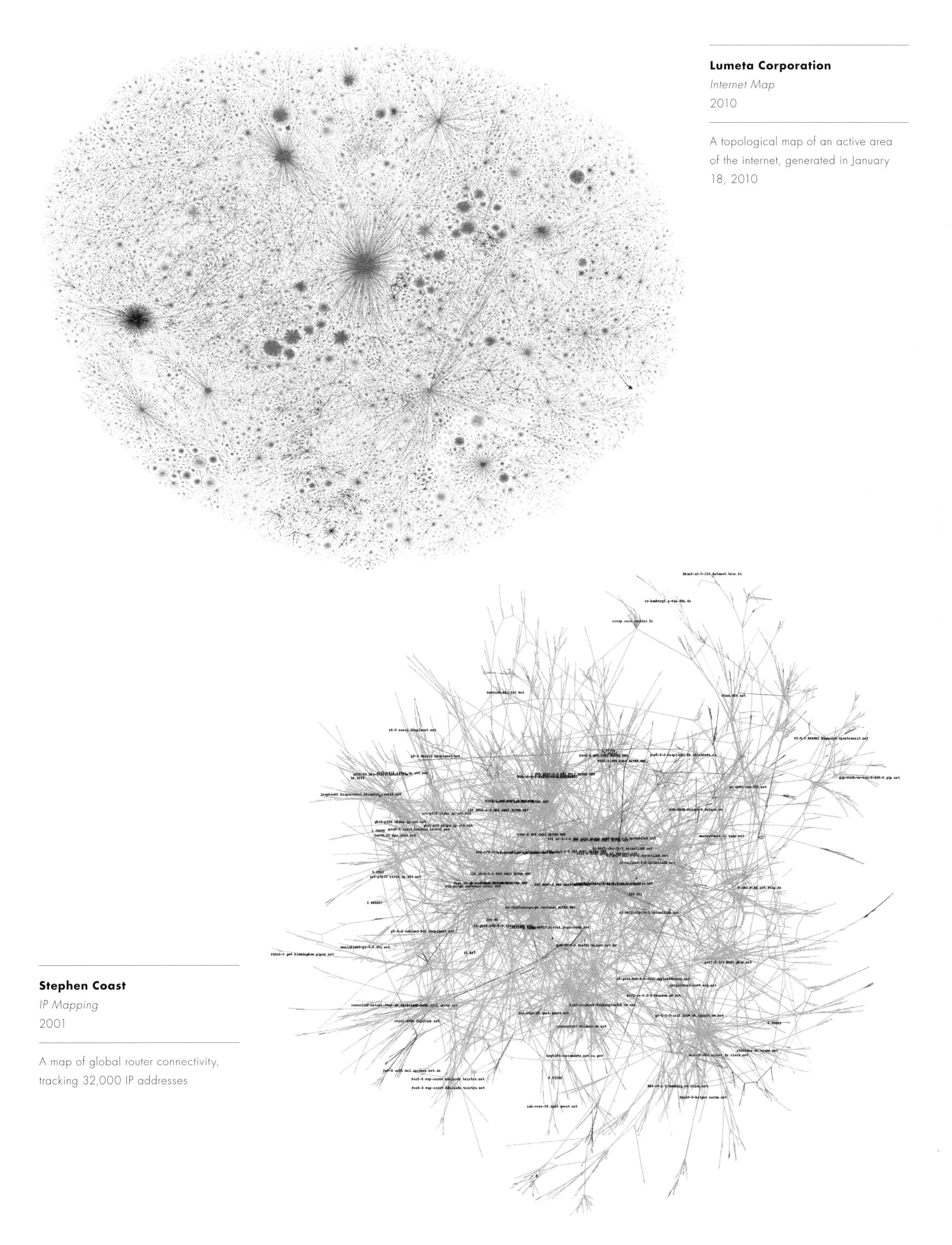

Lumeta Corporation

Internet Map

2010

A topological map of an active area of the internet, generated in January 18, 2010

Stephen Coast

IP Mapping

2001

A map of global router connectivity, tracking 32,000 IP addresses

Literature

Any text presents an intricate mesh of relationships between its many lines of words. From mapping the occurrence of particular sentences and words in a single book to cross-citations of particular topics, or even ties between books, literature is a growing subject matter for network visualization. Many visualization initiatives analyze large quantities of text in order to expose relevant insights within the narrative structure and the author's style, or hidden associations between discussed subjects.

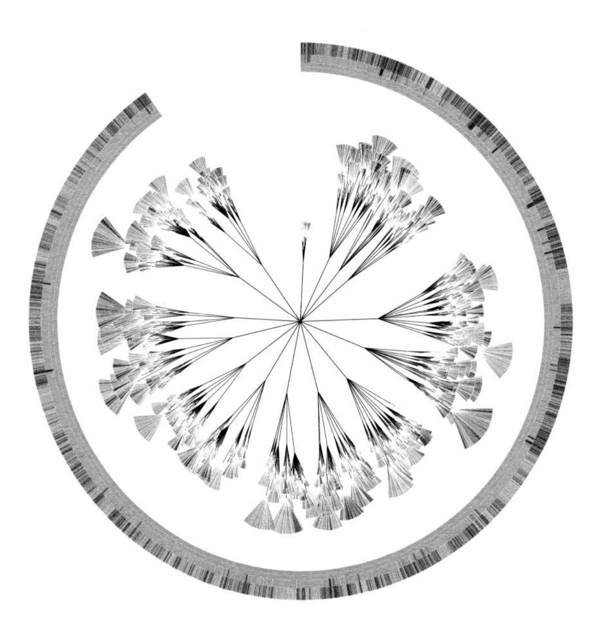

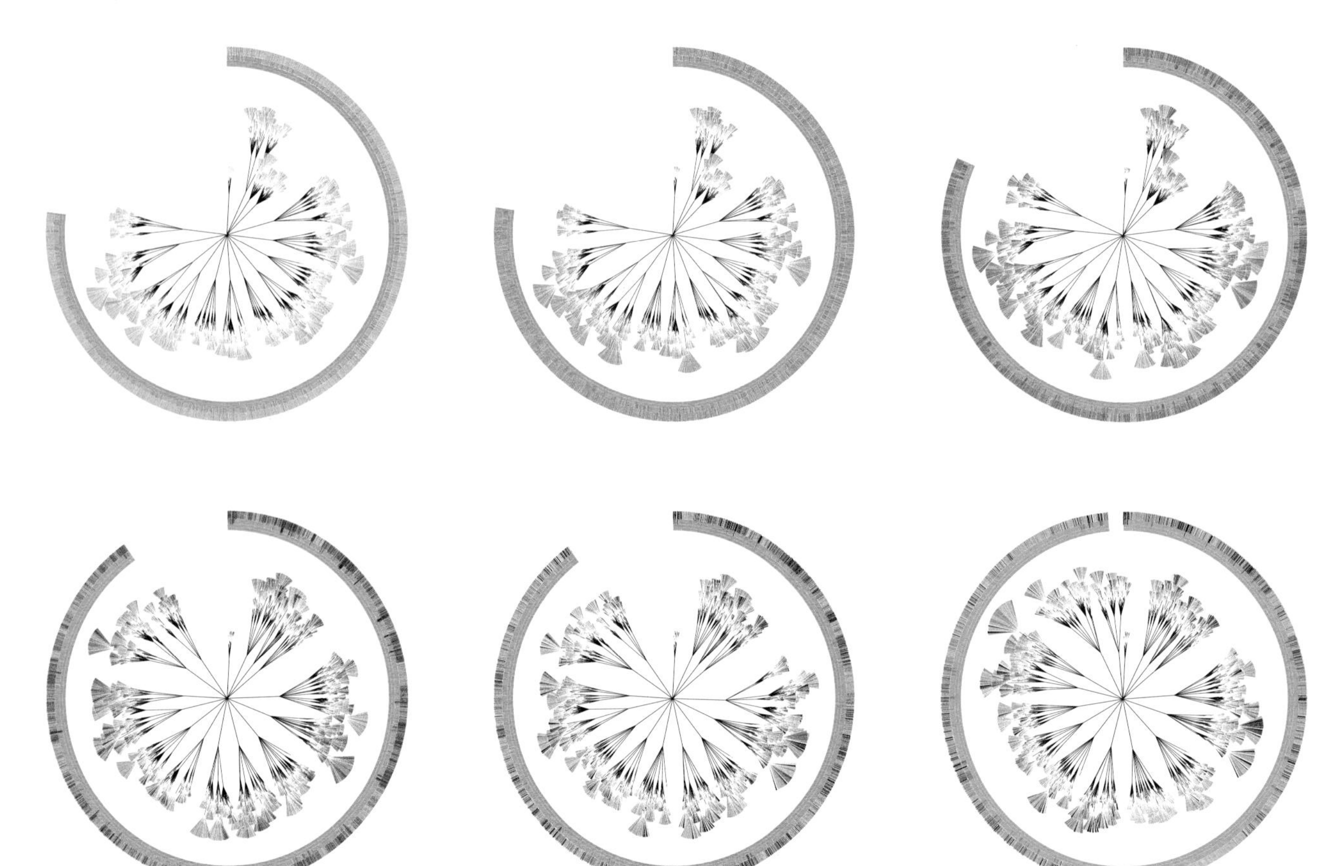

Stefanie Posavec and Greg McInerny
The Evolution of The Origin of Species
2009

Top right: A map of the chapter structure of the 1866 fourth edition of Charles Darwin's *The Origin of Species* (1859). In the inner tree, each splitting of the branch into progressively smaller sections parallels the organization of the content from chapters to subchapters, paragraphs, and sentences (represented as small wedge-shaped leaflets at the very end of the branches). The color of each sentence indicates whether the sentence survived to the fifth edition (green) or whether it was deleted from the third edition (orange). Center row, from left to right: The chapter structures of the first, second, and third editions of *The Origin of Species*. Bottom row, from left to right: The chapter structures of the fourth, fifth, and sixth editions of *The Origin of Species*.

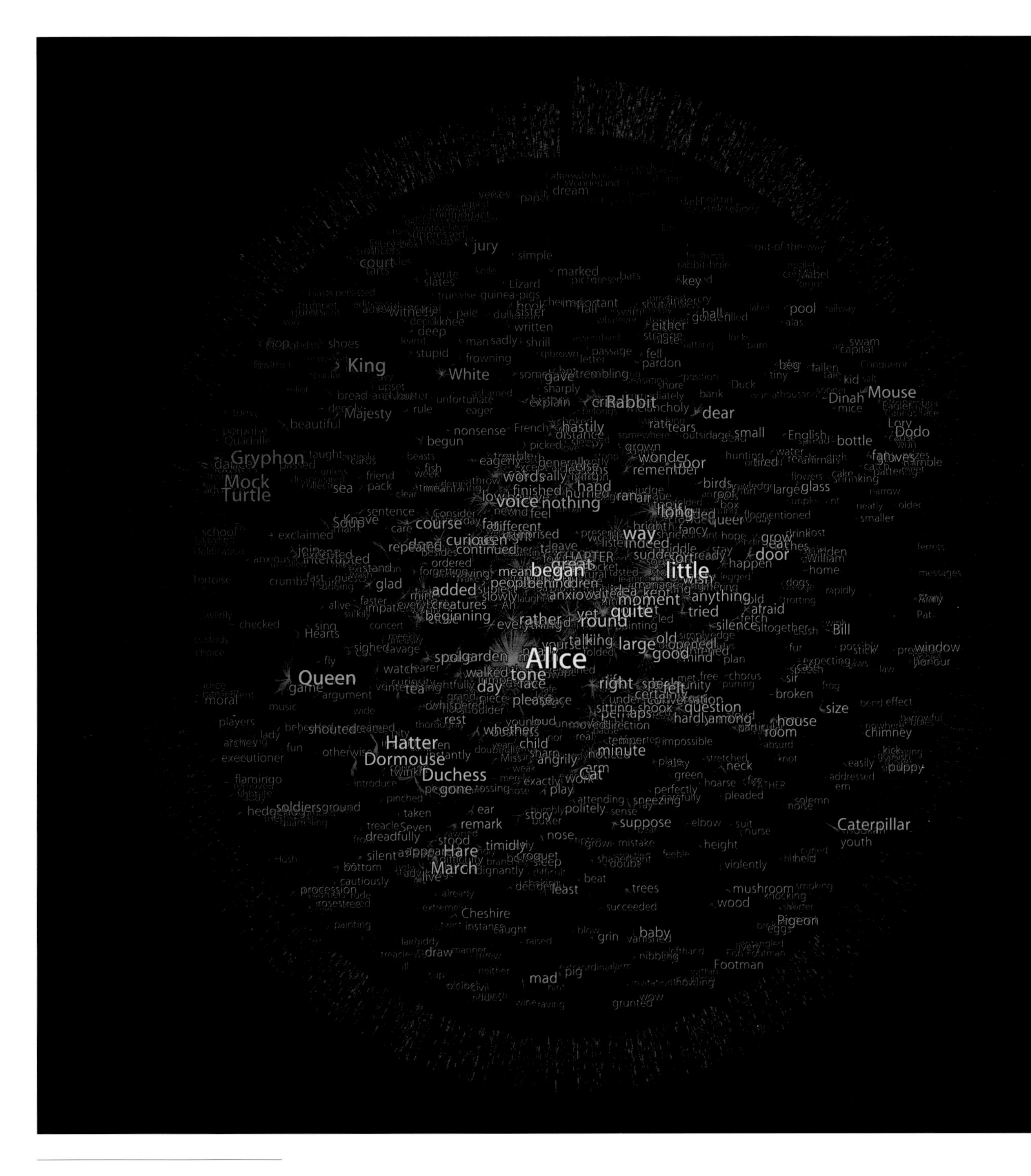

W. Bradford Paley

TextArc: Alice in Wonderland

2009

A map of word frequency and associations in Lewis Carroll's *Alice in Wonderland* (1865). TextArc is a text-analysis tool that shows the distribution of words in texts that have no metadata descriptions, such as a table of contents or an index. It first draws the entire text, sentence by sentence, in the shape of an ellipse. Every word is then drawn at its position in the book, just inside the ellipse next to the sentence in which it appears. Words that appear near one another in the book share a similar color and are brighter if they appear more frequently. The selection of any word (e.g., *Alice*) sprouts lines that immediately show the distribution of its usage (where it appears in the book).

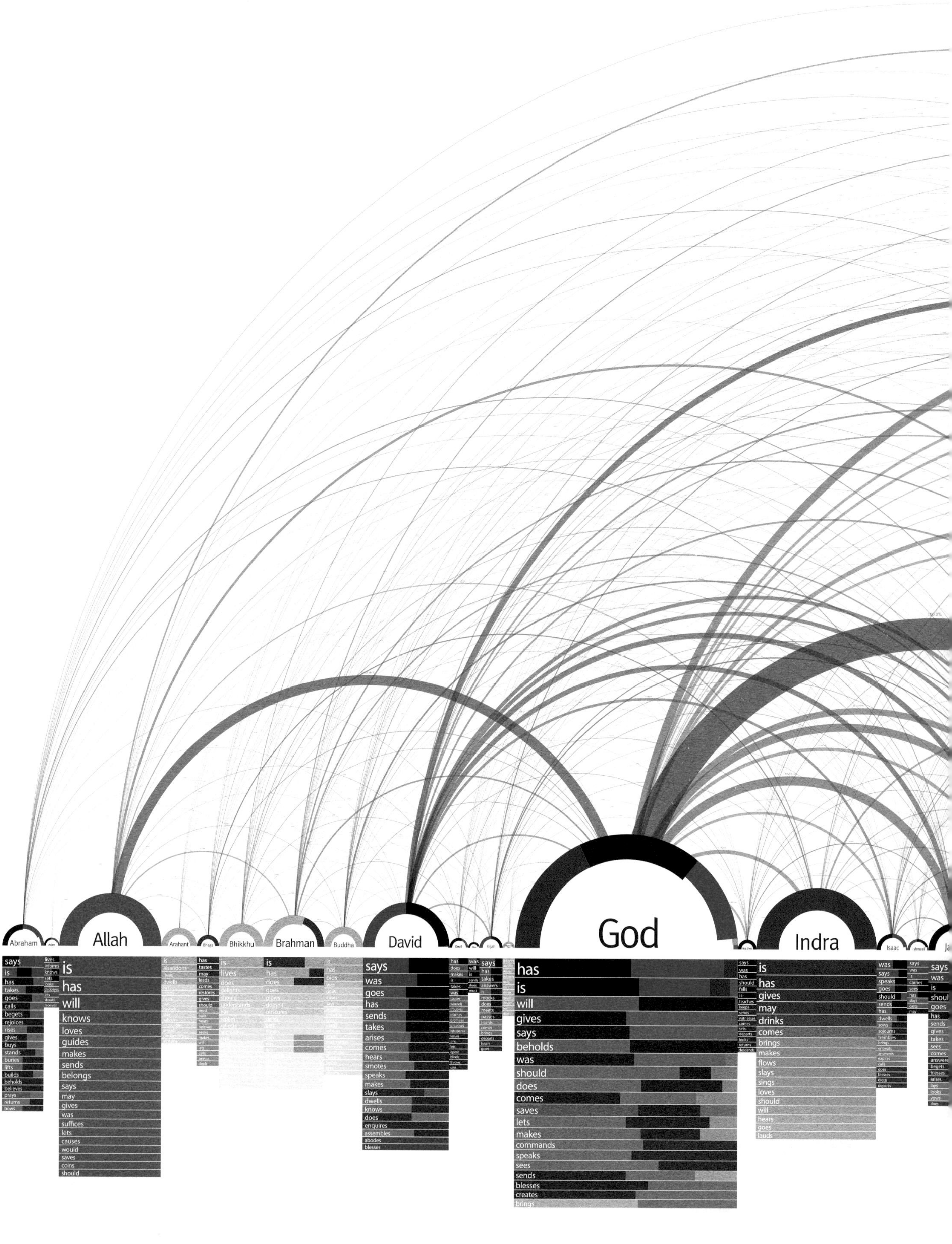

Abraham
says
is
has
takes
goes
calls
begets
rejoices
rises
gives
buys
stands
buries
lifts
builds
beholds
believes
prays
returns
bows

Allah
is
has
will
knows
loves
guides
makes
sends
belongs
says
may
gives
was
suffices
lets
causes
would
saves
coins
should

Arahant

Bhaga

Bhikkhu

Brahman

Buddha

David
says
was
goes
has
sends
takes
arises
comes
hears
smotes
speaks
makes
slays
dwells
knows
does
enquires
assembles
abodes
blesses

Elijah

God
has
is
will
gives
says
beholds
was
should
does
comes
saves
lets
makes
commands
speaks
sees
sends
blesses
creates
brings

Indra
is
has
gives
may
drinks
comes
brings
makes
flows
slays
sings
loves
should
will
hears
goes
lauds

Isaac
was
says
speaks
goes
should
sends
has
dwells

Ishmael

Philipp Steinweber and Andreas Koller

Similar Diversity

2007

A visualization of the similarities and differences between the holy books of five world religions: Christianity, Islam, Hinduism, Buddhism, and Judaism

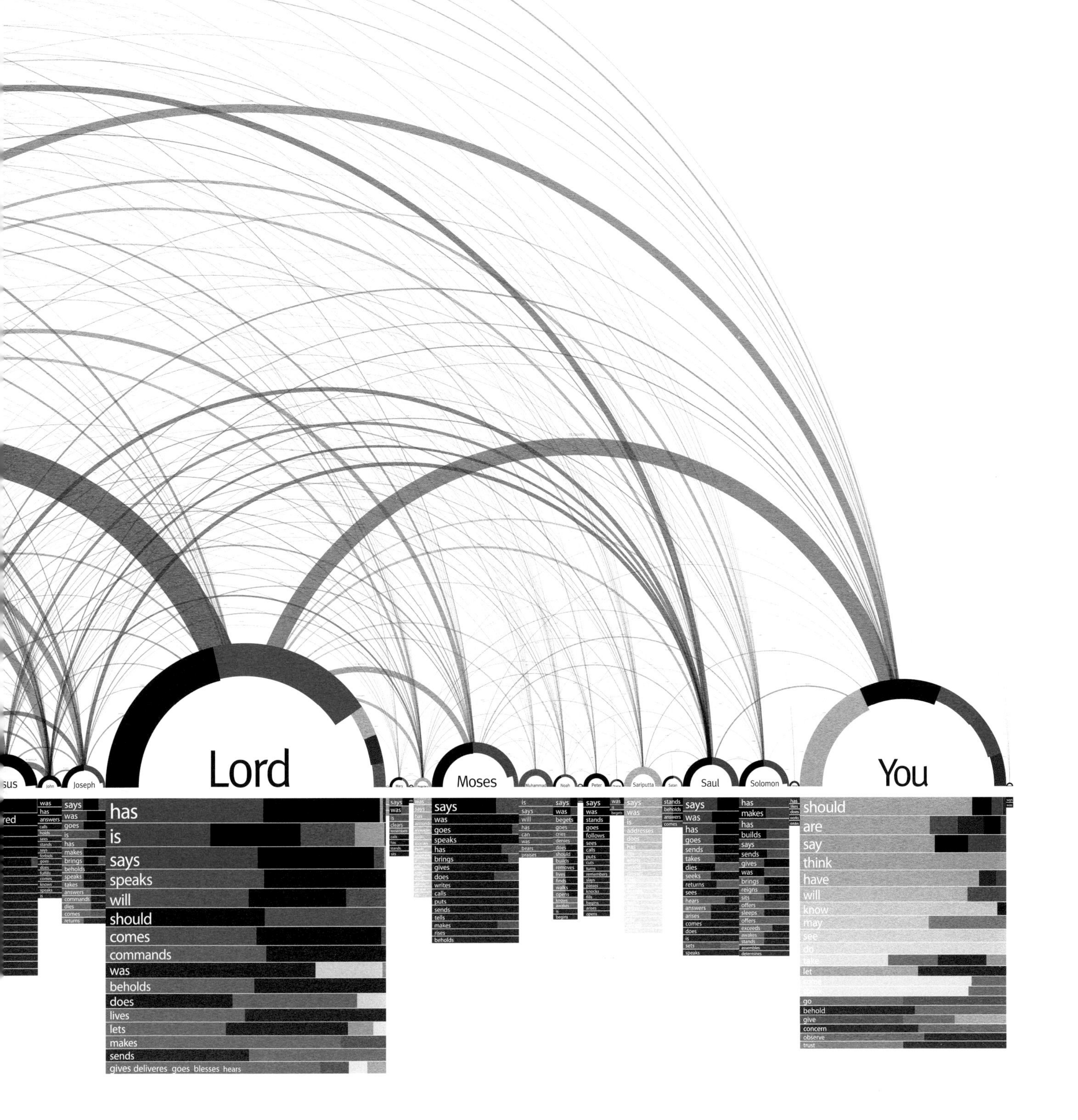

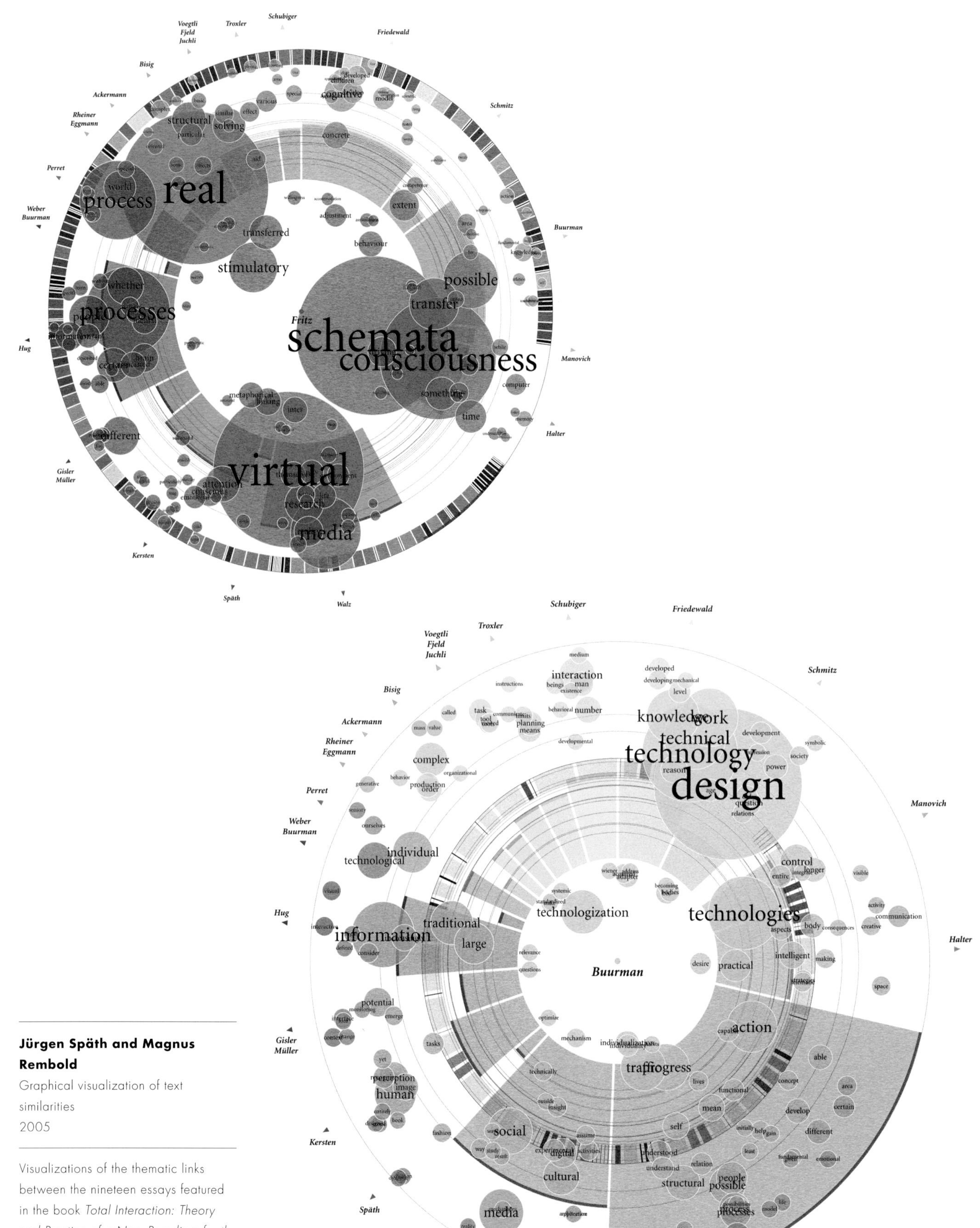

Jürgen Späth and Magnus Rembold
Graphical visualization of text similarities
2005

Visualizations of the thematic links between the nineteen essays featured in the book *Total Interaction: Theory and Practice of a New Paradigm for the Design Disciplines* (2004), edited by Gerhard M. Buurman

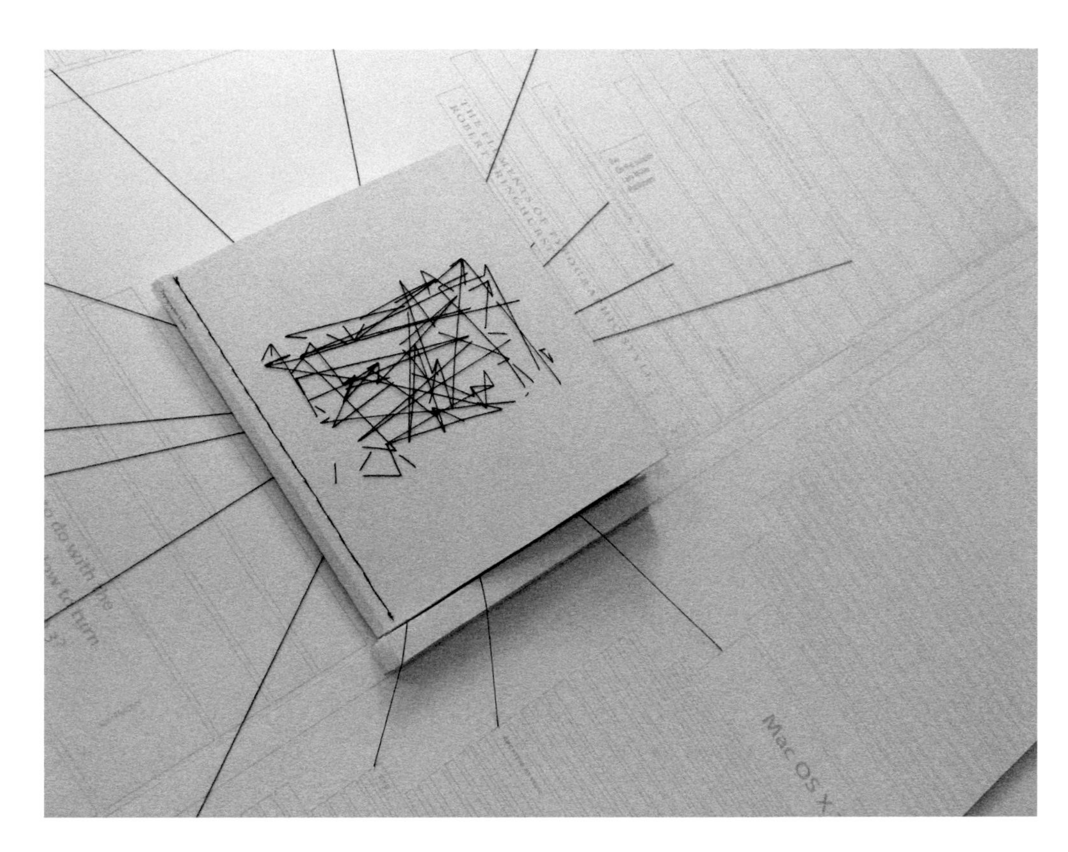

Dan Collier

Typographic Links

2007

A hand-sewn, three-dimensional hyperlink structure that guides the reader through the pages of a book (above); detail (right)

Music

One of the most interesting and recent themes for network visualization is music. Either by creating a visual metaphor for the notes of a song or by mapping similarities and differences between artists across extensive data sets, music is an enthralling emergent topic. A significant object of study in this context has been the popular music website Last.fm. Apart from the familiar social features, Last.fm's application programming interface (API) has been used by many visualization authors to better understand music affinities, personal playing habits, and overall community structure.

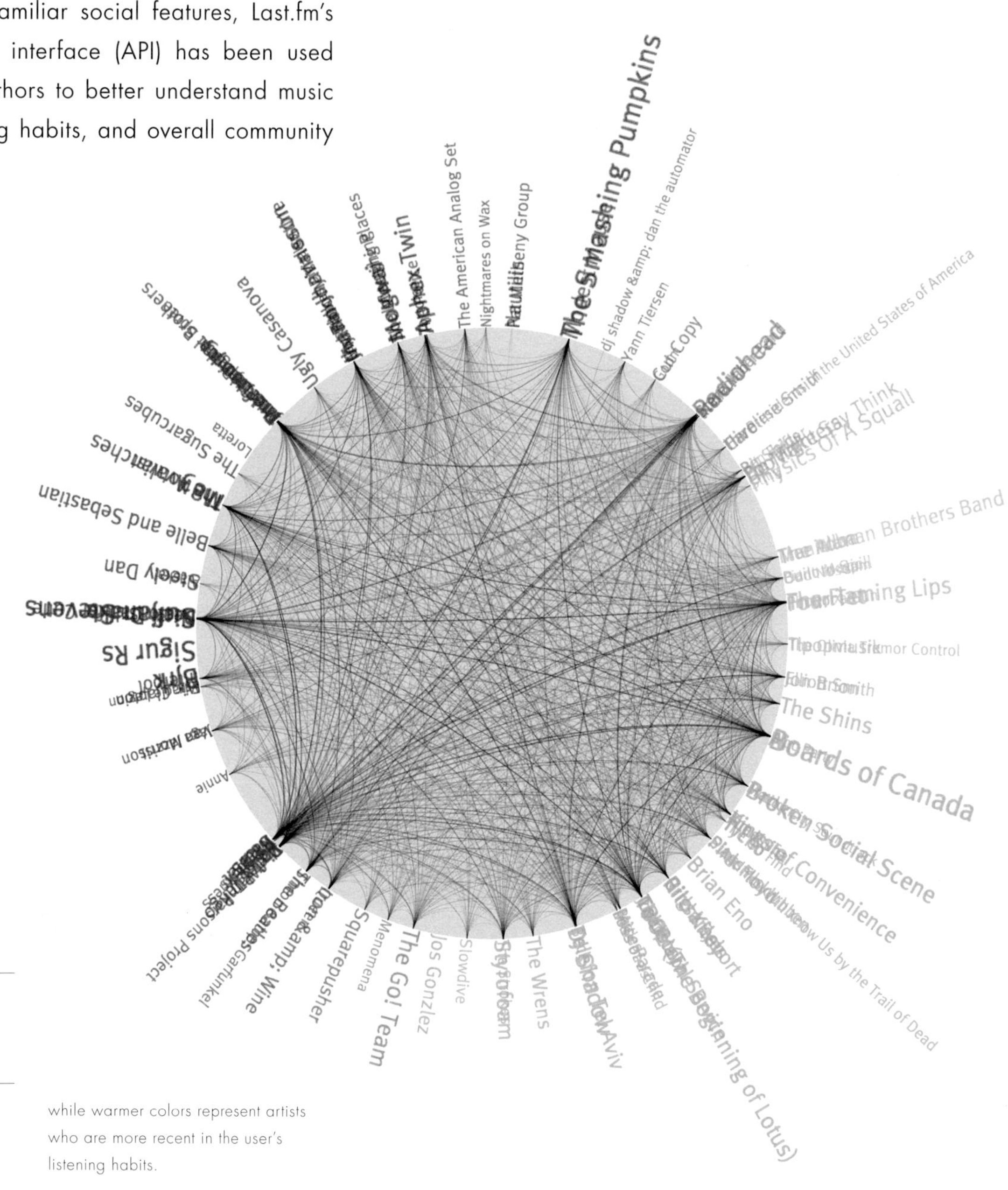

Lee Byron
Listening History
2007

A visualization of a Last.fm user's listening patterns over an eighteen-month period. The names of artists, progressing in a clockwise movement through the eighteen-month span, increase in type size linearly with greater listening frequency. Cooler colors represent artists who have been listened to for a long period of time, while warmer colors represent artists who are more recent in the user's listening habits.

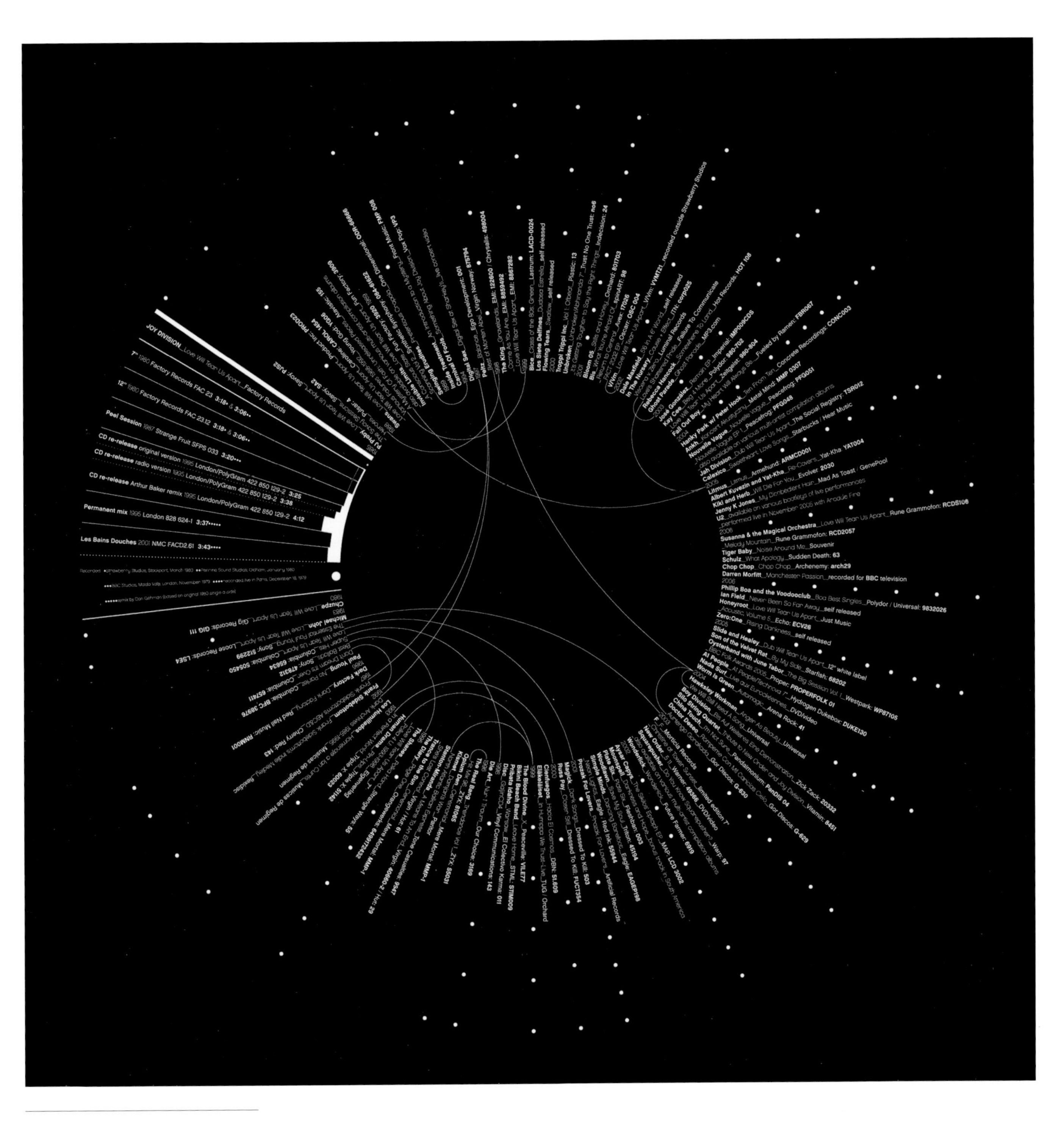

Peter Crnokrak
Love Will Tear Us Apart Again
2007

A map of eighty-five recorded covers of Joy Division's "Love Will Tear Us Apart," showing time since original recording, recording artist, release name, release date, and label

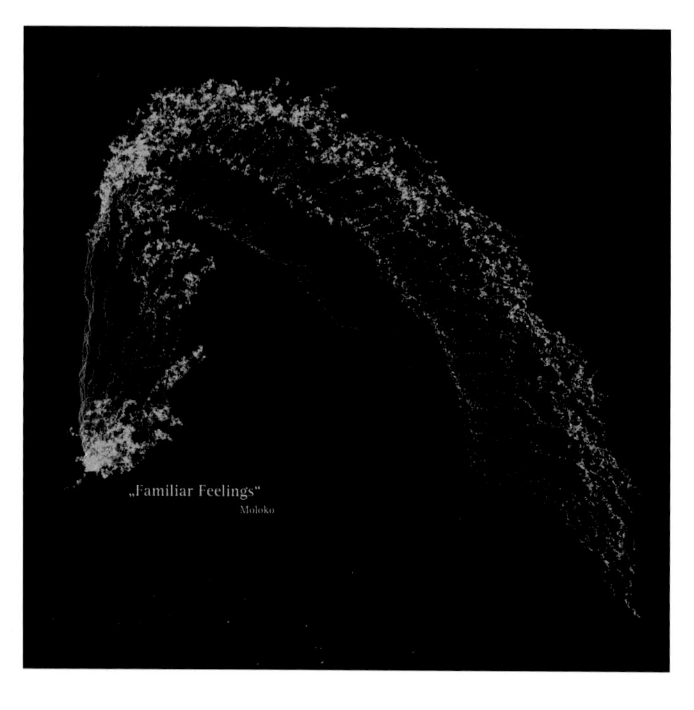

Matthias Dittrich (left)
Narratives 2.0
2008

A chart of Moloko's "Familiar Feelings" music track segmented into single channels depicted as fanlike lines moving away from the center with time. The angle of the line changes according to the frequency of the channel; and at high frequency levels, the channel is highlighted in orange.

Marco Quaggiotto, Giorgio Caviglia, and Adam Leibsohn
Visual i/zer
2009

An interactive visualization that uncovers syntax ties between various song lyrics

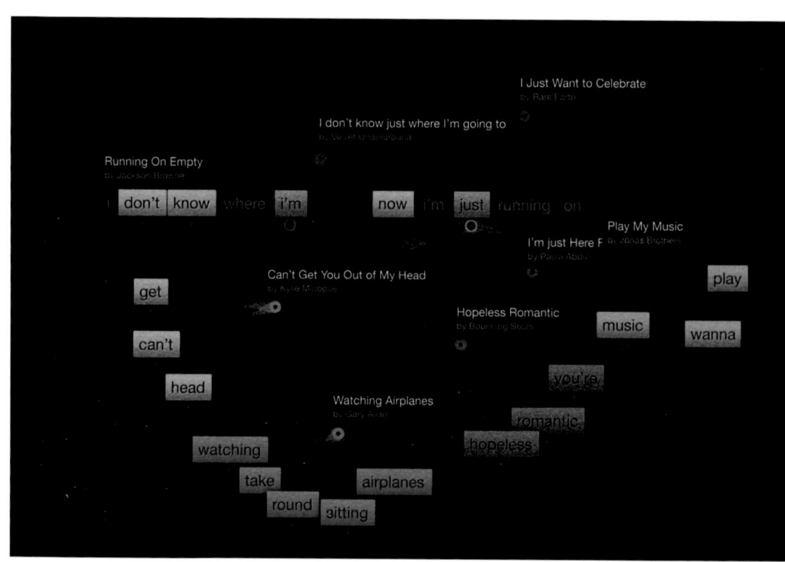

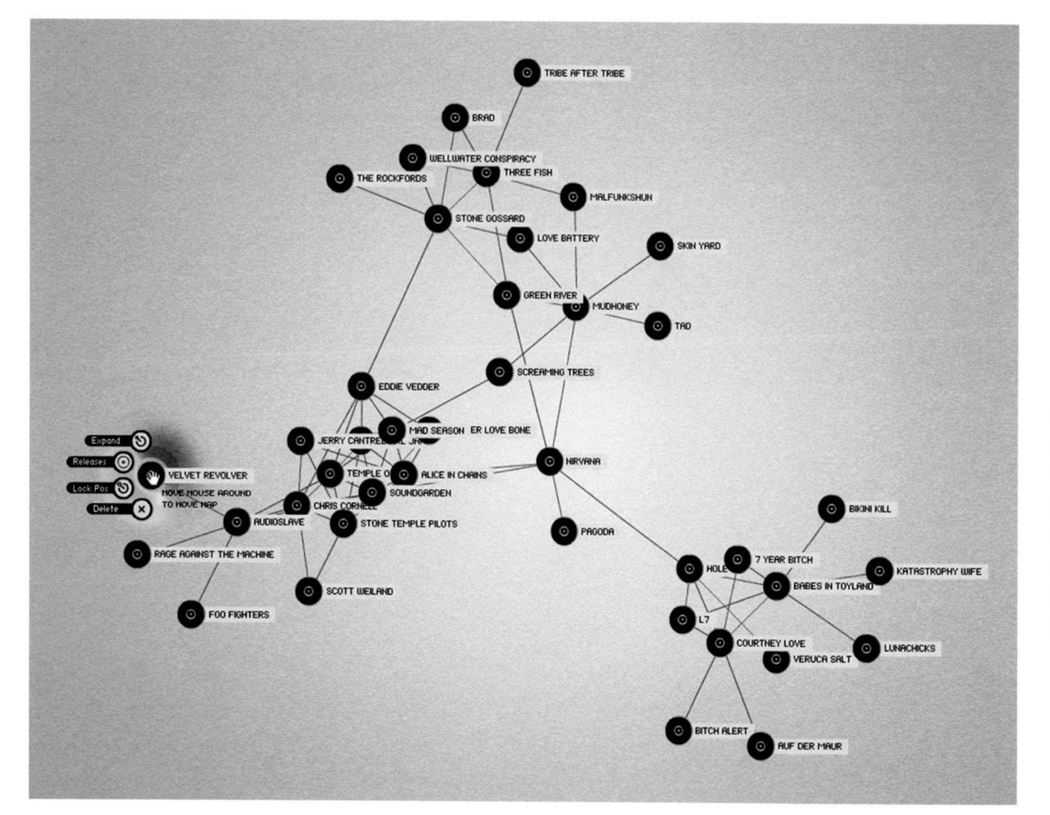

Onyro
TuneGlue
2006

An interactive graph of music artists connected by genre, based on Last.fm and Amazon.co.uk databases

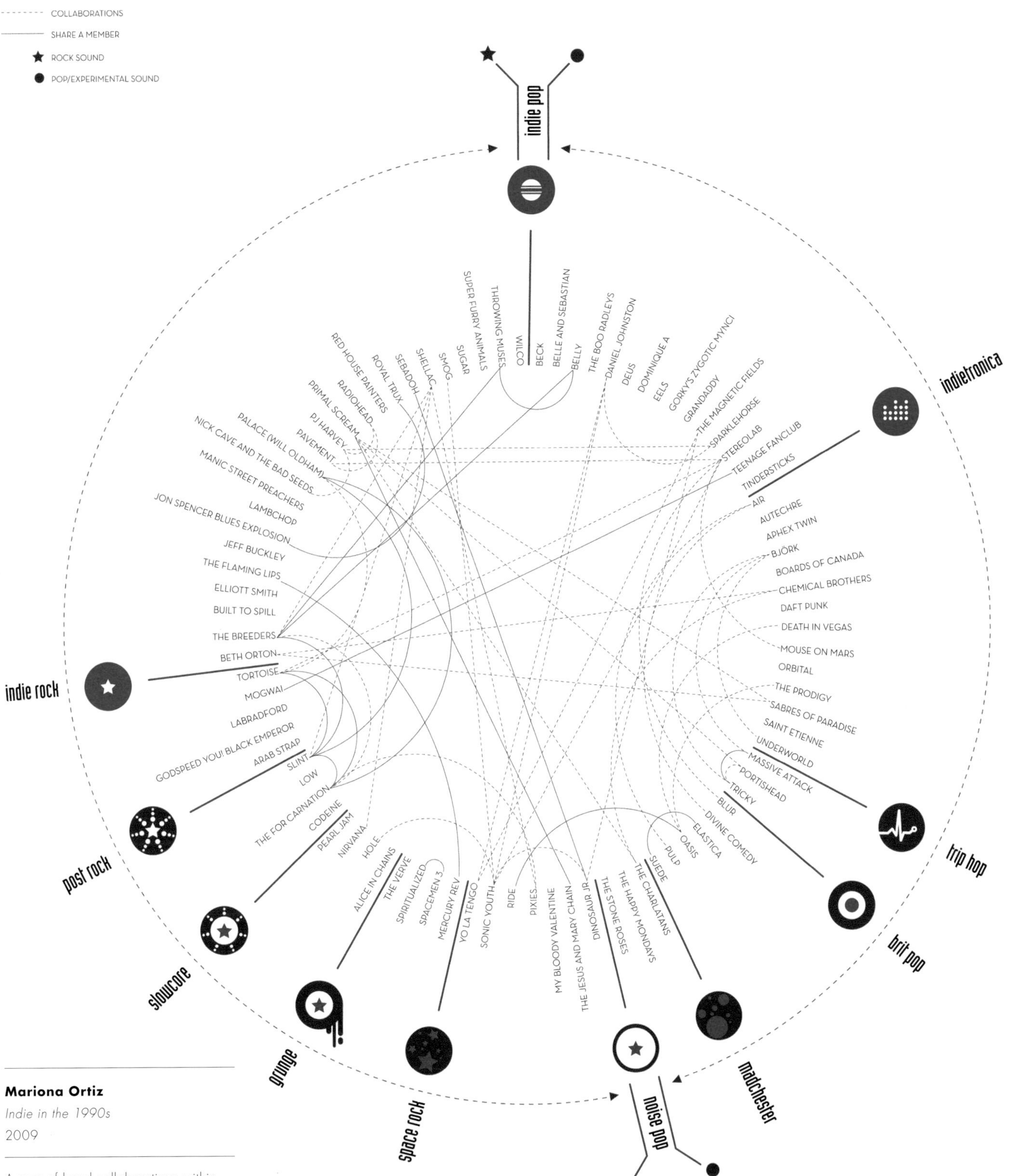
COLLABORATIONS
SHARE A MEMBER
ROCK SOUND
POP/EXPERIMENTAL SOUND
indie pop
indietronica
indie rock
post rock
slowcore
grunge
space rock
noise pop
madchester
brit pop
trip hop
SUPER FURRY ANIMALS
THROWING MUSES
WILCO
BECK
BELLE AND SEBASTIAN
BELLY
THE BOO RADLEYS
DANIEL JOHNSTON
DEUS
DOMINIQUE A
EELS
GORKY'S ZYGOTIC MYNCI
GRANDADDY
THE MAGNETIC FIELDS
SPARKLEHORSE
STEREOLAB
TEENAGE FANCLUB
TINDERSTICKS
AIR
AUTECHRE
APHEX TWIN
BJORK
BOARDS OF CANADA
CHEMICAL BROTHERS
DAFT PUNK
DEATH IN VEGAS
MOUSE ON MARS
ORBITAL
THE PRODIGY
SABRES OF PARADISE
SAINT ETIENNE
UNDERWORLD
MASSIVE ATTACK
PORTISHEAD
TRICKY
BLUR
DIVINE COMEDY
ELASTICA
OASIS
PULP
SUEDE
THE CHARLATANS
THE HAPPY MONDAYS
THE STONE ROSES
DINOSAUR JR
THE JESUS AND MARY CHAIN
MY BLOODY VALENTINE
PIXIES
RIDE
SONIC YOUTH
YO LA TENGO
MERCURY REV
SPACEMEN 3
SPIRITUALIZED
THE VERVE
ALICE IN CHAINS
HOLE
NIRVANA
PEARL JAM
CODEINE
THE FOR CARNATION
LOW
SLINT
ARAB STRAP
GODSPEED YOU! BLACK EMPEROR
LABRADFORD
MOGWAI
TORTOISE
BETH ORTON
THE BREEDERS
BUILT TO SPILL
ELLIOTT SMITH
THE FLAMING LIPS
JEFF BUCKLEY
JON SPENCER BLUES EXPLOSION
LAMBCHOP
MANIC STREET PREACHERS
NICK CAVE AND THE BAD SEEDS
PALACE (WILL OLDHAM)
PAVEMENT
PJ HARVEY
PRIMAL SCREAM
RADIOHEAD
RED HOUSE PAINTERS
ROYAL TRUX
SEBADOH
SHELLAC
SMOG
SUGAR

Mariona Ortiz

Indie in the 1990s

2009

A map of band collaborations within the 1990s independent music scene

News

Mass media is a central player in contemporary information-packed society. Having to reinvent themselves in order to adapt to a new digital landscape, many media channels are now incorporating alternative methods of dealing with the hefty amounts of daily news. Most projects in this category are either novel visualization approaches presented by the media source itself or original concepts by individual authors pursuing a better grasp of the underlying relationships within the provided data sets.

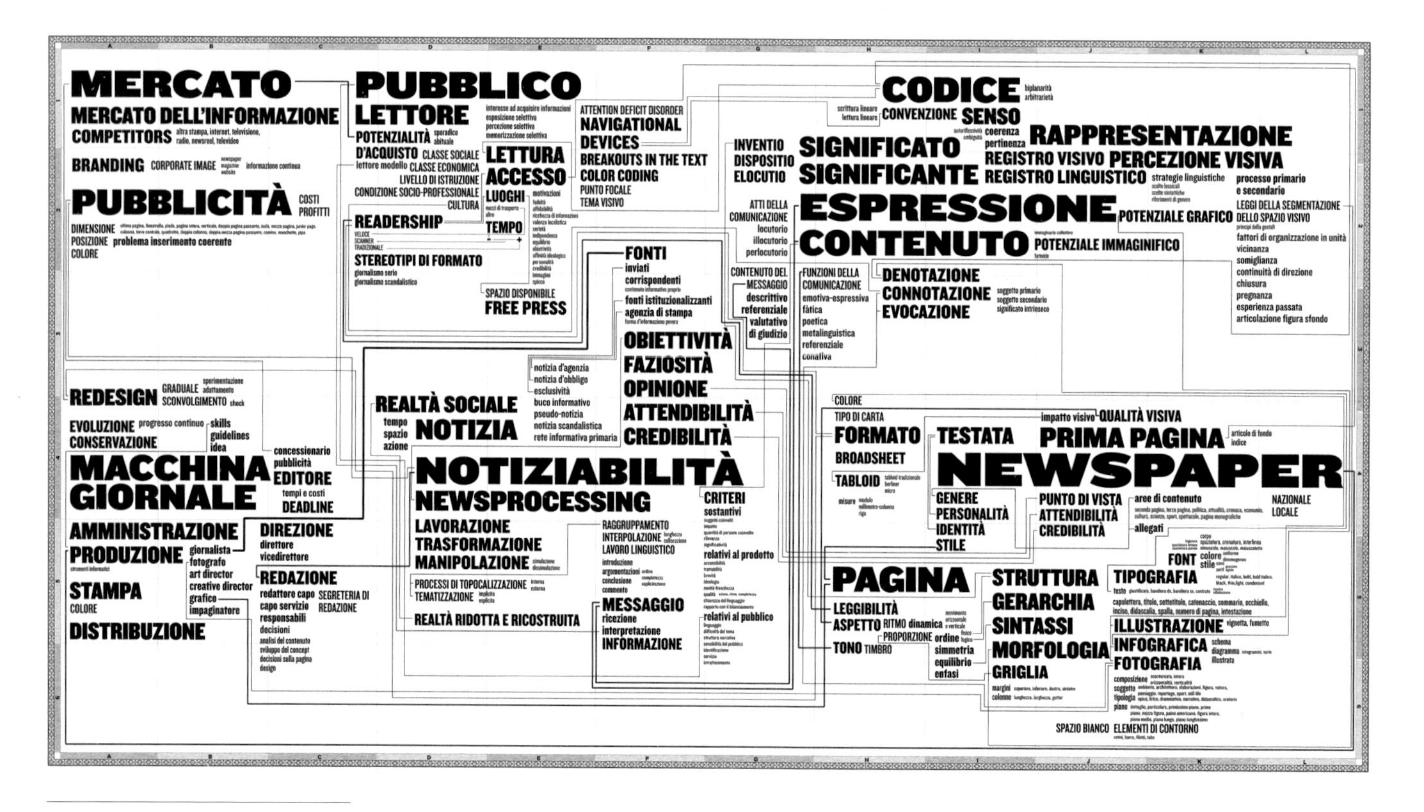

Francesco Franchi

Biologia della Pagina (Newspaper map)

2005

A typographical map of the complex network that underlies the production of a daily newspaper

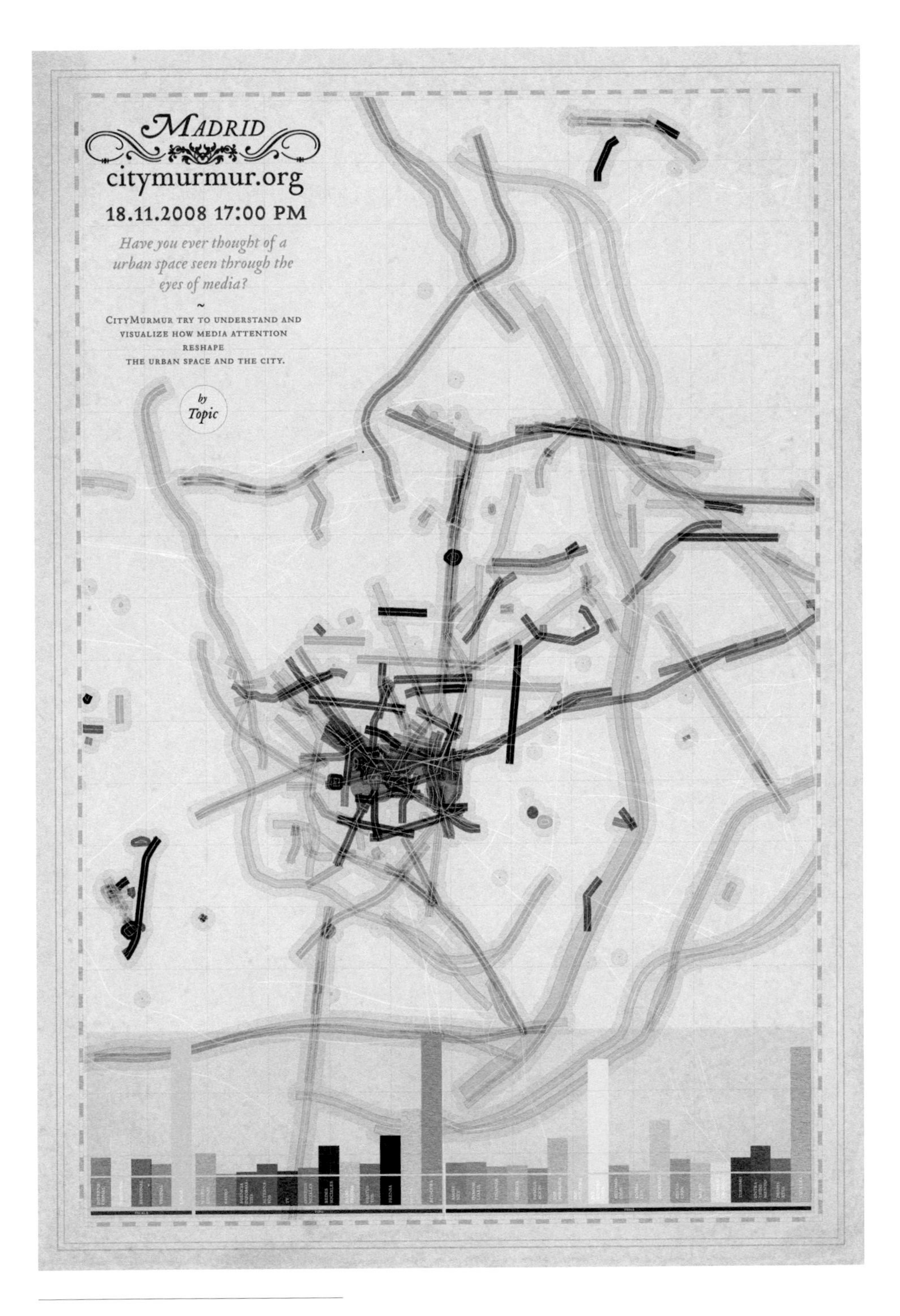

Giorgio Caviglia, Marco Quaggiotto, Donato Ricci, Gaia Scagnetti, Michele Graffieti, Samuel Granados Lopez, and Daniele Guido
City Murmur
2008

From a large RSS feed of 733 online sources, *City Murmur* tracks every time a street, place of interest, or district in Madrid is mentioned in the Spanish media, creating a time-based visual narrative of the urban space.

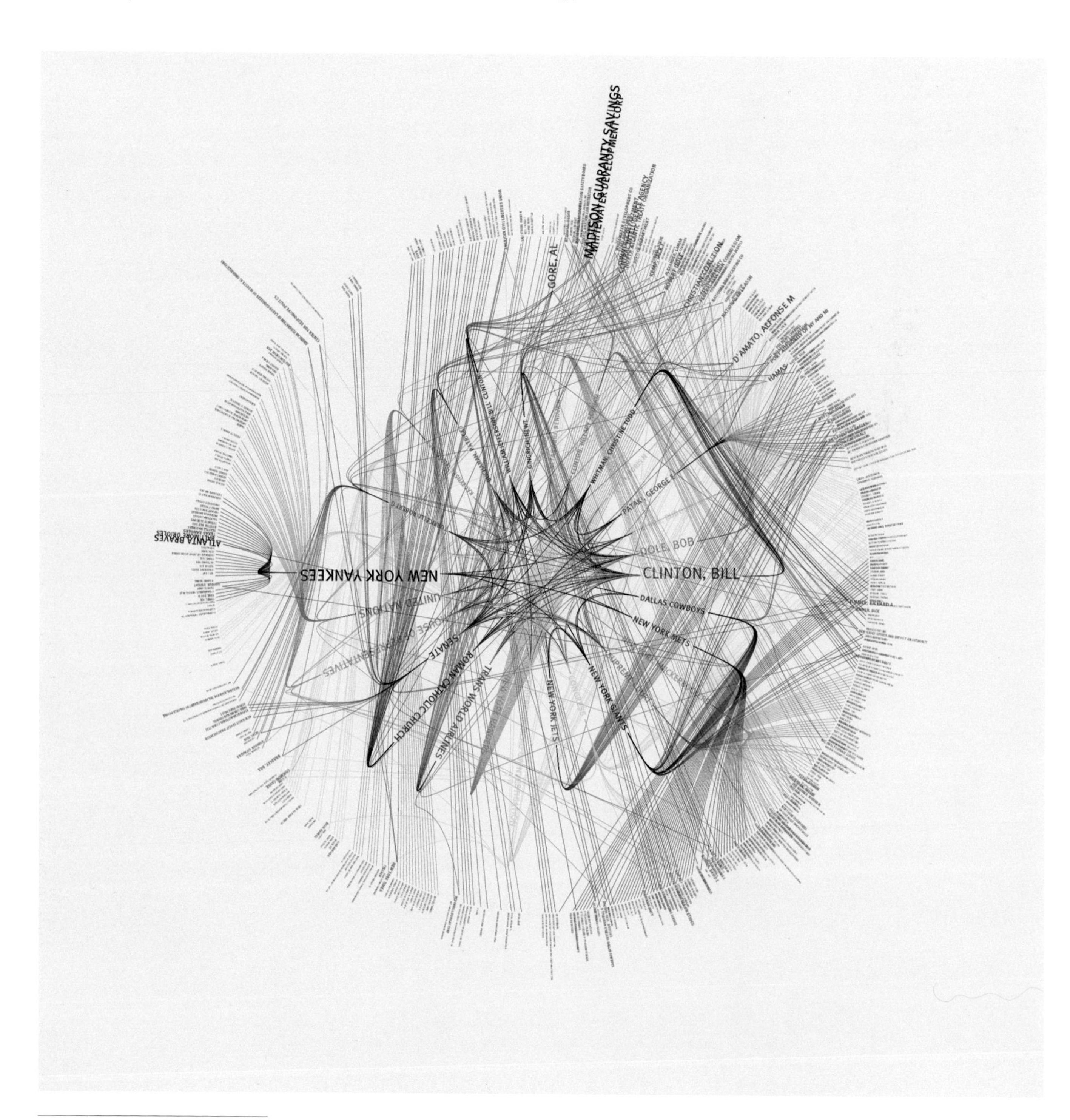

Jer Thorp
NYTimes: 365/360
2009

A visualization showing ties between the top organizations and personalities mentioned in the *New York Times* in 1996. A part of a large series of visualizations produced every year from 1985 to 2001, using the *New York Times* Article Search API.

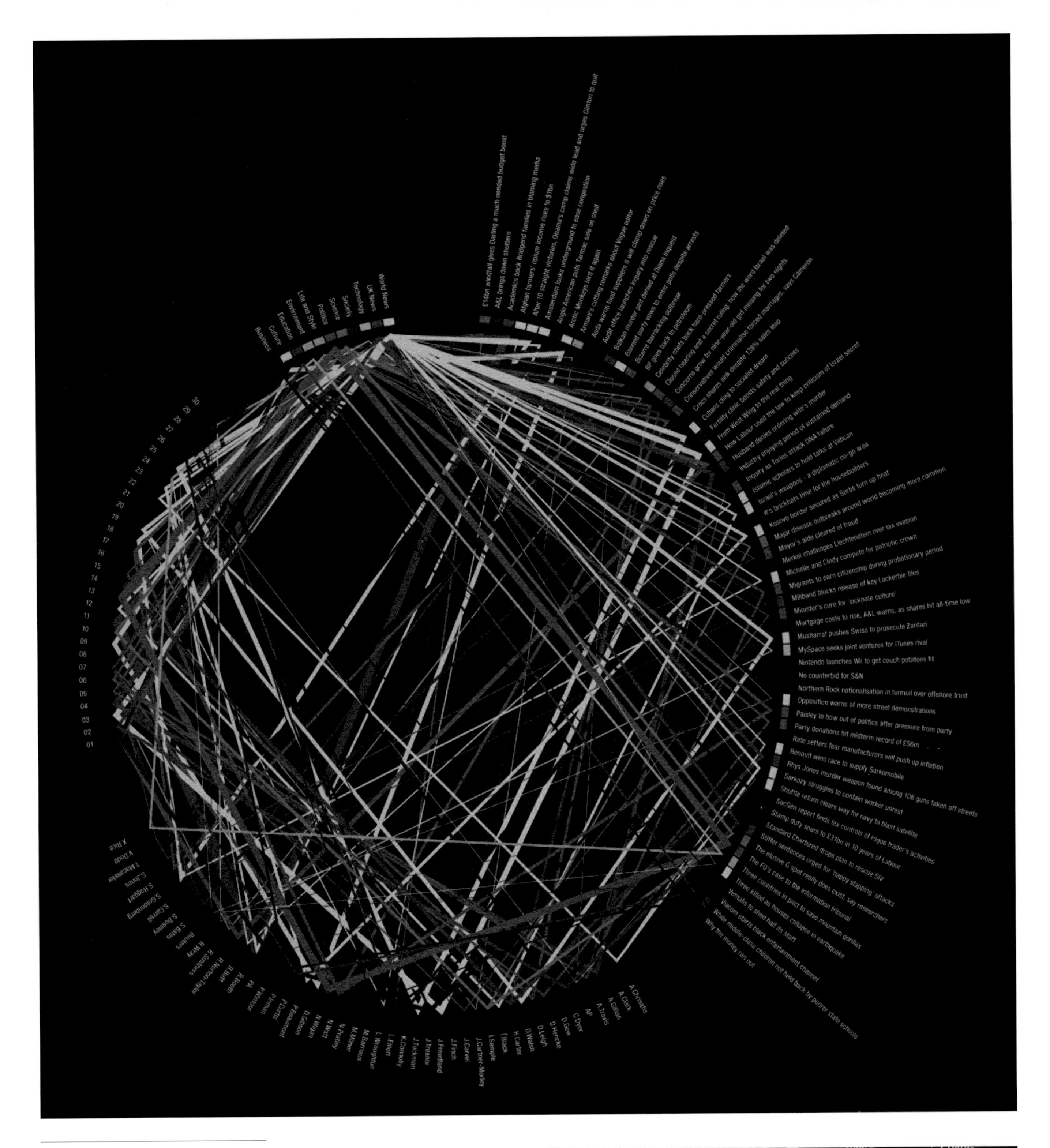

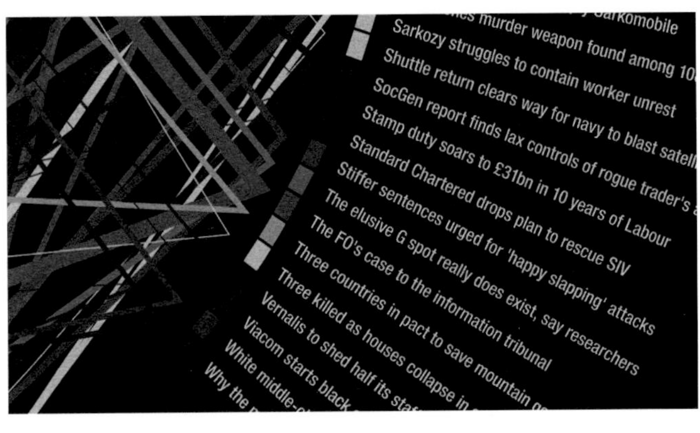

Dave Bowker

One Week of the Guardian

2008

A map of relationships between headlines, authors, page numbers, and categories in the *Guardian* newspaper in one single day. Prominent stories, depicted by lines, are color coded by category (e.g., science, politics), and their width is proportionate to their word count (above). Detail (right).

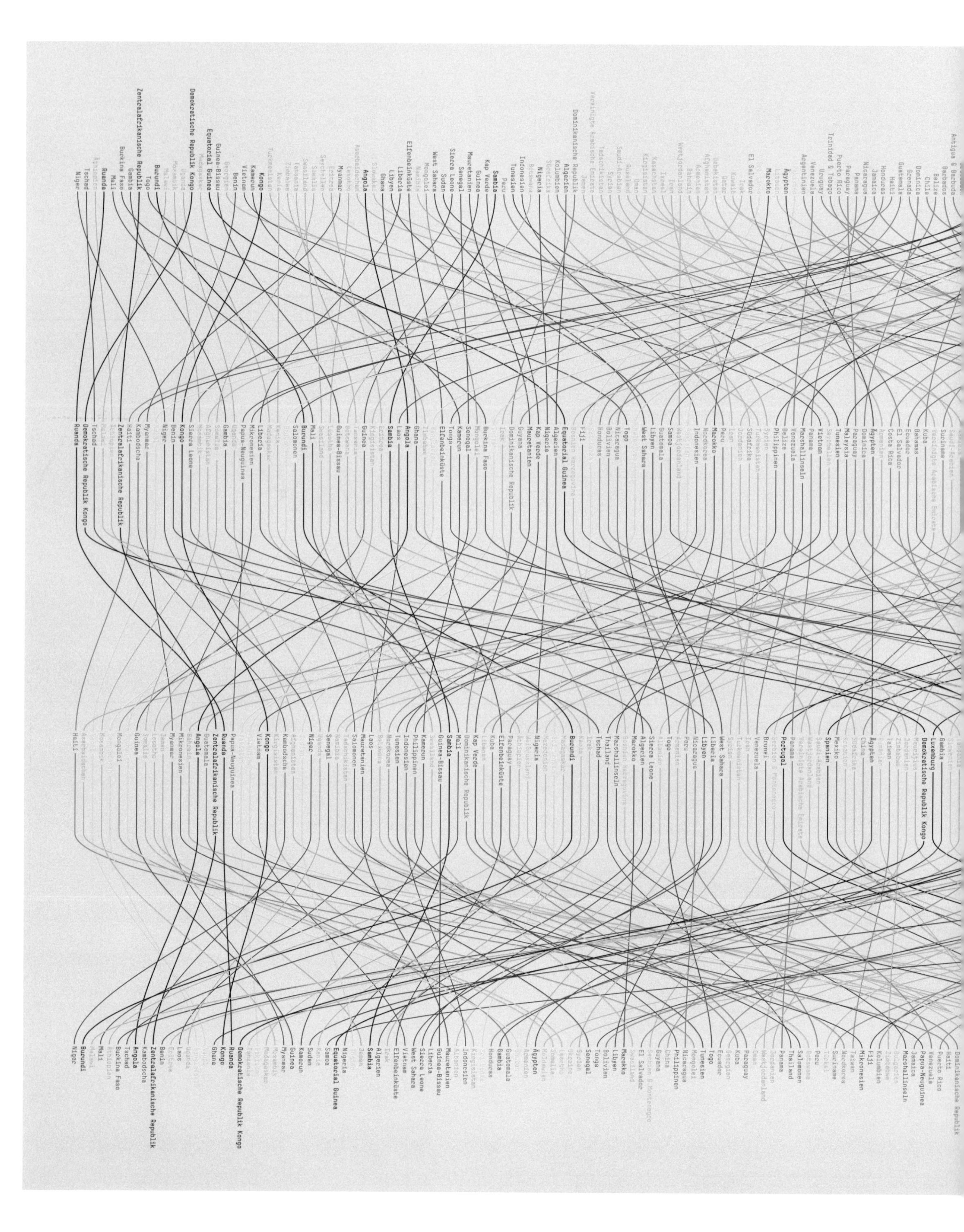

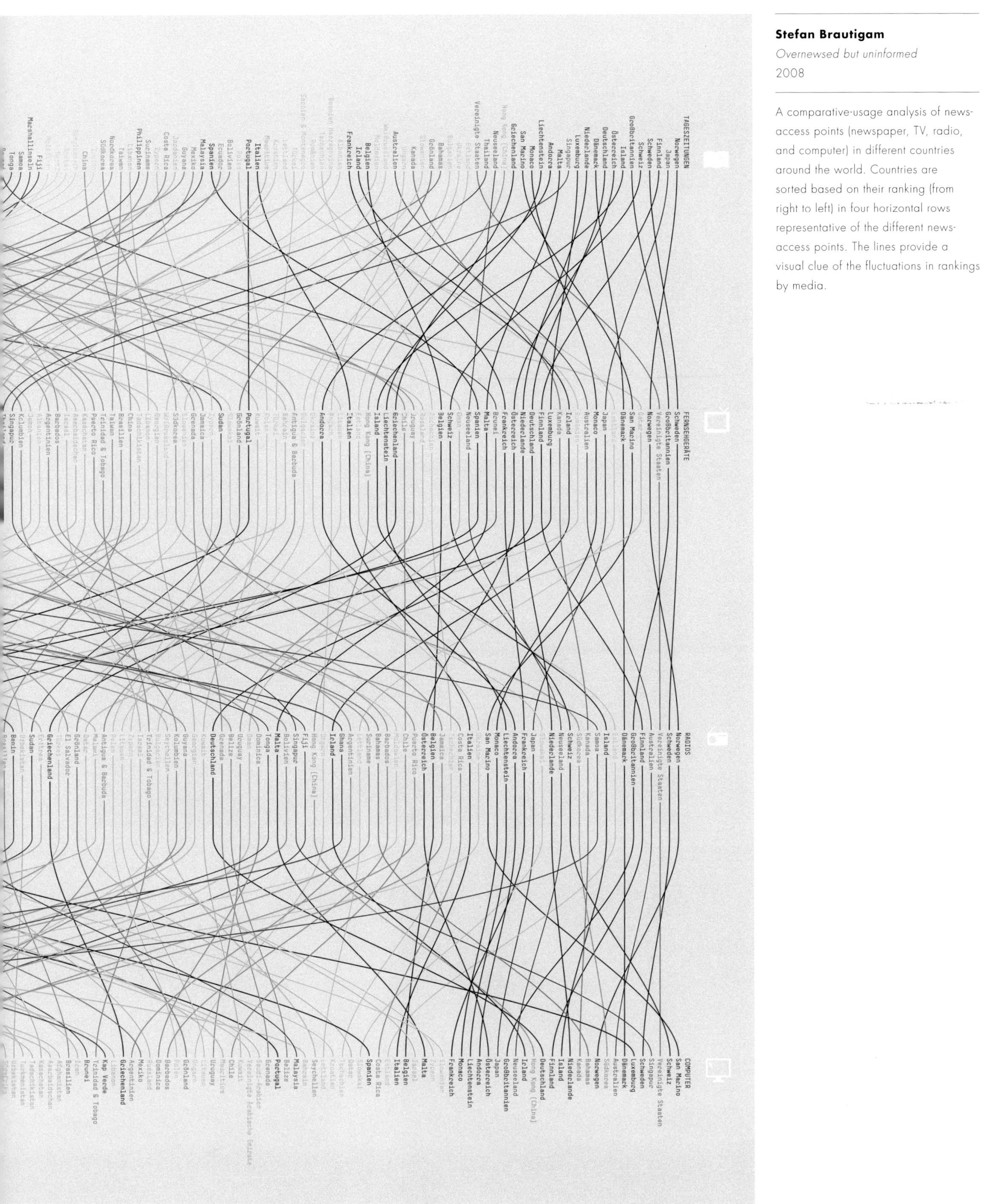

Stefan Brautigam
Overnewsed but uninformed
2008

A comparative-usage analysis of news-access points (newspaper, TV, radio, and computer) in different countries around the world. Countries are sorted based on their ranking (from right to left) in four horizontal rows representative of the different news-access points. The lines provide a visual clue of the fluctuations in rankings by media.

Proteins

Protein interaction networks have been one of the most popular subjects for network visualization and the chosen area of study for many biologists over the past decade. Interactions between proteins are crucial in almost every biological function and of fundamental importance for all processes in a living cell. The secrets behind many diseases reside within these intricate structures, and visualization has been an important tool for discovery and insight. The extracted knowledge from this exploratory process can prove essential in future therapeutic approaches.

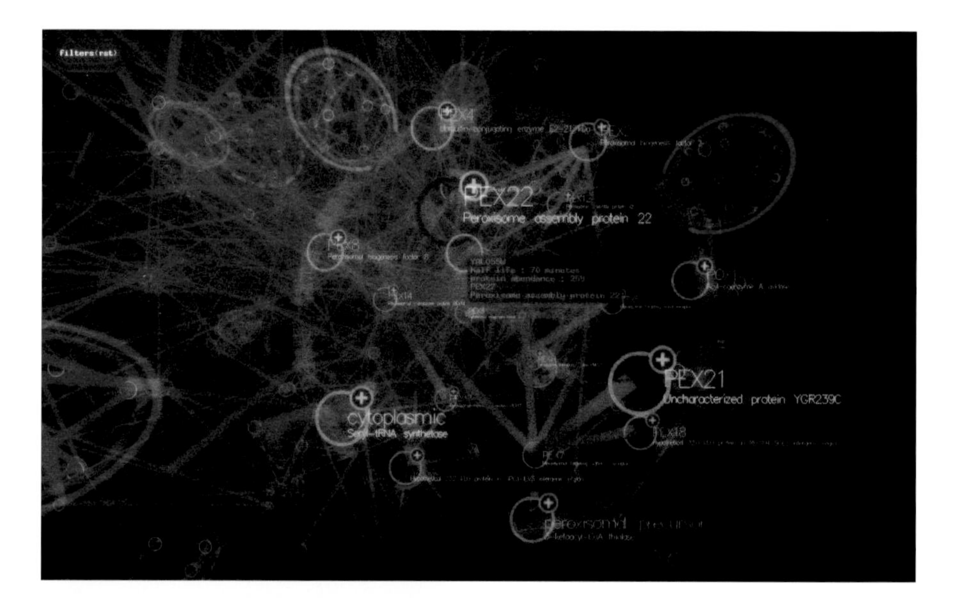

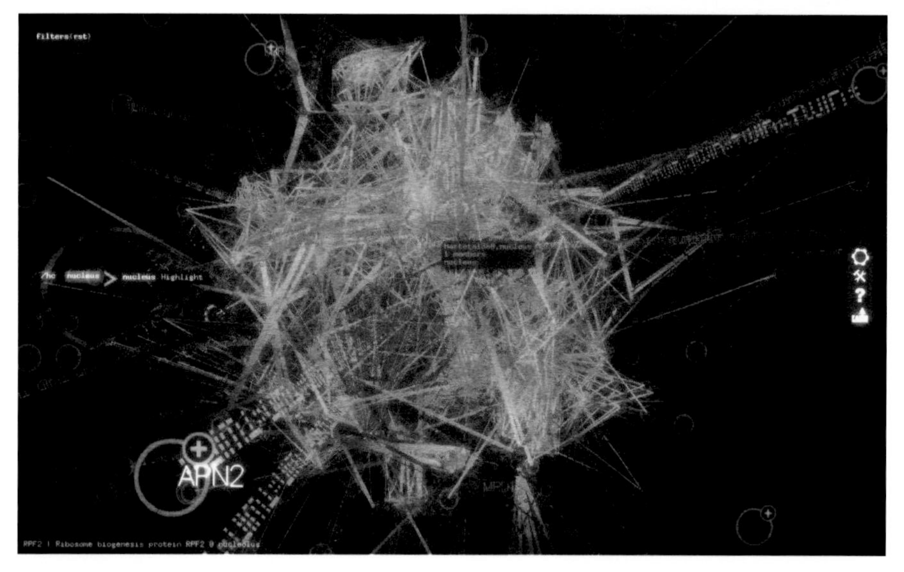

Yose Widjaja
The Interactorium
2009

A three-dimensional representation of the yeast interactome (all molecular interactions in cells) using Interactorium, a visualization platform that analyzes large interactome data sets. Developed in collaboration between Yose Widjaja and the School of Computer Science and Engineering at the University of New South Wales (above). Detail (left).

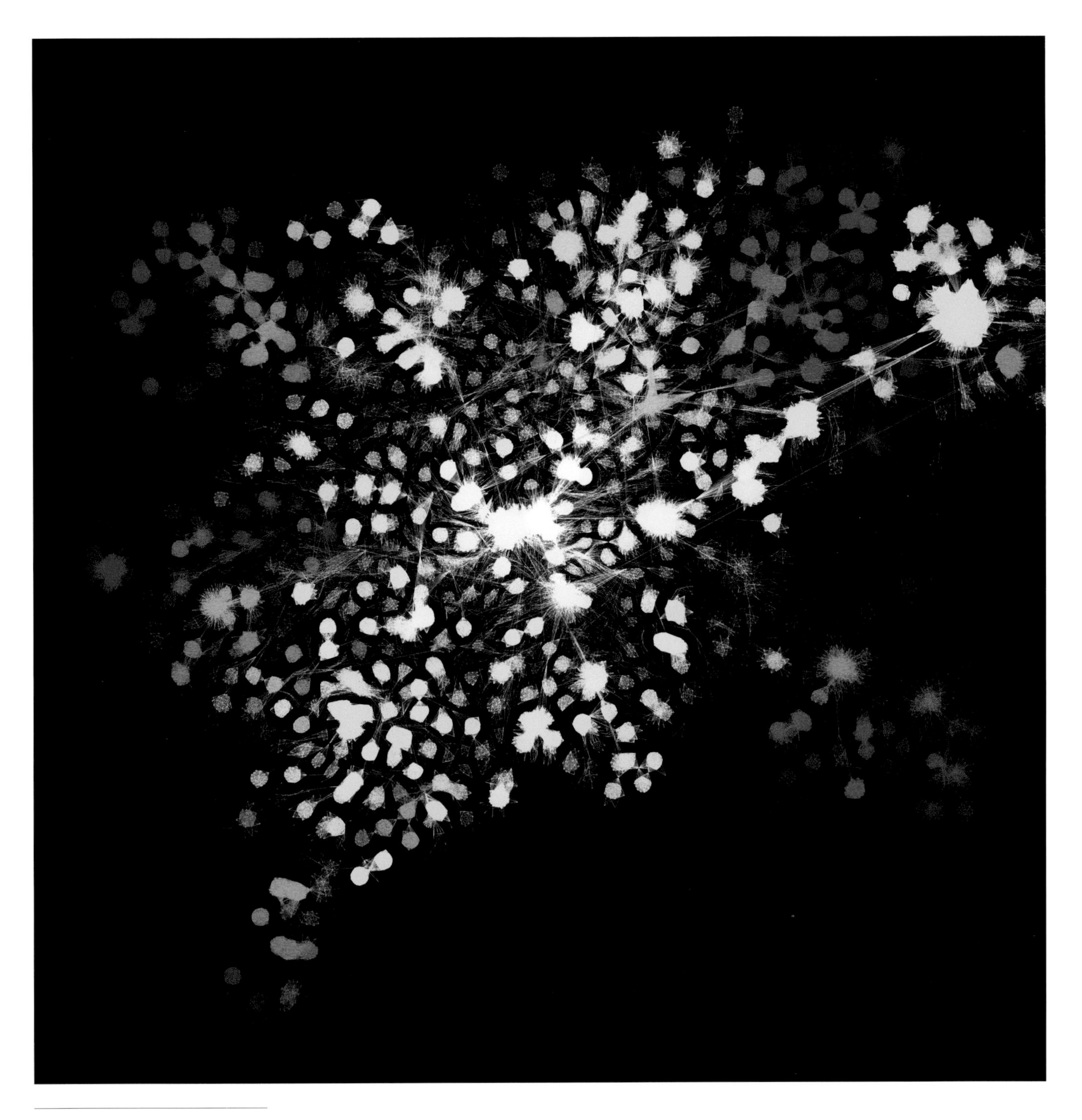

Alex Adai and Edward Marcotte
Protein Homology Network
2002

A protein-homology graph depicting
32,727 proteins and 1,206,654 edges

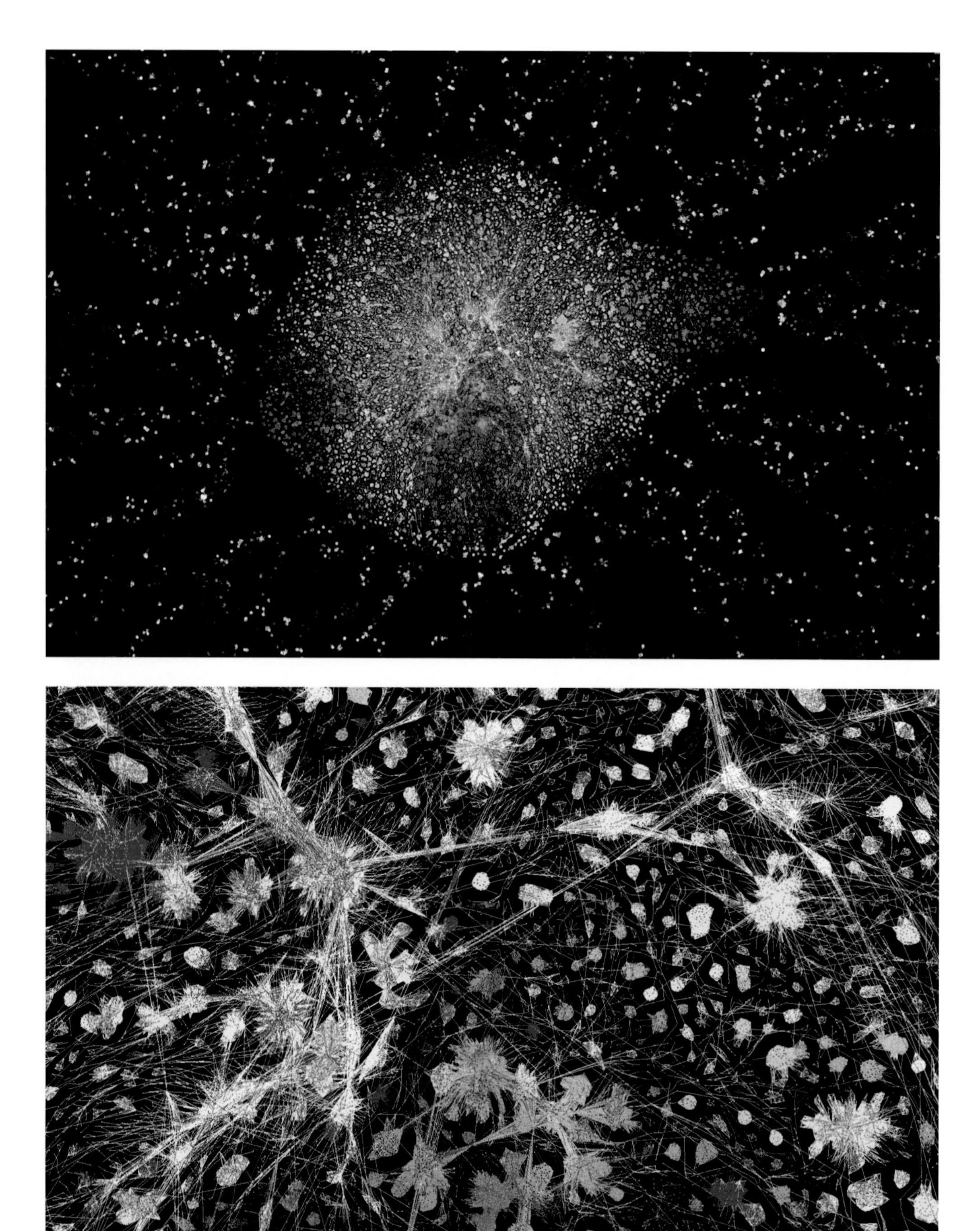

Alex Adai and Edward Marcotte
Minimum Spanning Protein Homology Tree
2002

A protein-homology map featuring 302,832 proteins. An edge is colored blue if it connects two proteins of the same species and red if from two different species. If no species information is available, edges are automatically colored based on a customized layout hierarchy. Proteins, on the other hand, are colored based on their Clusters of Orthologous Groups, a classification system determined by their genetic lineages (top). Detail (bottom).

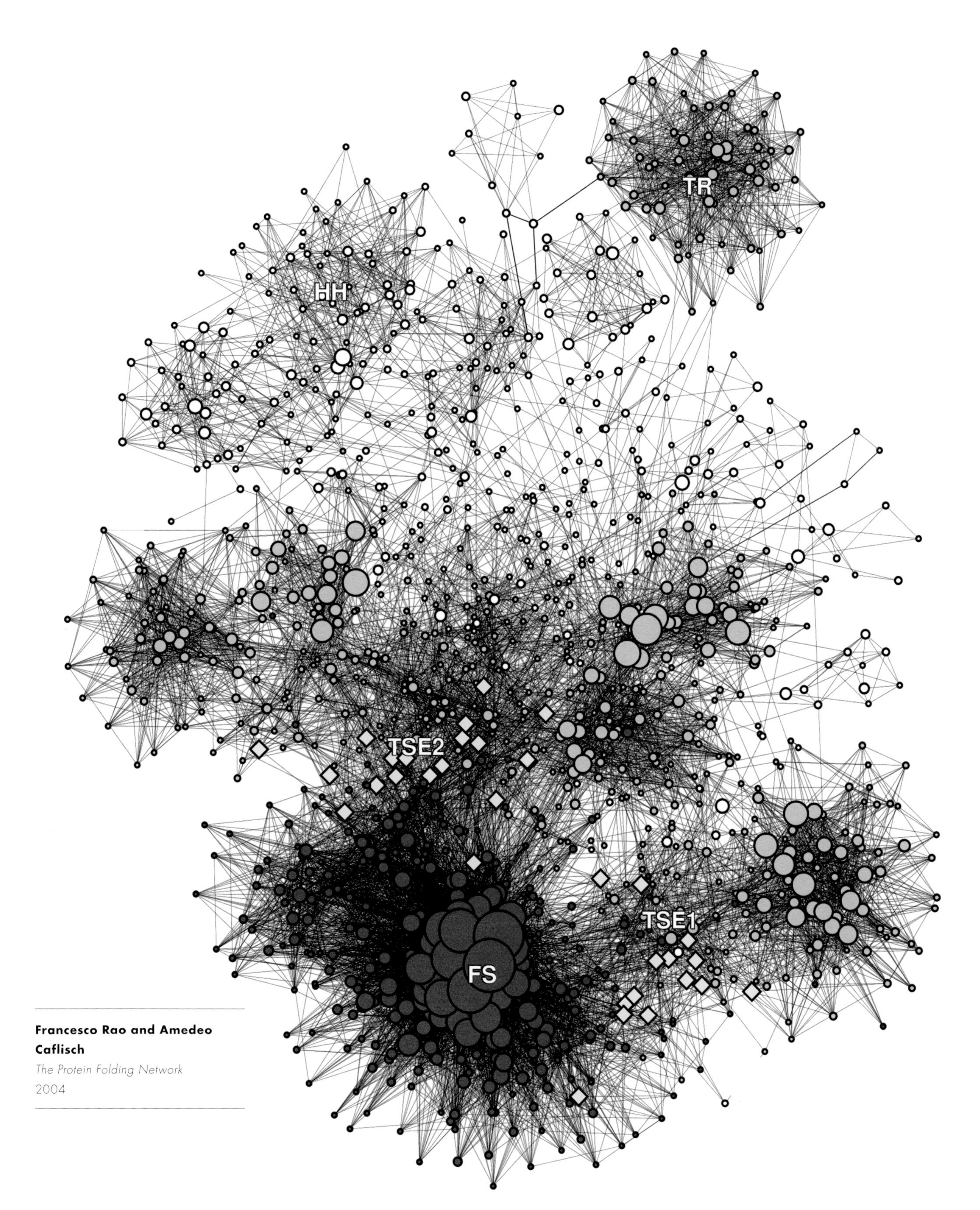

Francesco Rao and Amedeo Caflisch
The Protein Folding Network
2004

Terrorism

Characterized by idiosyncratic social structures and a loose hierarchy, terrorist cells are a challenging system to decipher. The unraveling of these decentralized organizations is an important pursuit for governments and military agencies across the world as well as for inquisitive citizens. Network visualization provides an important analytical tool by highlighting the ties between groups and individuals while also singling out prominent and influential people within the organizations.

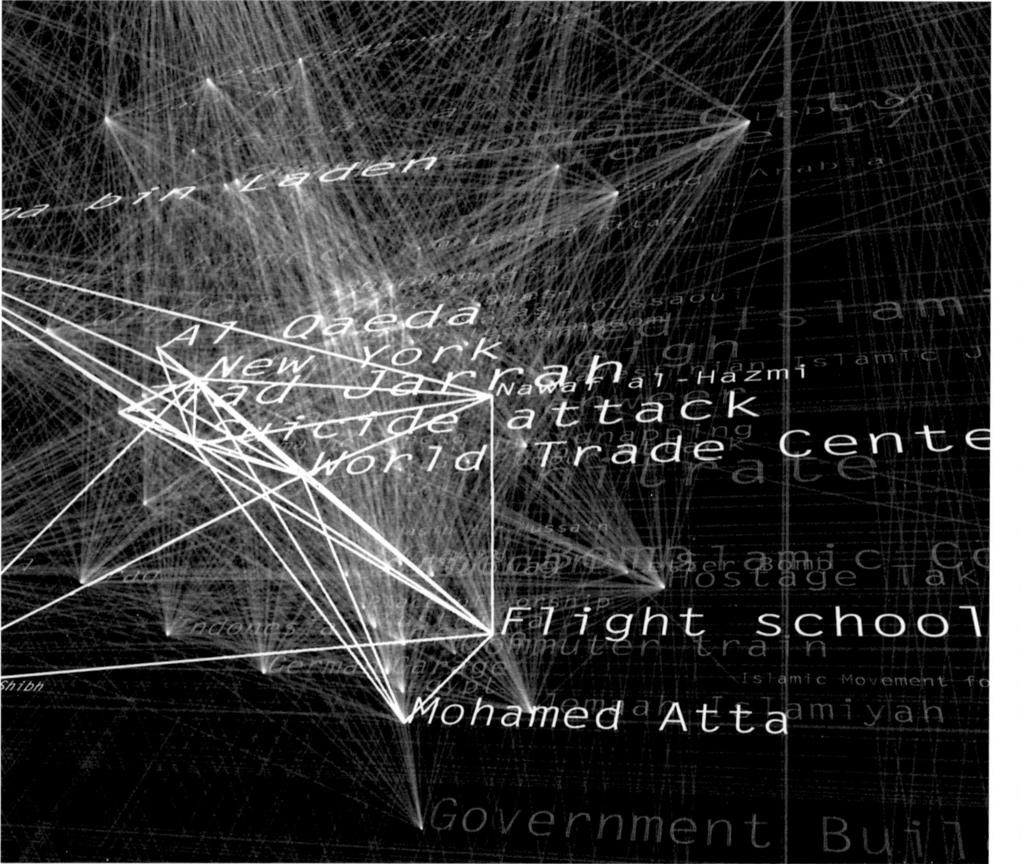

Lisa Strausfeld and James Nick Sears
Open-Source Spying
2006

A three-dimensional graph of associations between frequently searched words in a large counterterrorism database

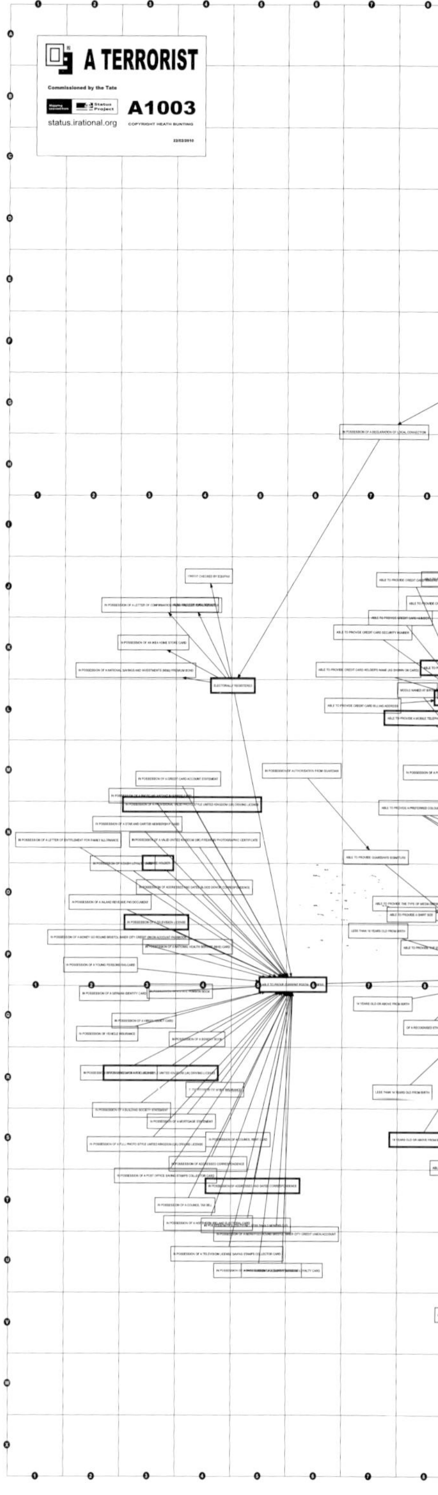

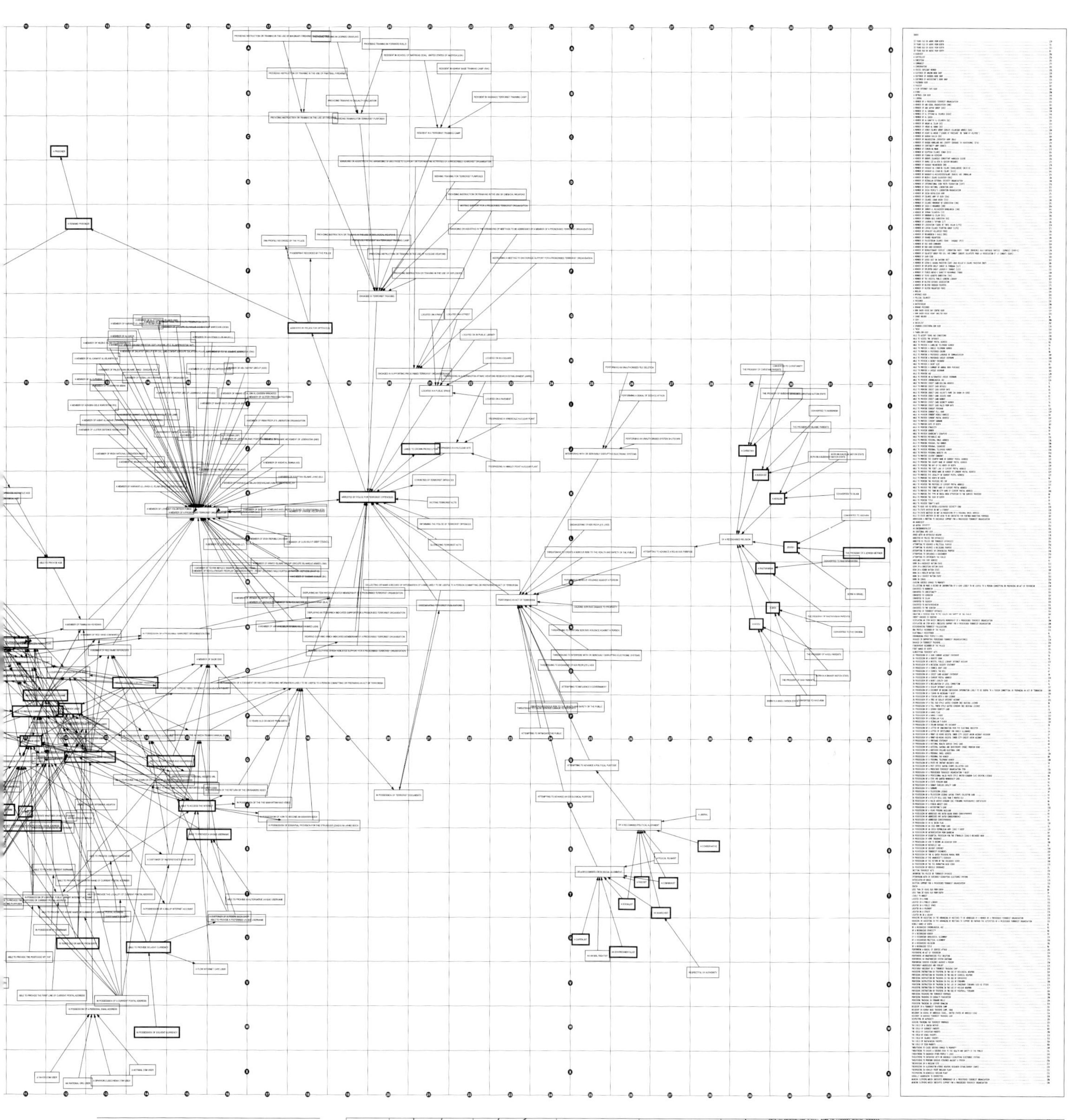

Heath Bunting

A Map of Terrorism

2008

A map of global activities and affiliations associated with terrorism. Commissioned by the Tate (above). Detail (right).

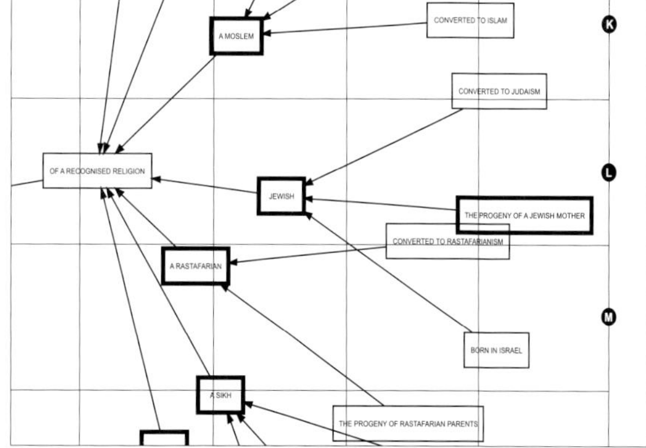

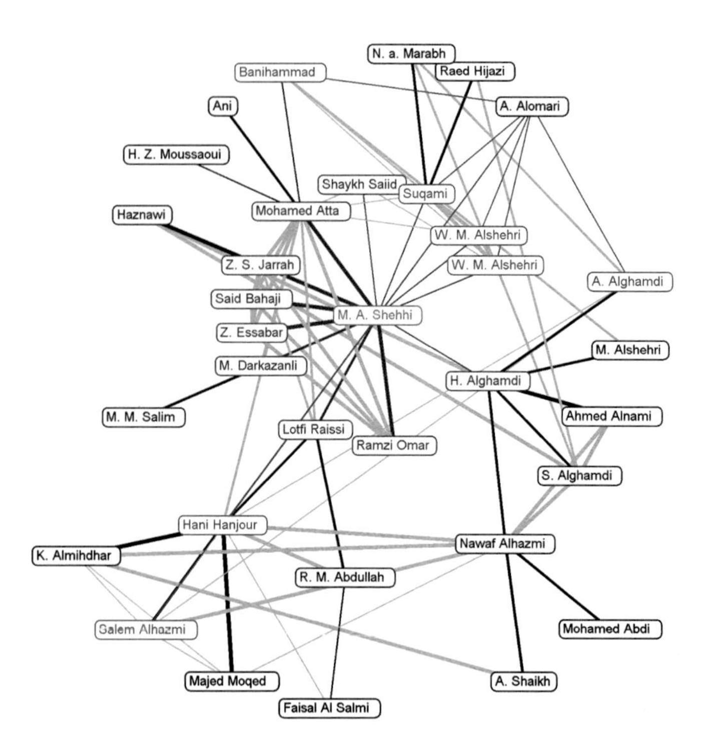

Jeffrey Heer and Alan Newberger
9/11 Terrorist Network
2004

A visualization of suspected connections between terrorists involved in the September 11 attacks

FMS Advanced Systems Group
TrackingTheThreat.com
2005

Screenshots of an interactive application that provides a graphical breakdown of the Al Qaeda network—part of an open-source database

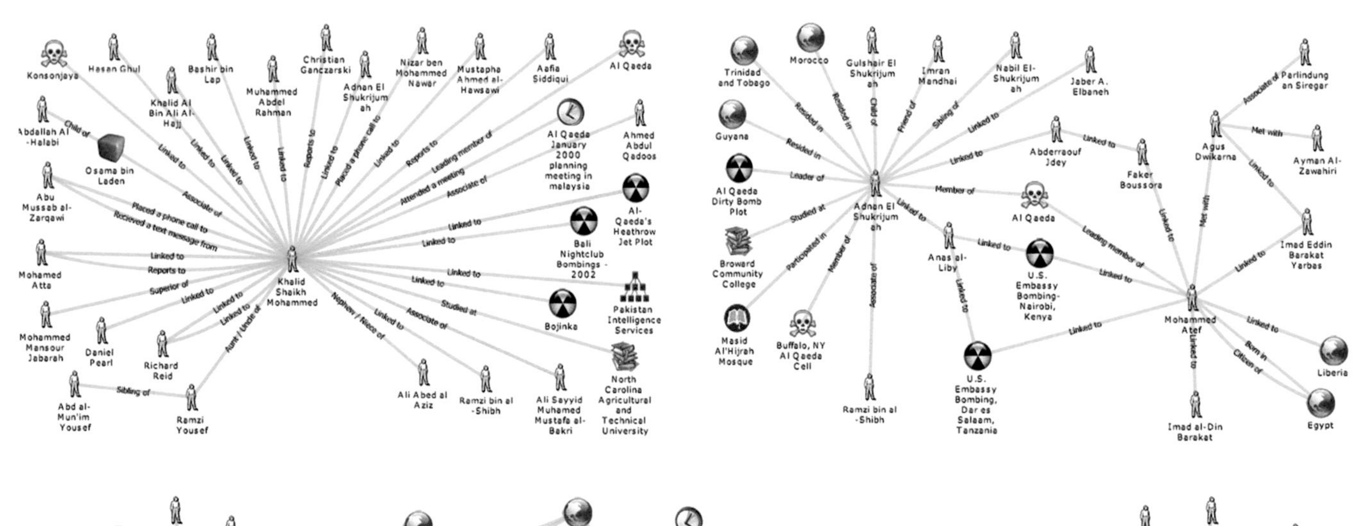

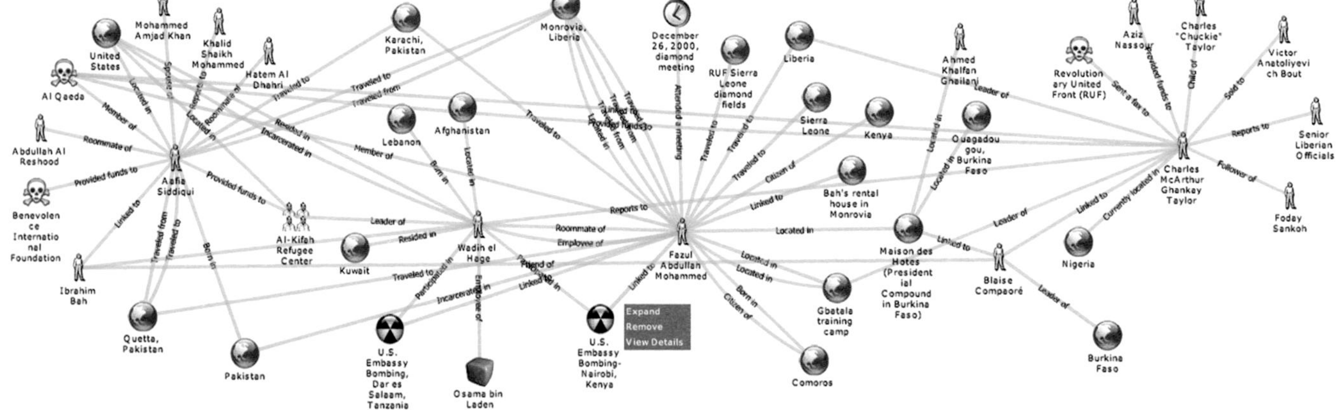

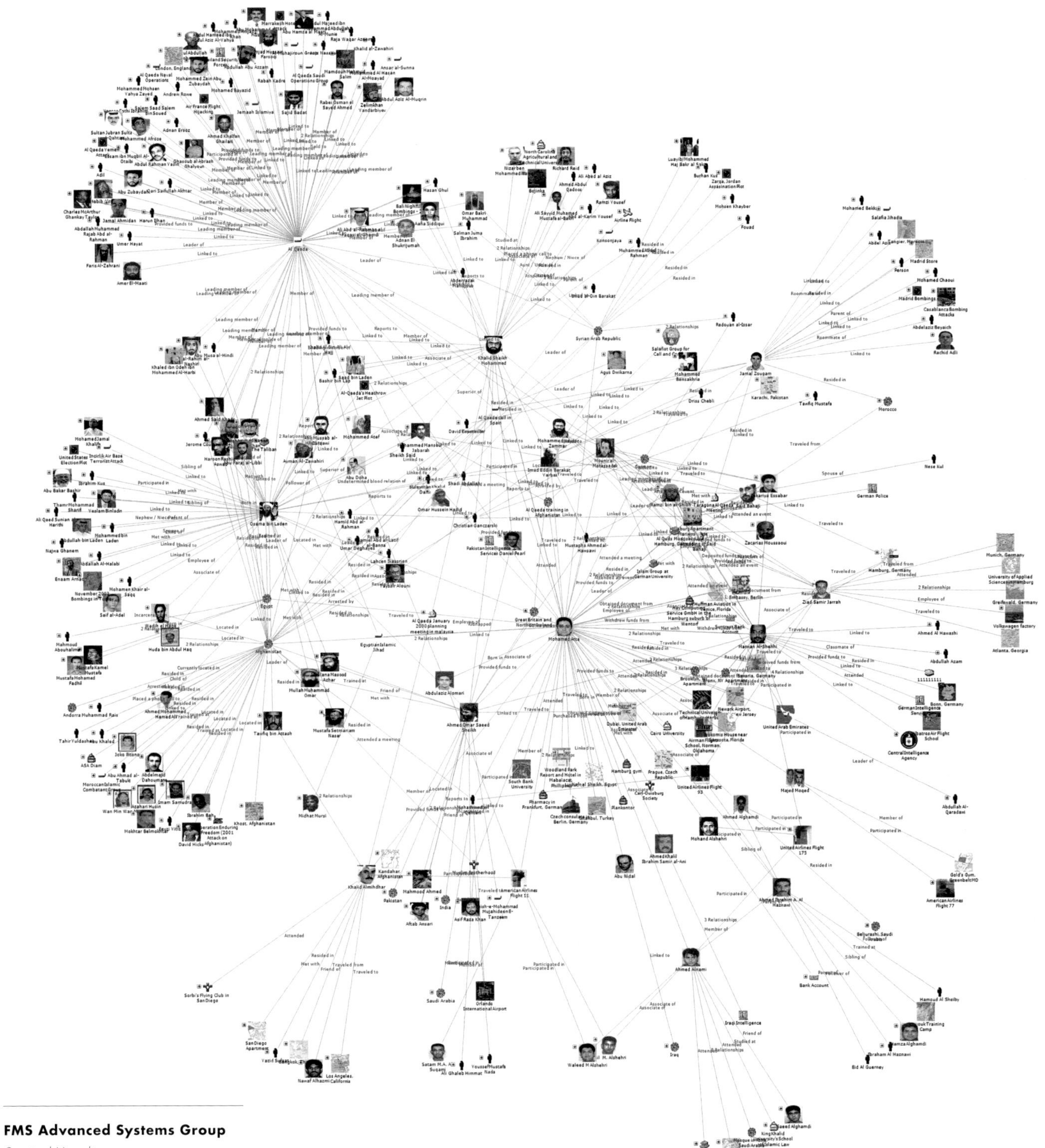

FMS Advanced Systems Group

Sentinel Visualizer

2008

A map of interconnections between members of the Al Qaeda terrorist network

Trajectories

Empowered by a wide range of cheap and accessible GPS and video-tracking devices, numerous authors are mapping an assortment of trails, paths, and movements within the physical environment, creating a dense lattice of individual networks. Social cartography is an emergent practice visible in multiple initiatives, from large collaborative projects like OpenStreetMap.org to smaller individual approaches aiming at exploring different facets of our own personal trajectories.

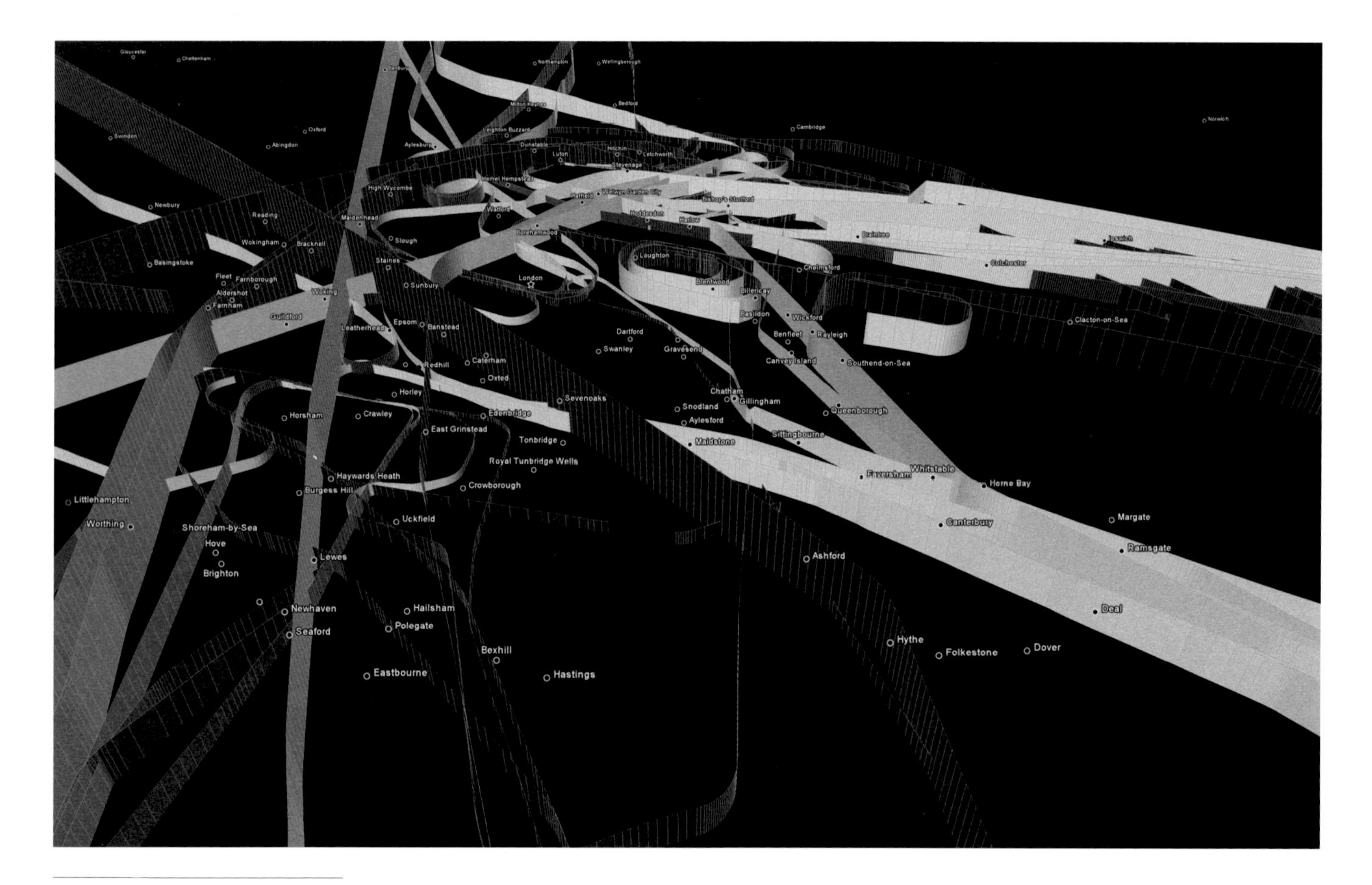

Jeremy Wood
Fly for Art (GPS Drawing)
2009

A map of commercial-airline tracks across London

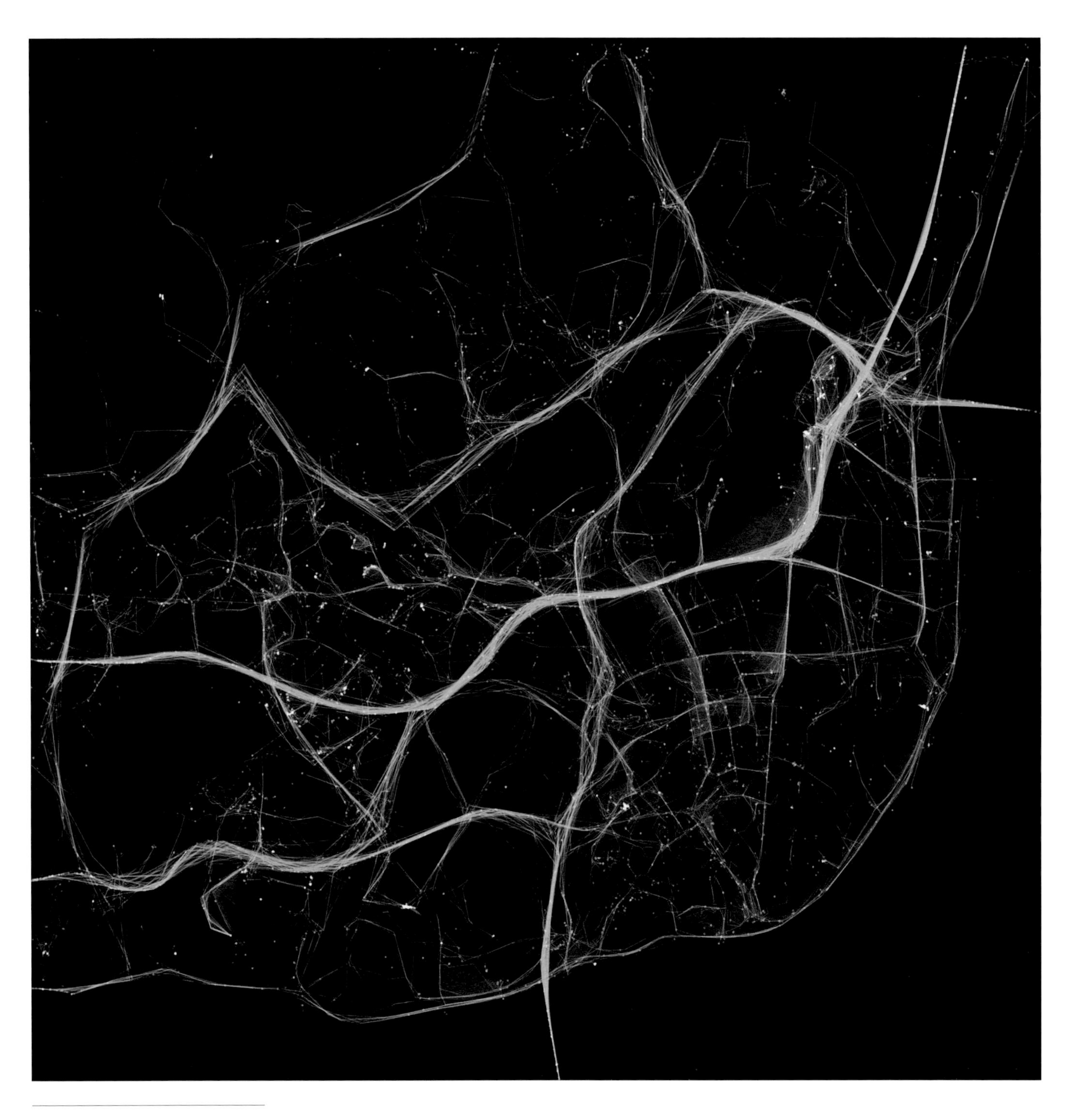

Pedro Miguel Cruz, Penousal Machado, and João Bicker
Visualizing Lisbon's traffic
2010

A map of the GPS coordinates and the speed of 1,534 taxis circulating in Lisbon, Portugal, during a single day in October 2009. White dots symbolize circulating vehicles. Each trail produced by the vehicles constitutes a temporary route, with the thickness standing for traffic intensity and a specific color gradation indicating the average speed. Cool hues (green and blue) represent rapid transit arteries, while the sluggish ones are depicted by warmer ones (red and orange). Pure green represents average speeds of 37 mph (60 km/h). A collaboration between the Centre for Informatics and Systems of the University of Coimbra and the CityMotion Project at MIT Portugal.

Tom Carden and Steve Coast
London GPS Tracking Map
2005

User-generated GPS traces of London collected for the OpenStreetMap project, a large collaborative project to create a free, editable map of the world

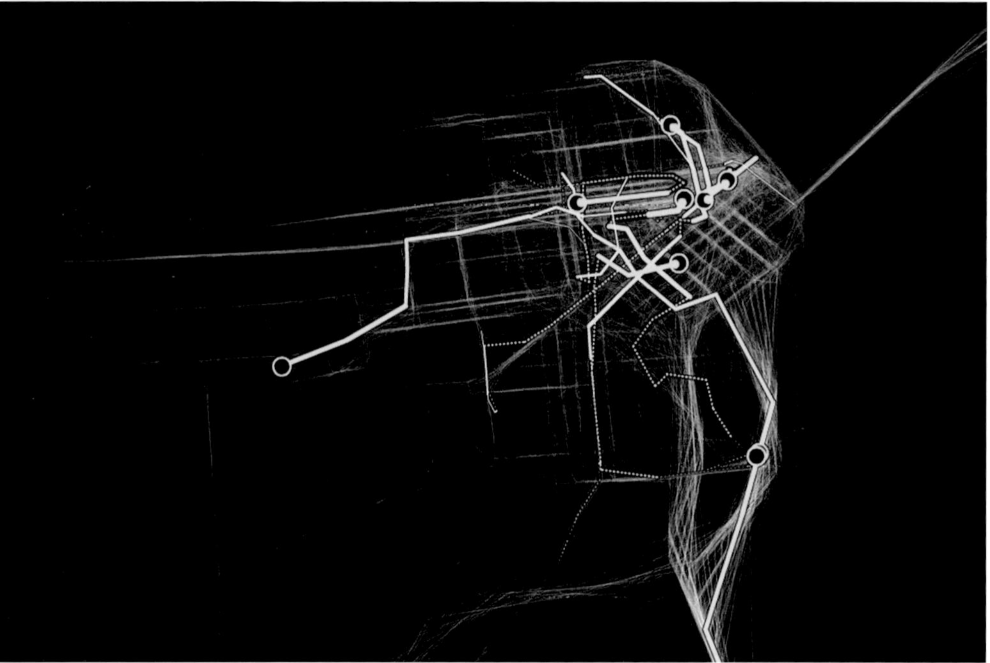

Stamen
Cabspotting
2006

A visualization of the GPS data generated by Yellow Cab taxis in San Francisco during a four-hour period

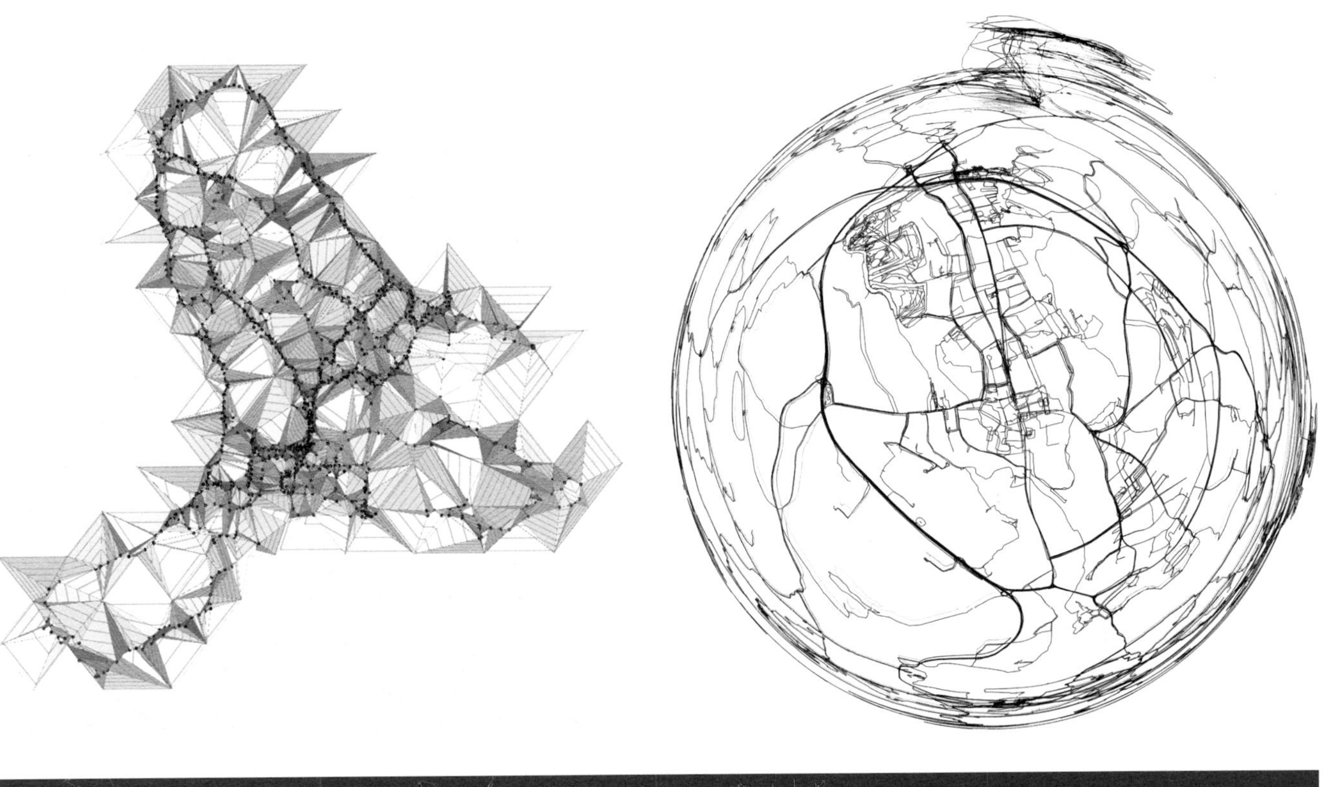

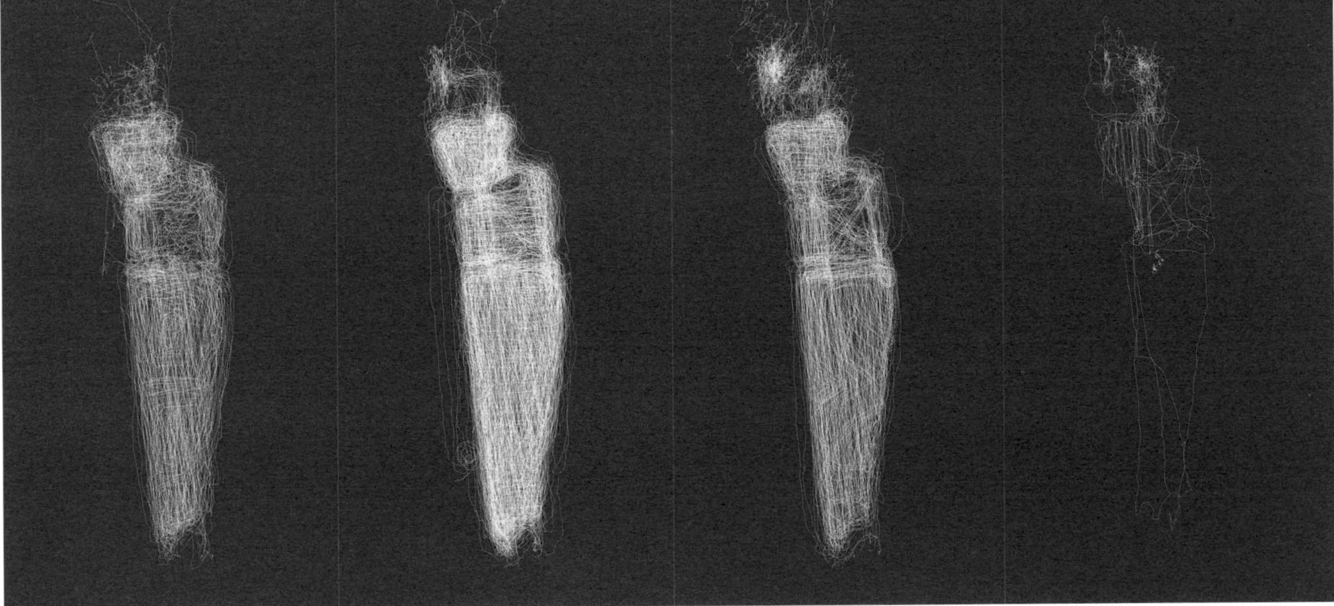

(top left)
Tom Carden
Biomapping Sketch
2006

A visualization of the emotion levels of people walking on the Greenwich Peninsula, London, calculated from GSR (Galvanic Skin Response) and GPS data. GSR is a simple indicator of emotional arousal. The height of the mesh is a measure of the GSR level.

(top right)
Jeremy Wood and Hugh Pryor
Oxford Fisheye (GPS Drawing)
2002

GPS traces in Oxford, England

(bottom)
Jeremy Wood
Lawn (GPS Drawing)
2008

The path generated by Jeremy Wood while mowing the lawn during different seasons throughout one year. From left to right: spring, summer, autumn, and winter.

Twitter

Twitter is one of the most popular microblogging social services, through which millions of people around the world communicate, using text messages of up to 140 characters, famously known as tweets. Twitter's underlying structure provides a great laboratory to investigate the behavioral traits of social groups and is an outstanding trend-analysis tool in tracking the fluctuations of public opinion in real time. In the context of network visualization, most authors try to disclose a range of relationships between Twitter users and the vast number of messages they regularly post.

Jer Thorp
Just Landed
2009

A visualization of the flight activity of Twitter users over a period of thirty-six hours. It finds tweets containing the phrase "Just landed in" and then marks the location where the user landed and the home location listed on their Twitter profile to generate the travel paths.

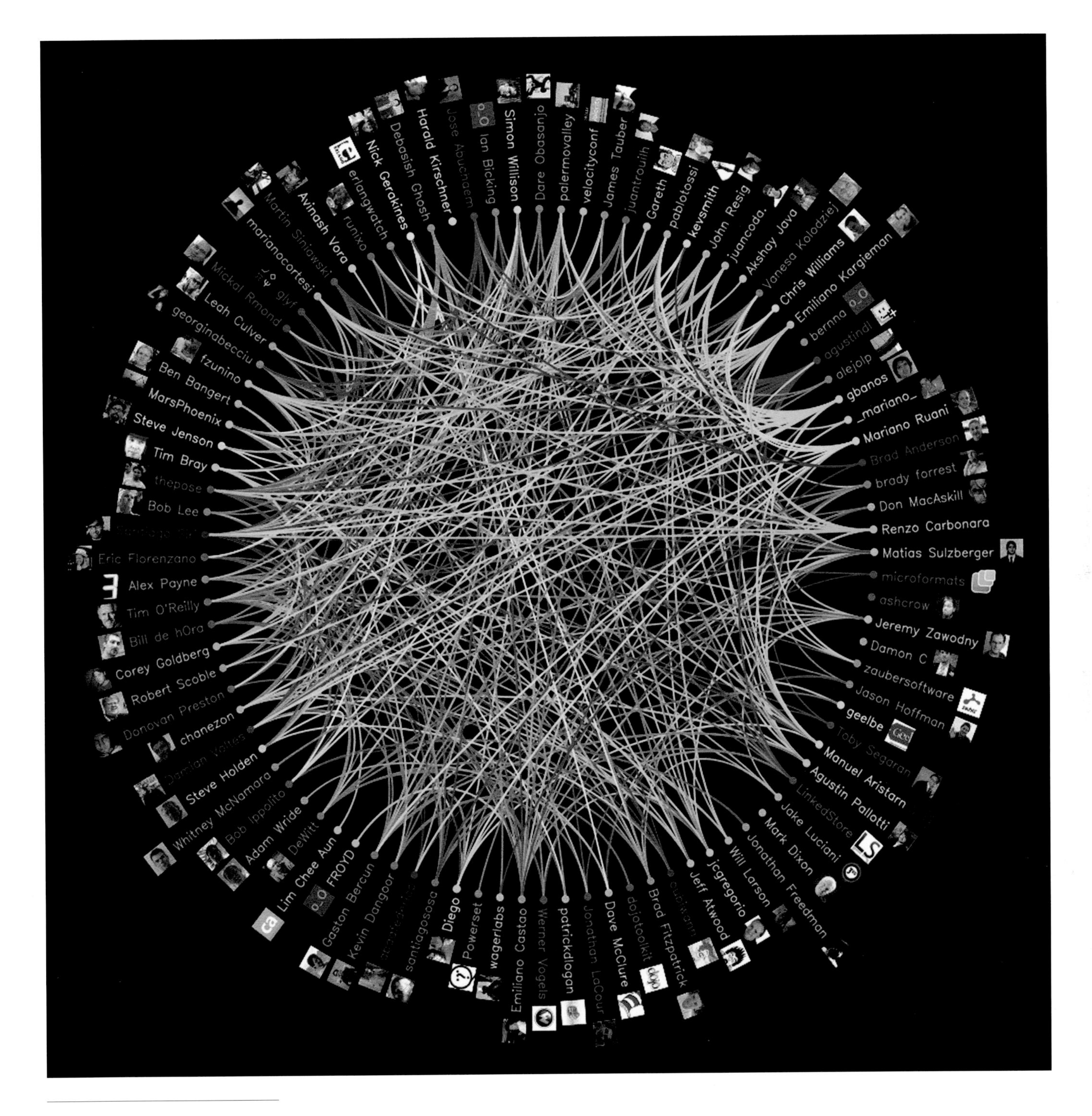

Augusto Becciu
TweetWheel
2008

A radial visualization of an individual's Twitter network and the connections between followers

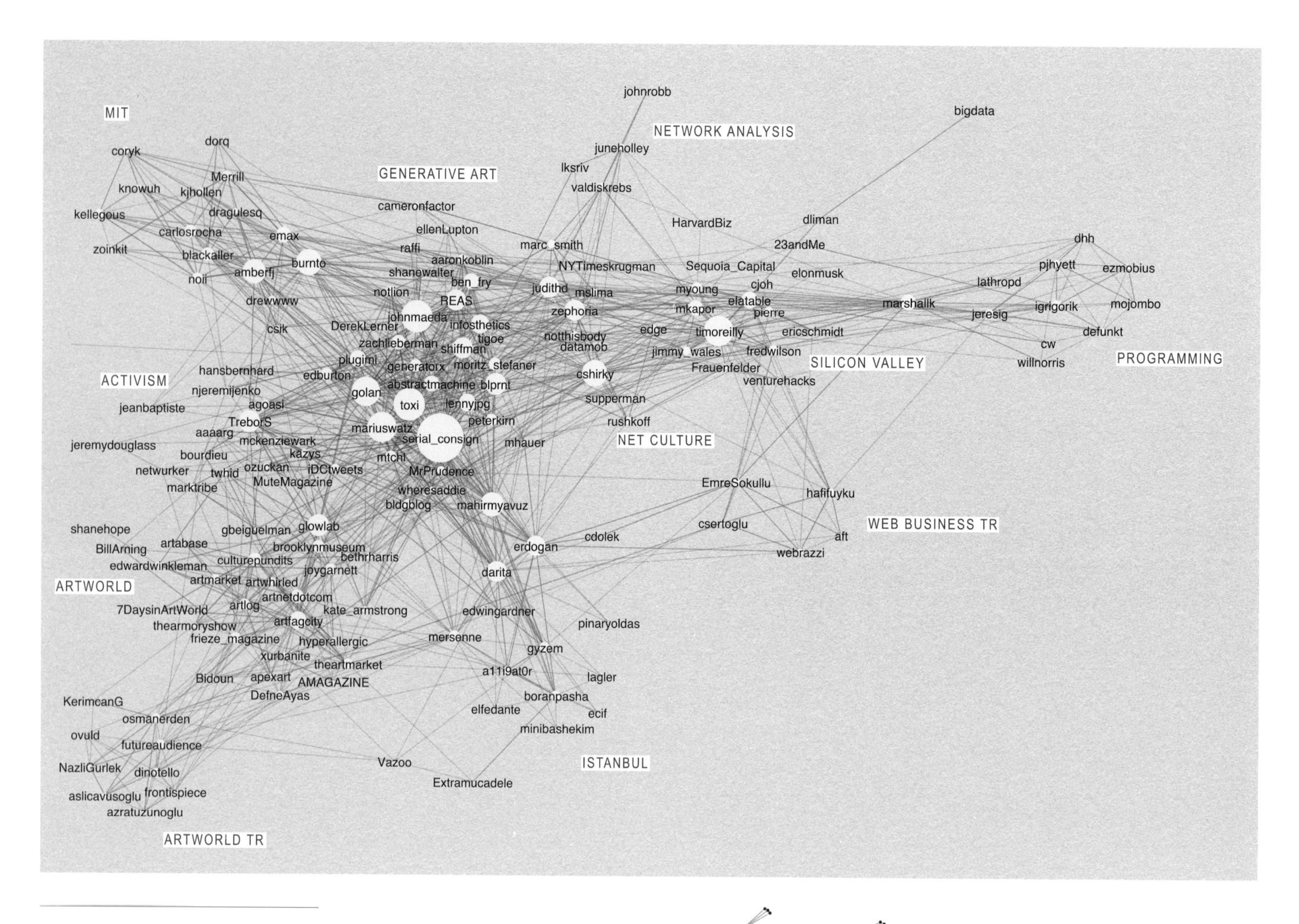

Burak Arikan

Growth of a Twitter Graph

2008

A map of connections within a user's Twitter network, grouped by topics of interest

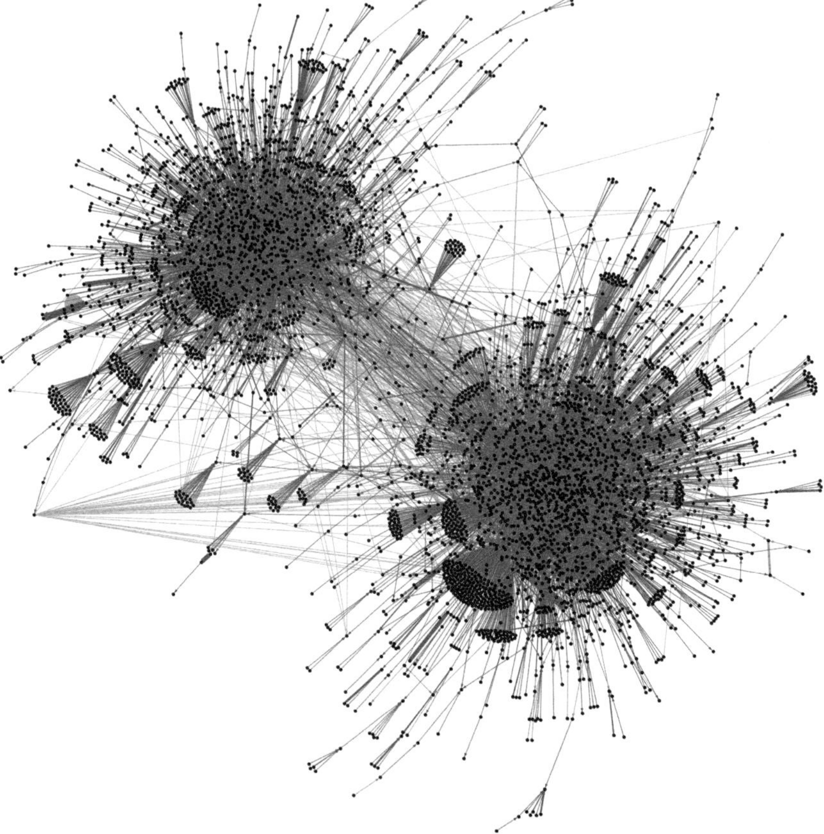

Jacob Ratkiewicz et al.

Truthy

2010

A diffusion network of the Twitter #GOP hashtag using Truthy, a system to analyze and visualize the diffusion of information on Twitter. Truthy evaluates thousands of tweets an hour to identify new and emerging bursts of activity around different memes.

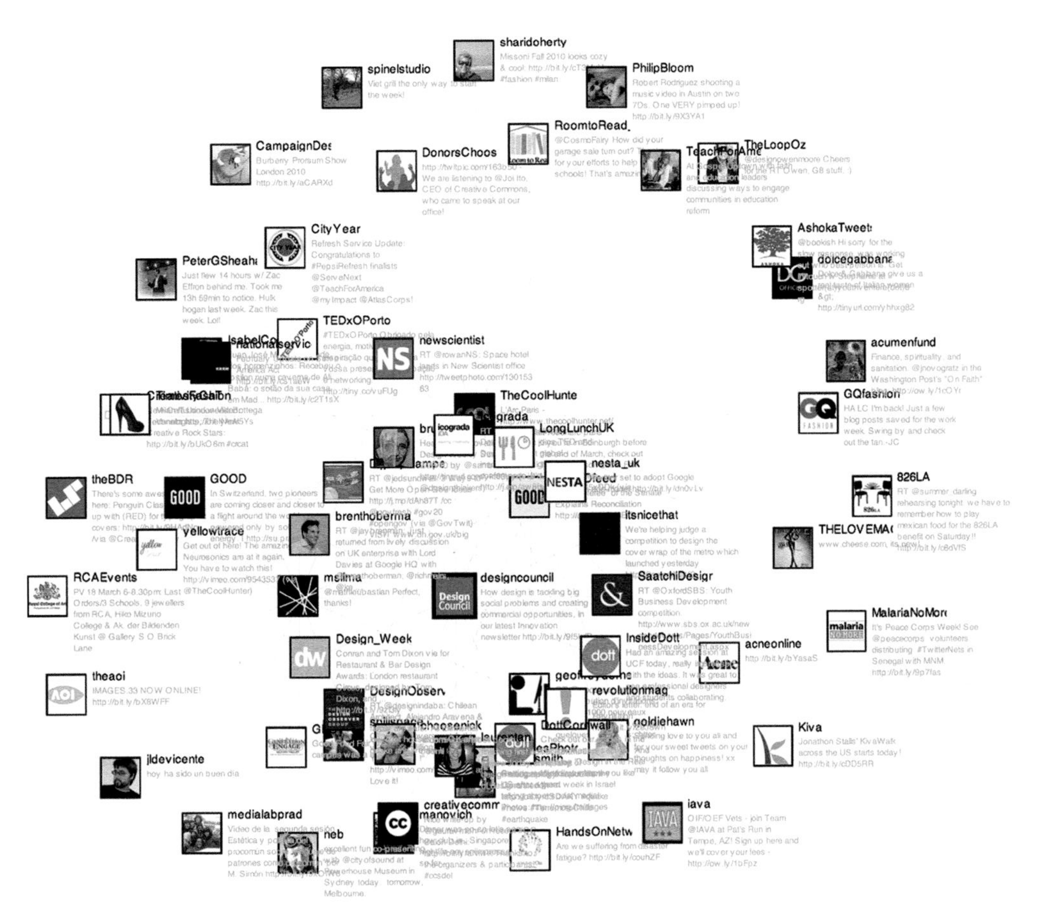

Kris Temmerman

Twitter Friends Browser

2008

A screenshot of an interactive application that allows anyone to explore their immediate Twitter followers as well as the followers' own followers

Daniel McLaren

Mentionmap

2009

A screenshot of a web application used to explore an individual's Twitter network. Links indicate a name mention of one user by another, with line thickness corresponding to number of mentions.

Wikipedia

With three million published articles by August 2009, in its English version alone, Wikipedia is the largest encyclopedia ever created in the history of humankind. This dense body of knowledge, connected by millions of hyperlinks, has been an intriguing subject for many who have felt compelled to uncover its intricate patterns. This fascination is in part due to Wikipedia's evolving nature and how it continues to redefine knowledge territories within its complex rhizomatic structure.

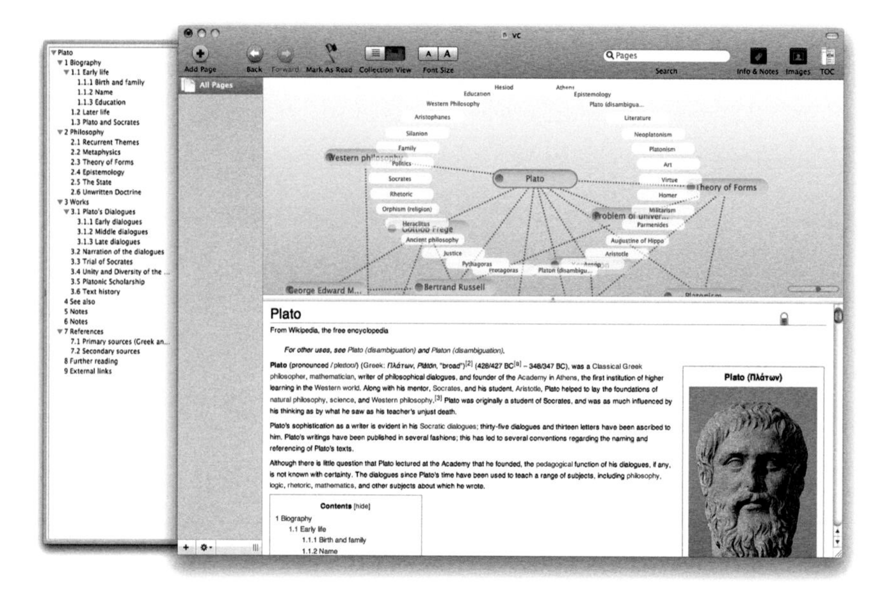

Dennis Lorson
Pathway
2006

A screenshot generated by Pathway, a free Mac application designed to help navigate Wikipedia—depicting the trail of article pages visited by a user (above); detail (right)

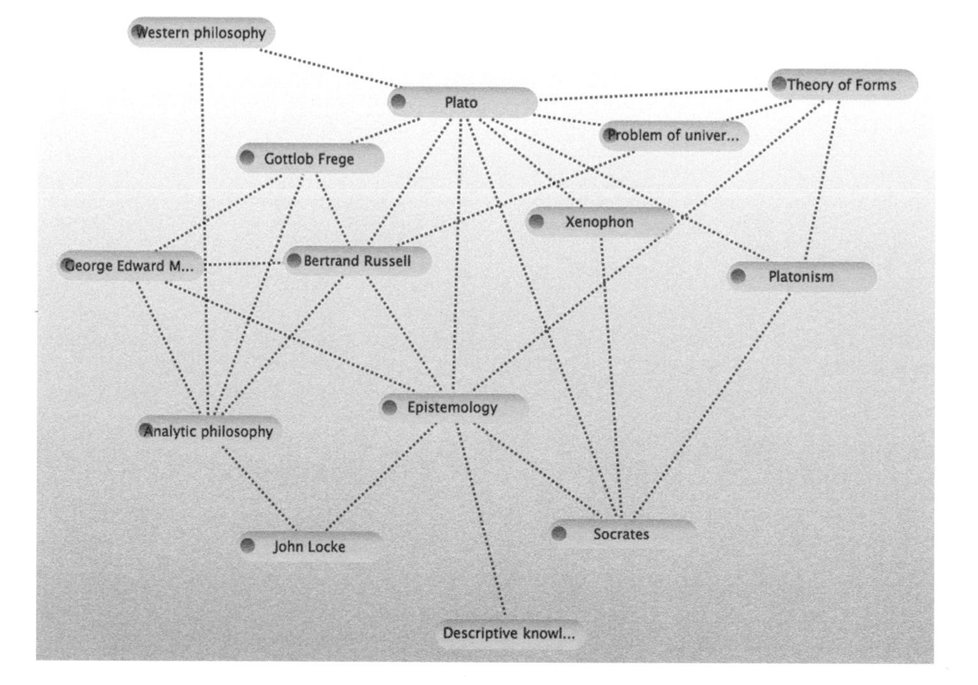

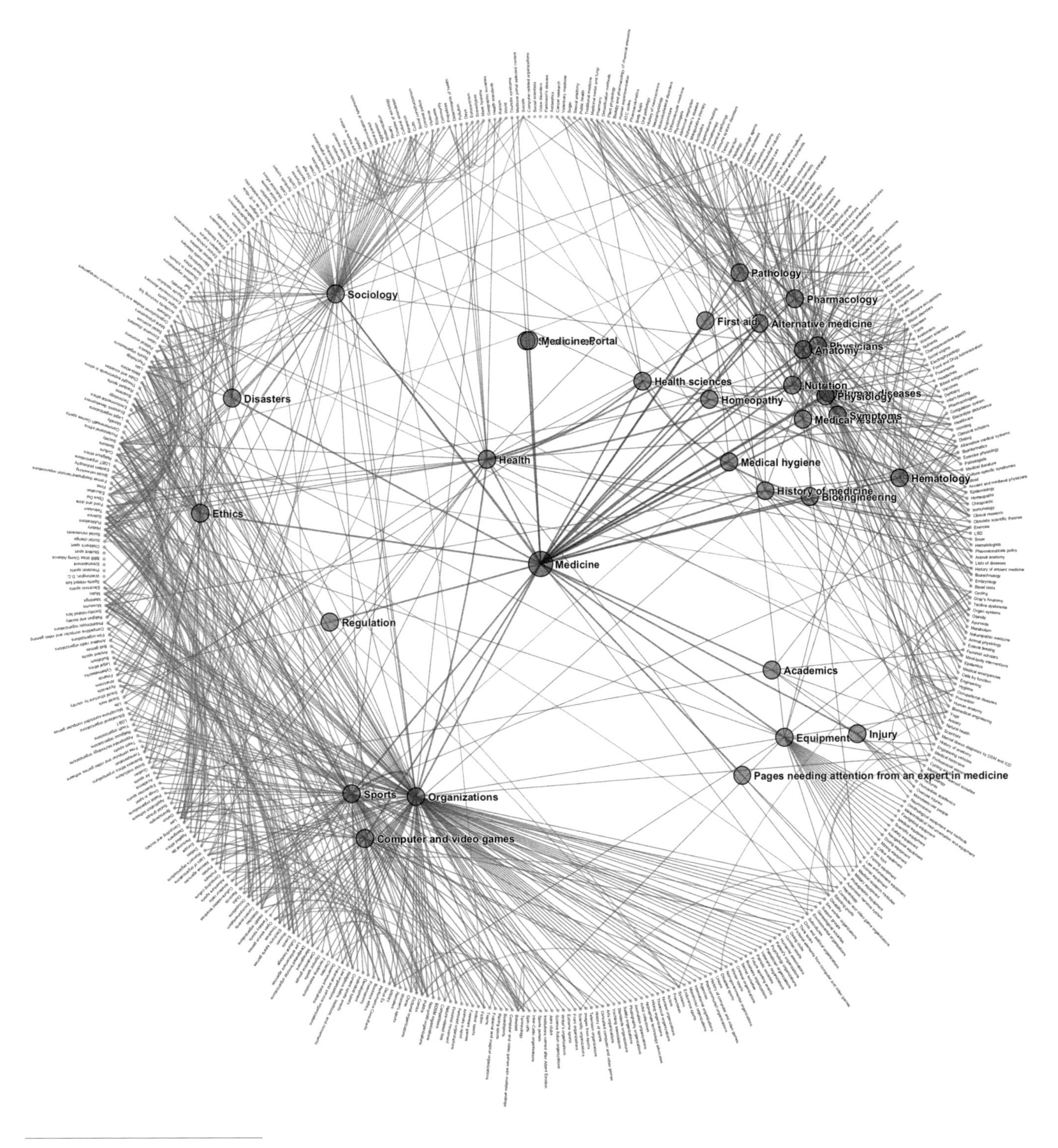

Chris Harrison

ClusterBall

2007

A chart of interconnections in subcategories of the Wikipedia *Medicine* category page

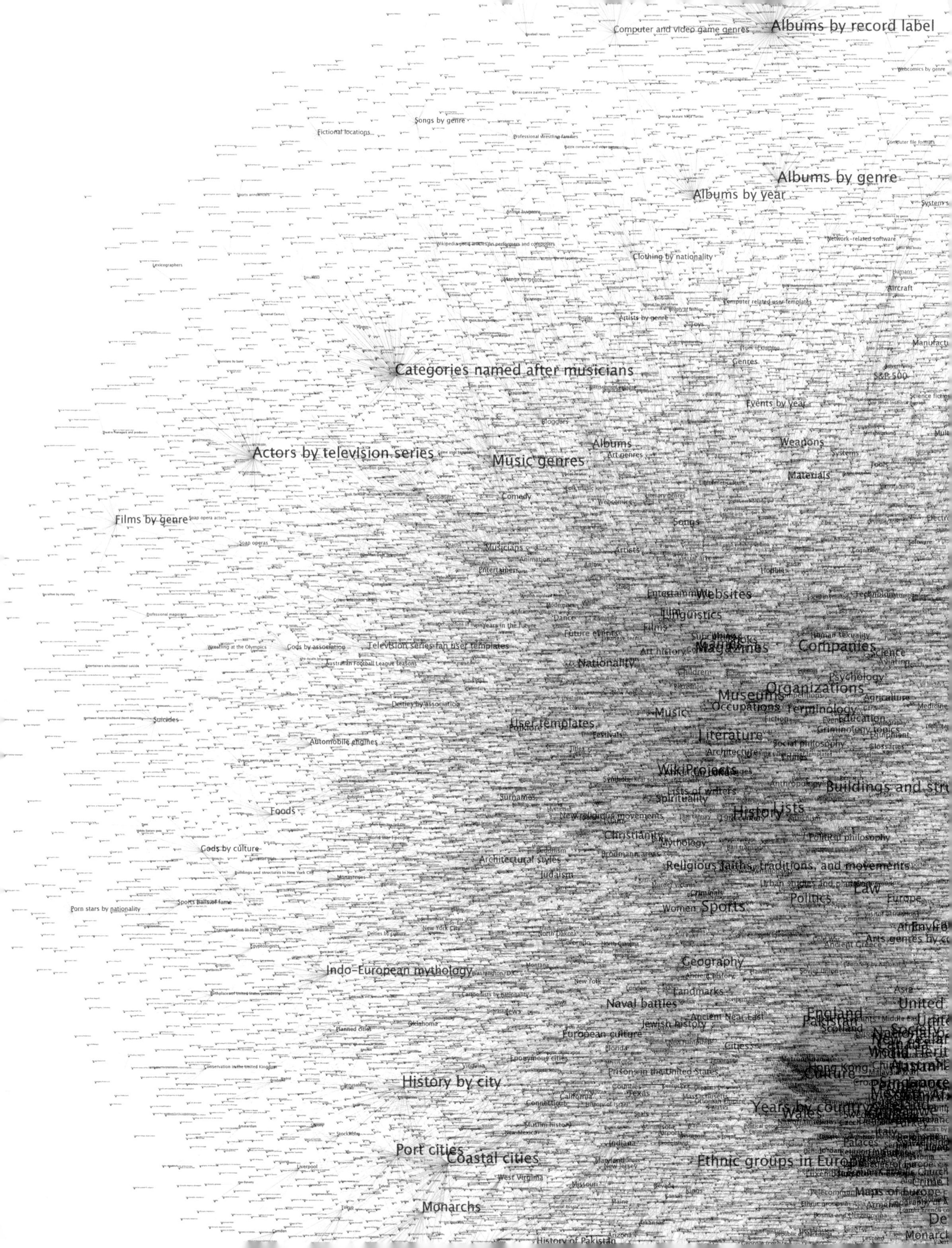

Computer and video game genres
Albums by record label
Webcomics by genre
Renaissance paintings
Songs by genre
Fictional locations
Teenage Mutant Ninja Turtles
Professional wrestling families
Computer file formats
Albums by genre
Albums by year
Sports announcers
System s
Anime by genre
Folk songs
Wikipedia good articles on performers and composers
Network-related software
Lexicographers
Clothing by nationality
Manga by genre
Humans
Aircraft
Computer related user templates
Artists by genre
Toys
Musicians by band
Manufactu
Genres
Advertising
Categories named after musicians
S&P 500
Science fictio
Events by year
Bloggers
Opera managers and producers
Albums
Weapons
Actors by television series
Music genres
Art genres
Systems
Television genres
Tools
Materials
Composers
Comedy
Folk music
Webcomics
Literary genres
Films by genre
Soap opera actors
Songs
Design
Punk
Electro
Software
Soap operas
Drama
Musicians
Artists
Animation
Visual arts
Cognition
Anime
Entertainers
Hobbies
Radio
Socialites by nationality
Crime television series
Entertainment
Websites
Technology
Dance
Linguistics
Photography
Comedy-drama television series
Professional magicians
Years in the future
Future events
Films
Human sexuality
Japanese terms
Subcultures
Books
Wrestling at the Olympics
Gods by association
Television series fan user templates
Magazines
Companies
Science
Art history
Aviation
Australian Football League seasons
Entertainers who committed suicide
Nationality
Children
Psychology
Museums
Organizations
Agriculture
Medicine
Occupations
Terminology
Deities by association
Suicides
Fiction
Education
Music
User templates
Criminology topics
Festivals
Automobile engines
Literature
Social philosophy
Glossaries
Architecture
Crime
WikiProjects
Anthropology
Buildings and str
Lists of writers
Surnames
Spirituality
Foods
New religious movements
19th century
20th century
History
Lists
Anarchism
Christianity
World War I people
Mythology
Political philosophy
Gods by culture
Buddhism
Brodmann areas
Architectural styles
Religious faiths, traditions, and movements
Judaism
Buildings and structures in New York City
Monasteries
Urban studies and planning
Murder victims
Criminals
Politics
Law
Europe
Porn stars by nationality
Sports halls of fame
Women
Sports
Communism
Visitor attractions
Transportation in New York City
Deities by culture
New York City
North Dakota
Africa
Egyptologists
Colorado
North Carolina
Cold War
Ancient Greece
Arts genres by c
Geography
Montana
Indo-European mythology
Washington, D.C.
Hawaii
Soviet Union
Ancient history
New York
Asia
Birthplaces of United States presidents
Cartoonists by nationality
Landmarks
United
Jews
Naval battles
London
Ancient Near East
England
Pakistan
Scotland
Oklahoma
Jewish history
Society
Planned cities
European culture
New Hampshire
Cities
Florida
Virginia
Metropolitan areas
Hong Kong
Culture
Malaysia
Eponymous cities
Conservation in the United Kingdom
Prisons in the United States
History by city
Jerusalem
Counties
Massachusetts
Connecticut
California
Texas
History of Israel
Years by country
Muslim history
Wales
Amsterdam
Minnesota
Washington
Sweden
Italy
Stockholm
New Mexico
Palaces
Indiana
Port cities
Coastal cities
Ethnic groups in Europe
Liverpool
Maryland
New Jersey
West Virginia
Missouri
Nevada
Maps of Europe
Maine
Ethnic groups in Asia
Bosnia and Herzegovina
Monarchs
Tokyo
Arizona
History of Pakistan
Monarch

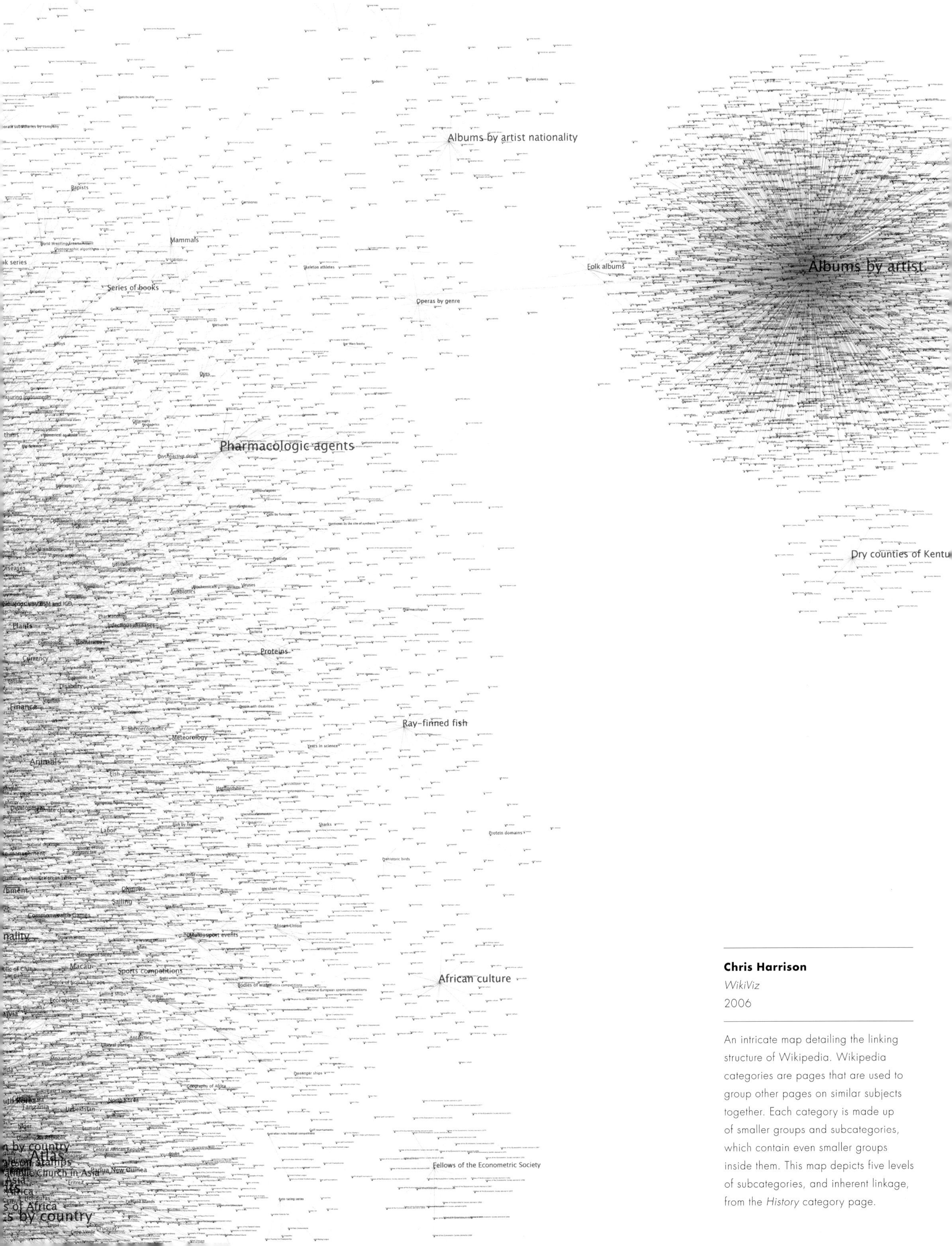

Chris Harrison

WikiViz

2006

An intricate map detailing the linking structure of Wikipedia. Wikipedia categories are pages that are used to group other pages on similar subjects together. Each category is made up of smaller groups and subcategories, which contain even smaller groups inside them. This map depicts five levels of subcategories, and inherent linkage, from the *History* category page.

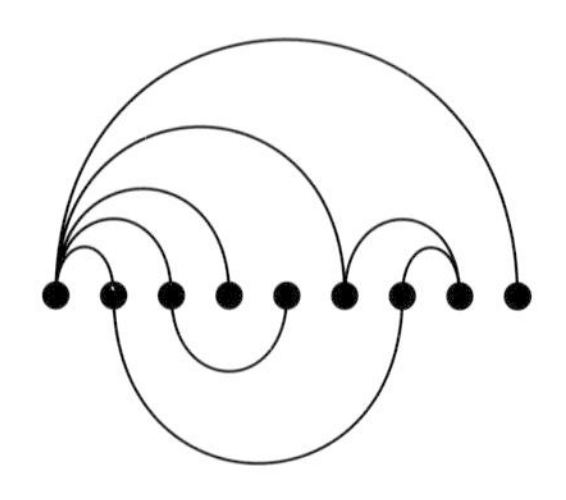

Arc Diagram

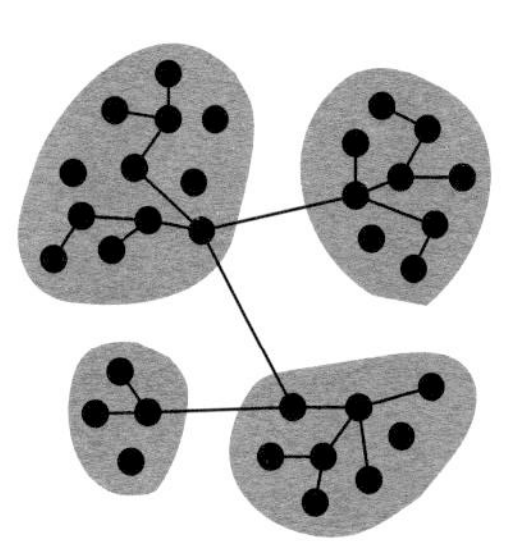

Area Grouping

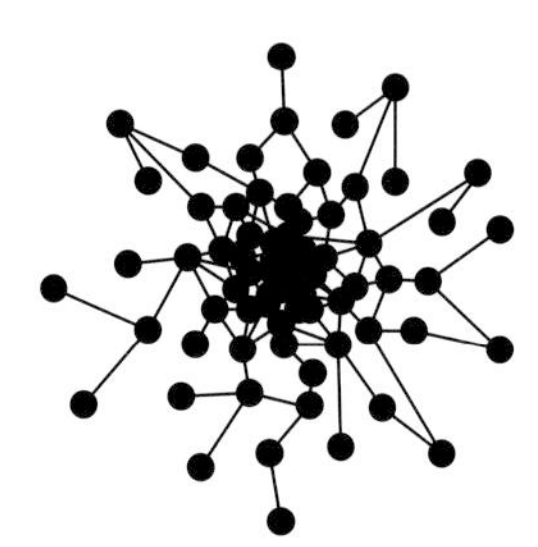

Centralized Burst

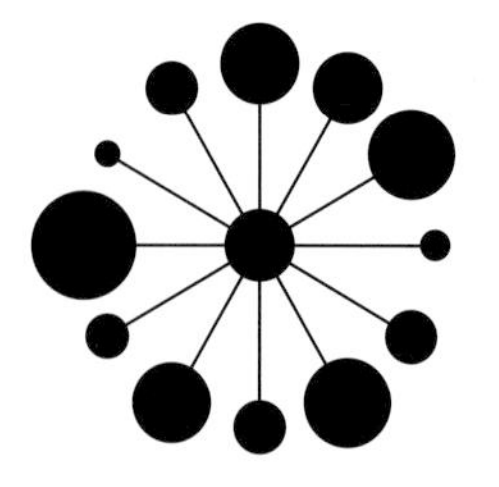

Centralized Ring

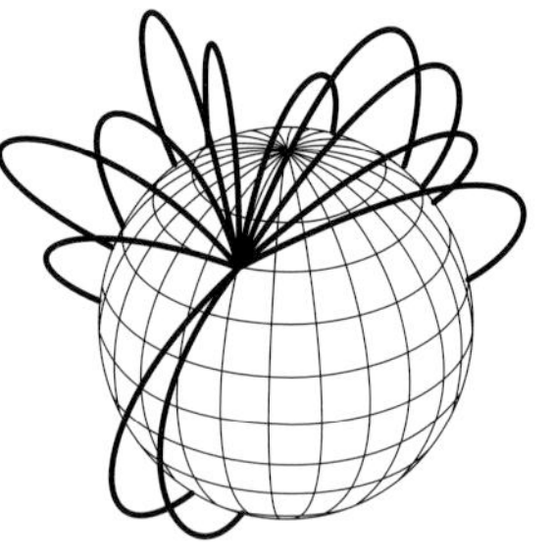

Circled Globe

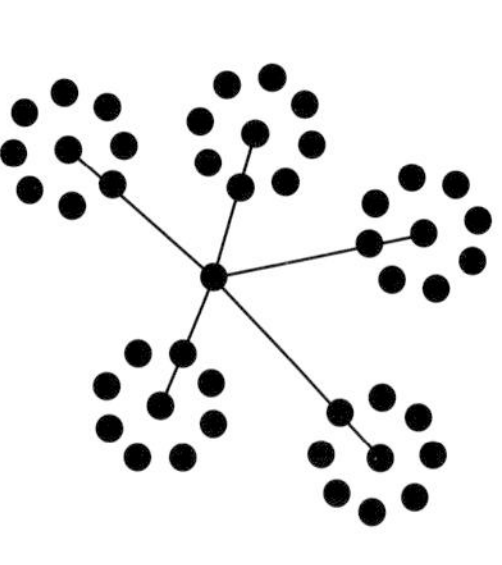

Circular Ties

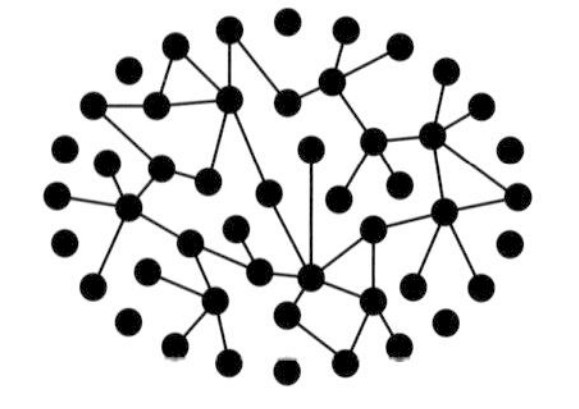

Elliptical Implosion

Flow Chart

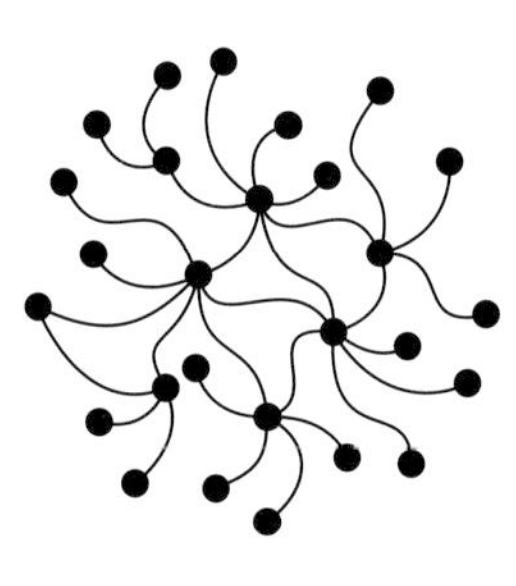

Organic Rhizome

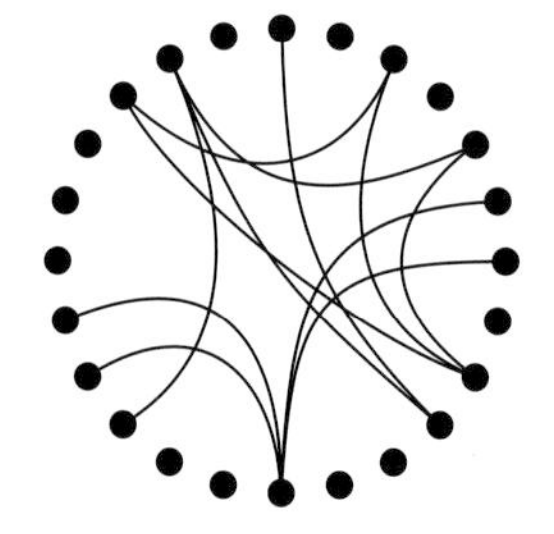

Radial Convergence

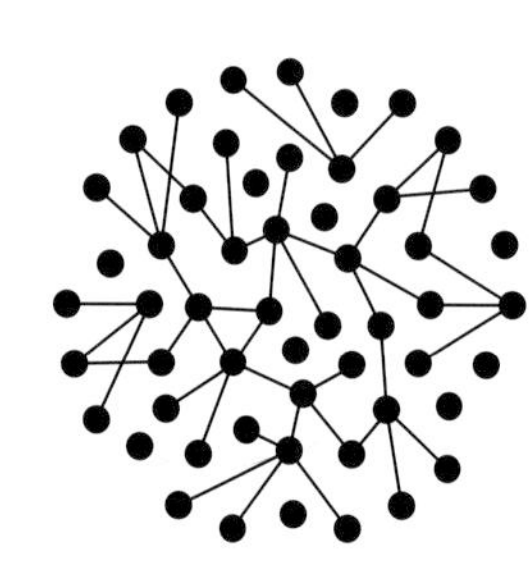

Radial Implosion

Ramification

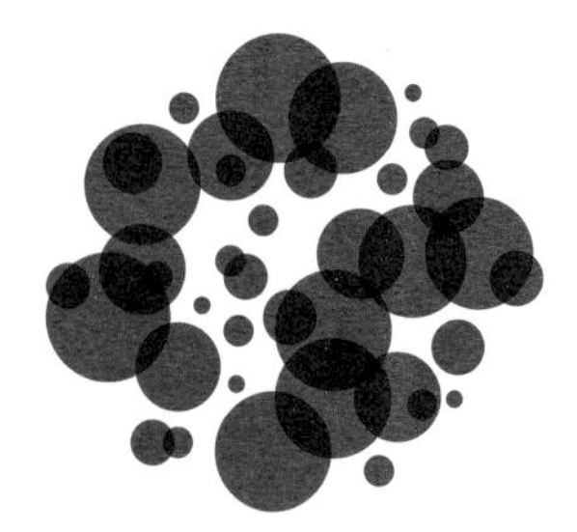

Scaling Circles

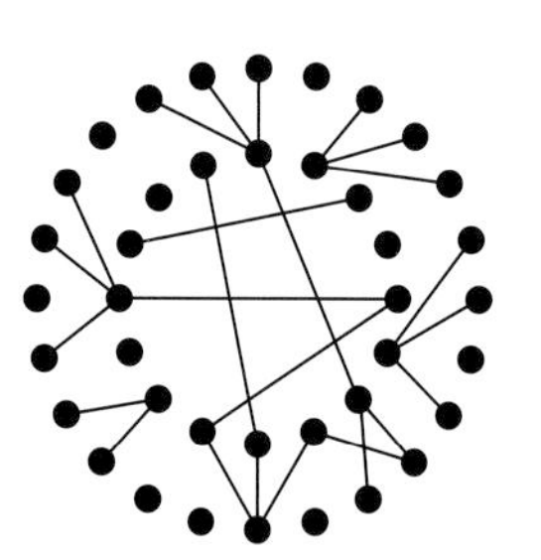

Segmented Radial Convergence

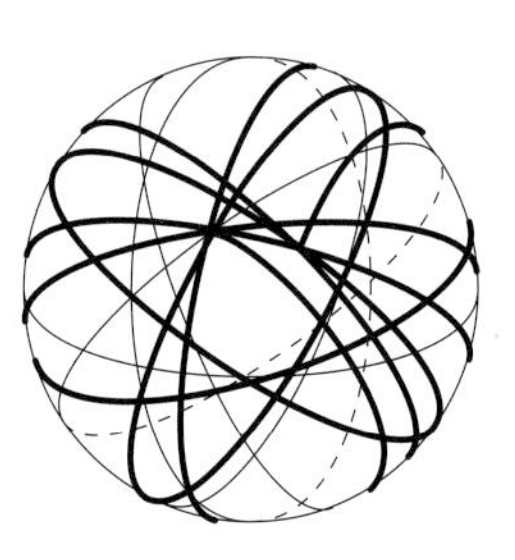

Sphere

05 | The Syntax of a New Language

Graphic representation constitutes one of the basic sign-systems conceived by the human mind for the purposes of storing, understanding, and communicating essential information. As a "language" for the eye, graphics benefits from the ubiquitous properties of visual perception.

—Jacques Bertin

Functional visualizations are more than innovative statistical analyses and computational algorithms. They must make sense to the user and require a visual language system that uses colour, shape, line, hierarchy and composition to communicate clearly and appropriately, much like the alphabetic and character-based languages used worldwide between humans.

—Matt Woolman

Looking at the range of network depictions produced in the last fifteen years, one cannot help but marvel at the diversity of topics and subjects being explored. But almost as staggering as the assortment of portrayed subjects is the variety of employed visual techniques. Frequently generated by computer algorithms and enhanced by interactive features, most projects showcase a broad palette of visual elements and variations that consider color, text, imagery, size, shape, contrast, transparency, position, orientation, layout, and configuration. Despite this rich graphical diversity, many projects tend to follow noticeable trends and common principles, which in turn result in a type of emergent taxonomy. This embryonic and evolving taxonomy provides a portrait of the current state of the practice and reveals the initial building blocks shaping a new visual language.

The two epigraphs to this chapter are drawn from Bertin, *Semiology of Graphics*, 2; and Woolman, *Digital Information Graphics*, 11.

Arc Diagram

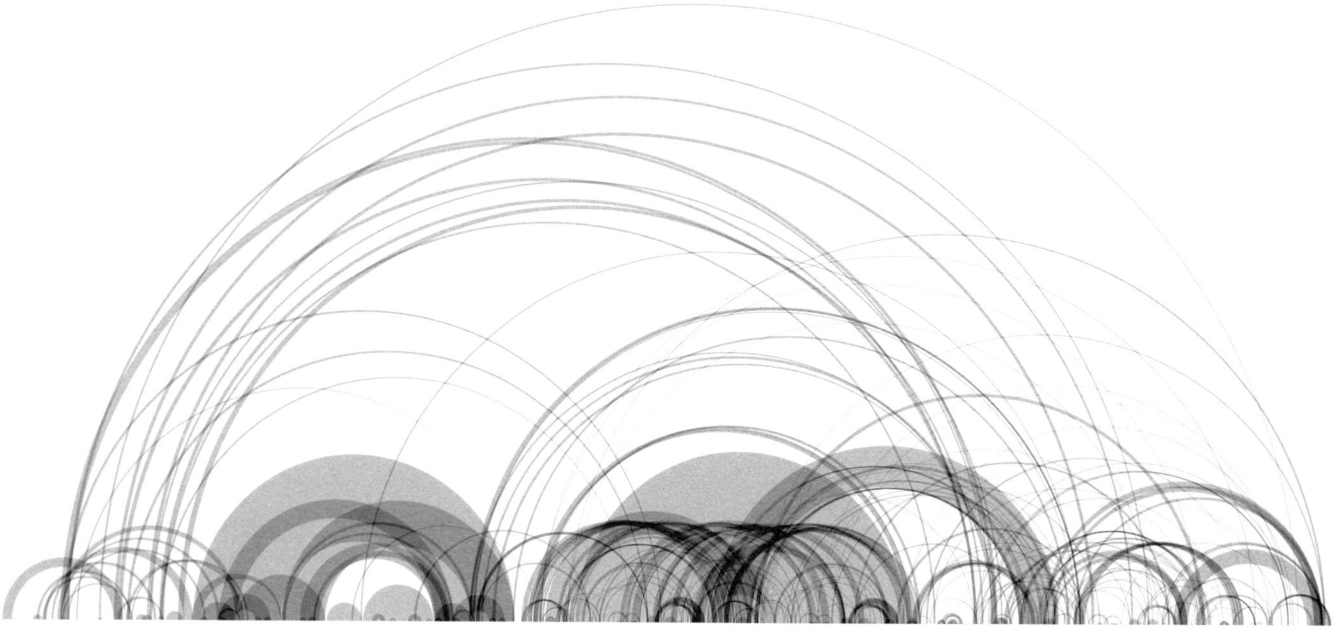

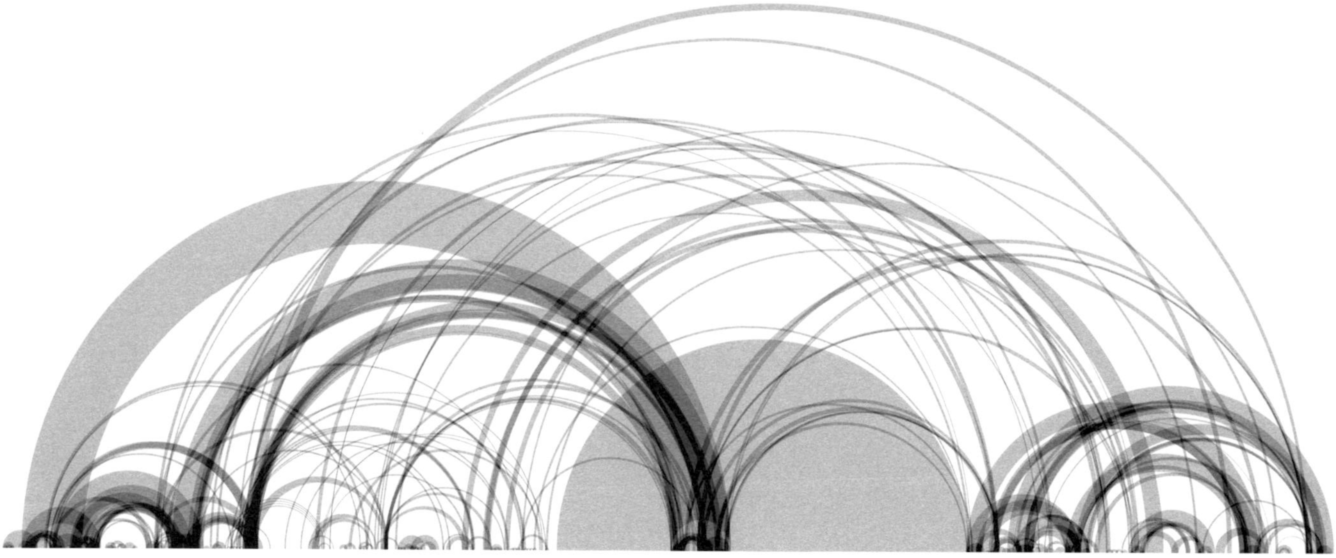

(top)
Martin Wattenberg
The Shape of Song
2001

A visualization of *Moonlight Sonata* by Ludwig van Beethoven. Each arc connects identical, repeated passages of the composition. With these passages as signposts, the diagram reveals the foundational structure of the music track.

(bottom)
Martin Wattenberg
The Shape of Song
2001

A visualization of *Four Seasons (Autumn)* by Antonio Vivaldi

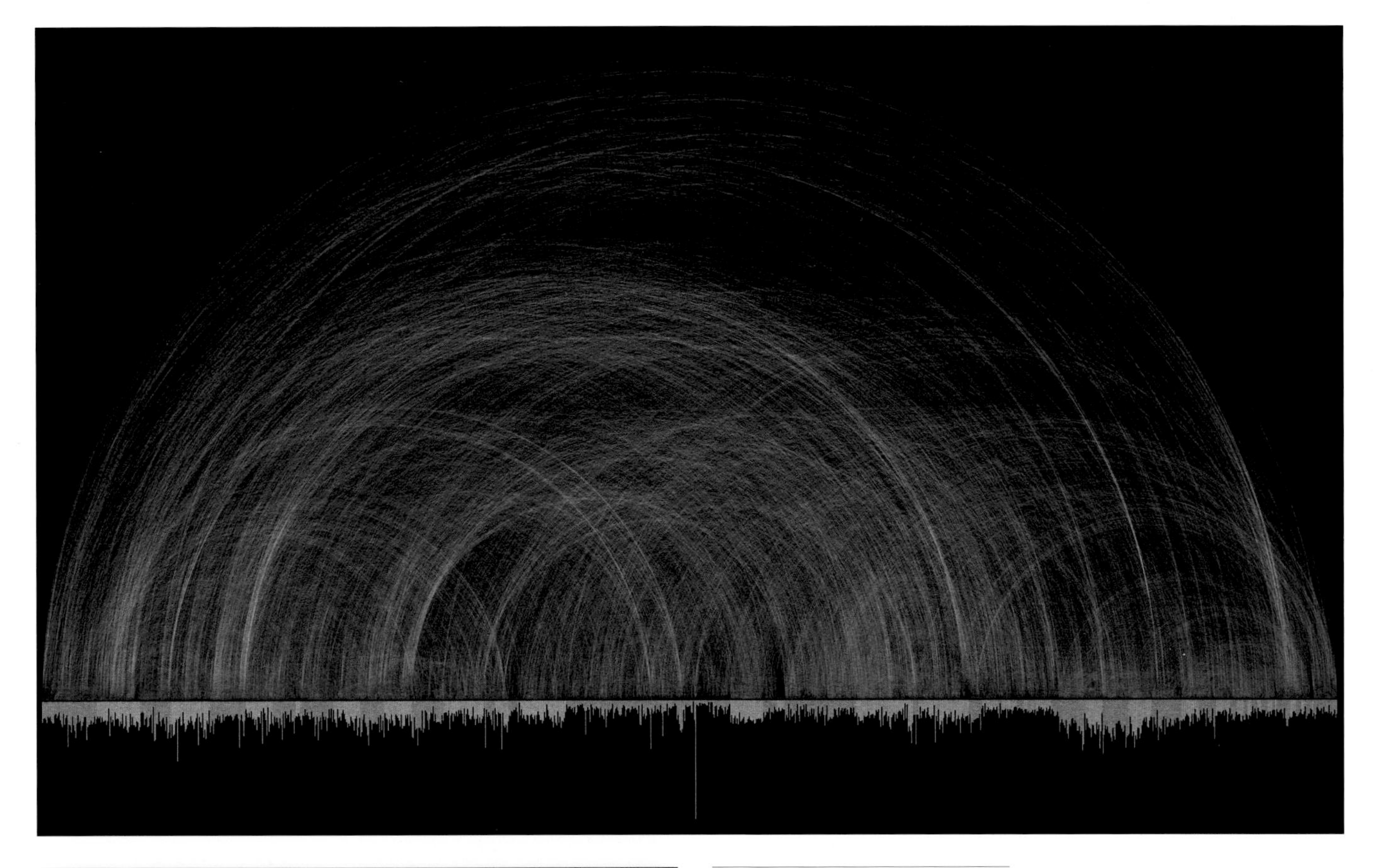

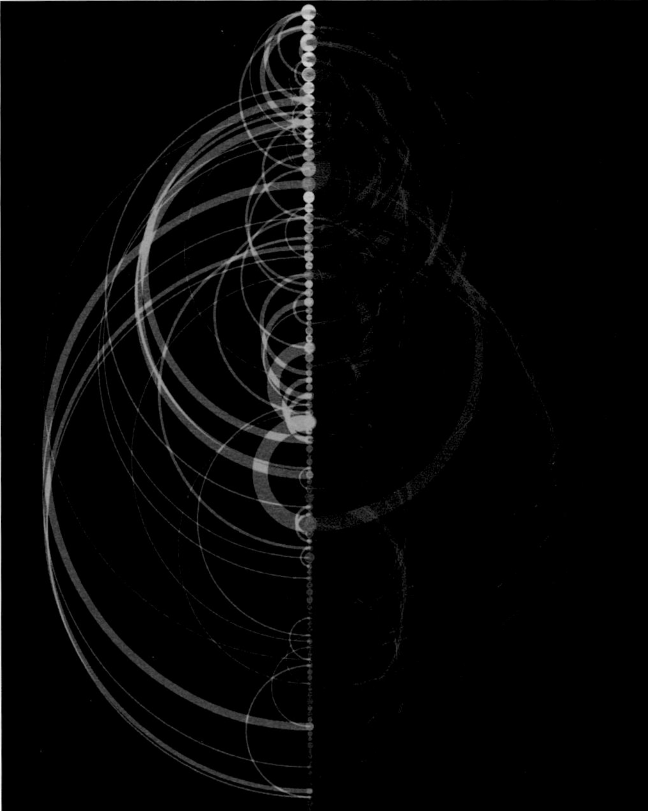

Chris Harrison
Visualizing the Bible
2007

A map of 63,779 cross-references found in the Bible. The bar graph on the bottom represents all of the books in the Bible, alternating between white and light gray for easy differentiation. The length of each bar, representative of a book's chapter and dropping below the datum, corresponds to the number of verses in that chapter.

Each arc represents a textual cross-reference (e.g., place, person), and the color denotes the distance between the two chapters where the reference appears—ultimately creating a rainbowlike effect.

Martin Dittus
Chart Arcs
2006

Map of a person's weekly chart in Last.fm. The vertical yellow points indicate chart positions, while arcs represent movement on the chart.

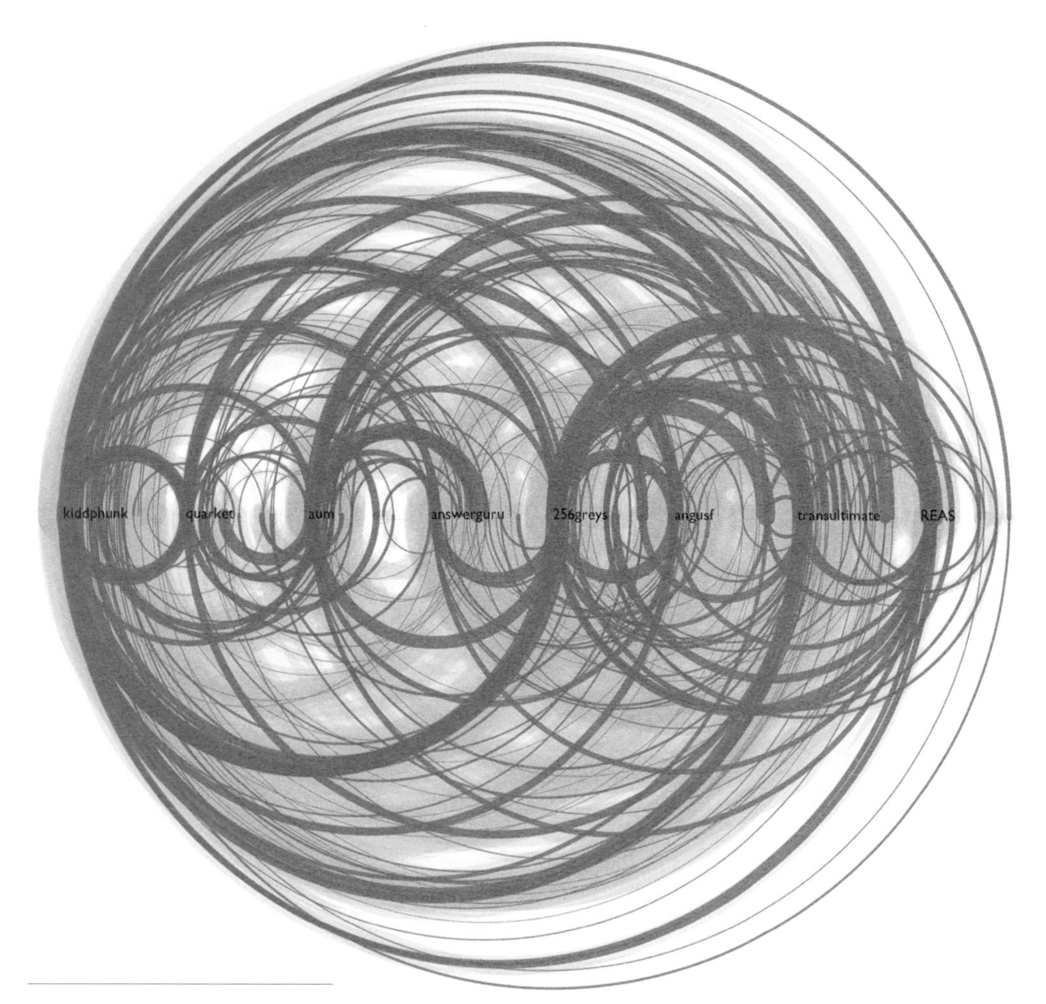

(opposite)

Felix Heinen

Data Visualisation of a Social Network

2007

A diagram, part of a larger composition, depicting the diverse activity of members from an online-social-network service similar to Facebook. It portrays the prevalence of popular features (e.g., blog, chat, search), each represented by a different arc color, across common demographical information (e.g., age, education, marital status) pulled from member profiles.

Ian Timourian

del.icio.us.discover

2006

A comparative analysis of tagging behavior of different Del.icio.us users

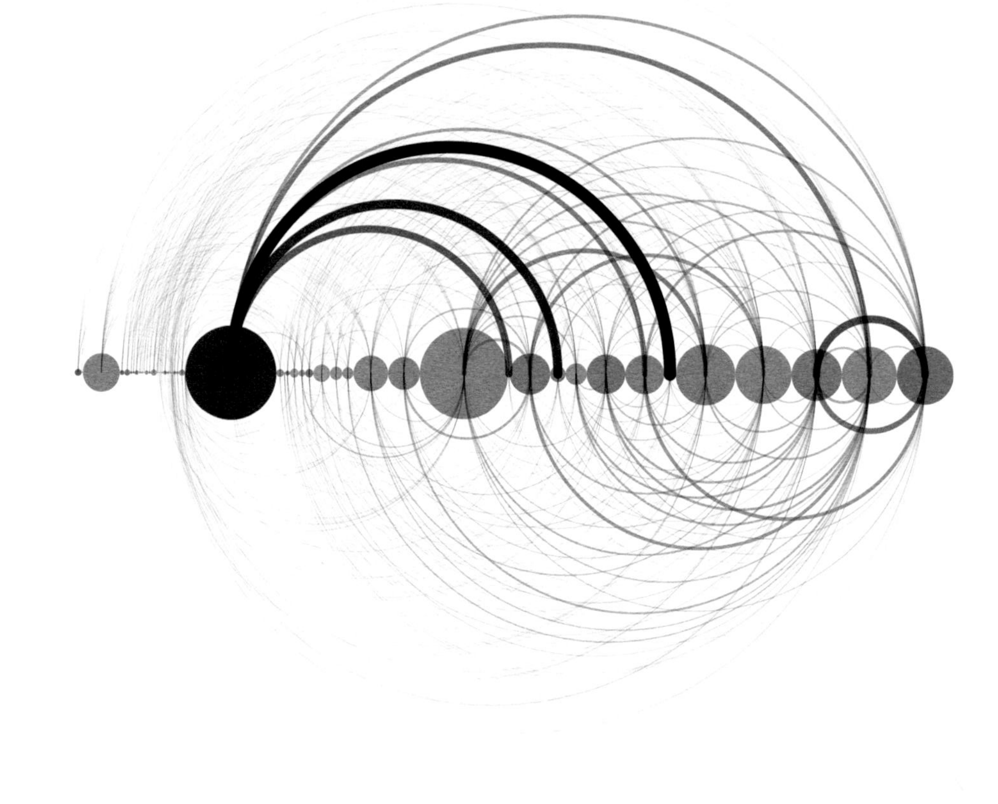

Martin Dittus

IRC Arcs

2006

A communication chart of an Internet Relay Chat (IRC) channel. Circles represent users, and arcs symbolize references, which are messages from a user containing the name of another user. The arcs are directional and are drawn clockwise. Arc strength (thickness) corresponds to the number of references from the source to the target.

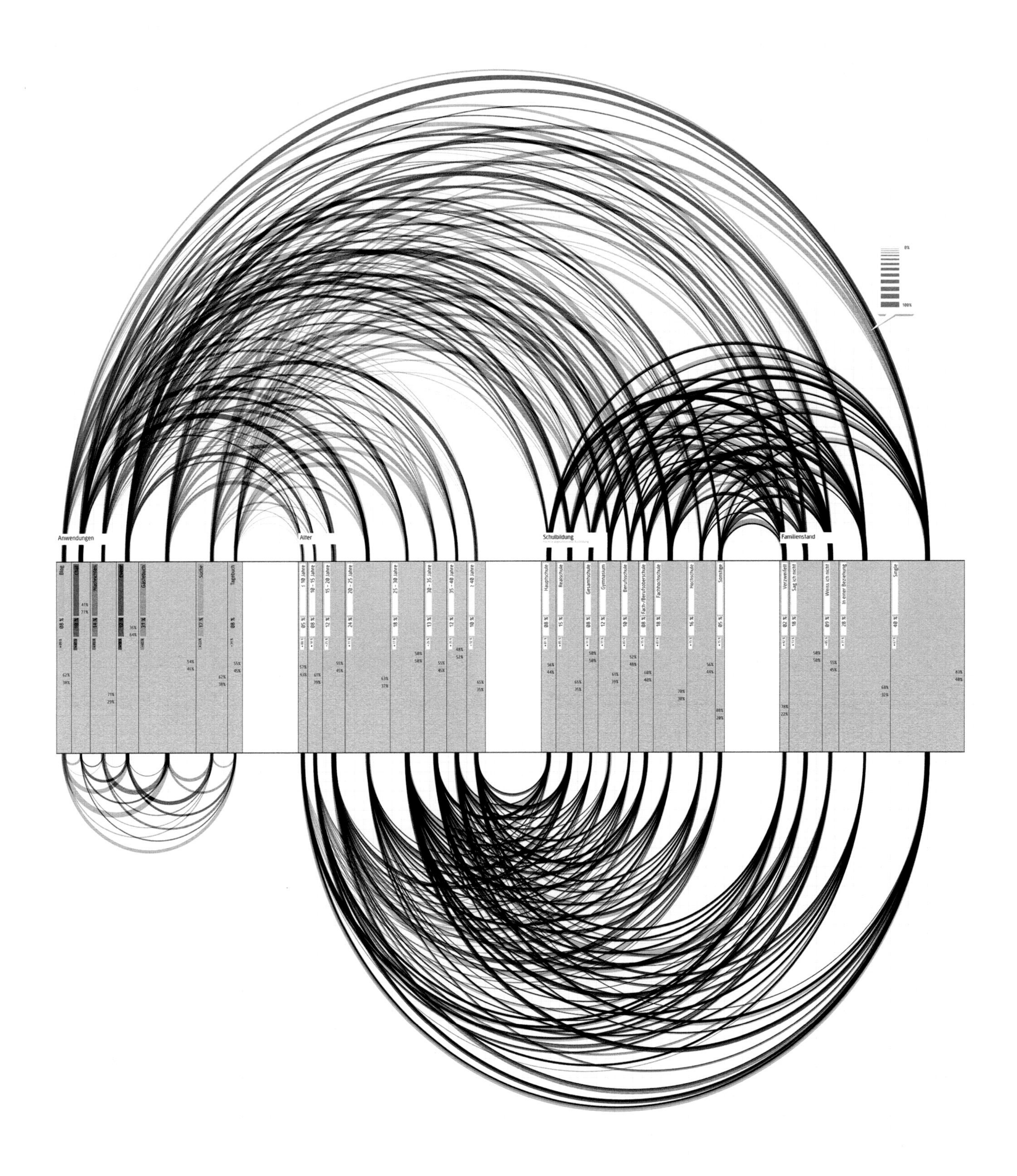
0%
100%
Anwendungen
Blog
08 %
Chat
18 %
41%
77%
Nachrichten
14 %
Event
22 %
36%
64%
Gästebuch
31 %
Suche
17 %
Tagebuch
08 %
62%
38%
71%
29%
54%
46%
62%
38%
55%
45%
Alter
≤ 10 Jahre
05 %
10 - 15 Jahre
08 %
15 - 20 Jahre
12 %
20 - 25 Jahre
24 %
25 - 30 Jahre
18 %
30 - 35 Jahre
13 %
35 - 40 Jahre
12 %
≥ 40 Jahre
10 %
57%
43%
61%
39%
55%
45%
63%
37%
50%
50%
55%
45%
48%
52%
65%
35%
Schulbildung
Hauptschule
08 %
Realschule
15 %
Gesamtschule
08 %
Gymnasium
12 %
Berufsschule
10 %
Fach-/Berufsoberschule
08 %
Fachhochschule
18 %
Hochschule
16 %
Sonstige
05 %
56%
44%
65%
35%
50%
50%
61%
39%
52%
48%
60%
40%
70%
30%
56%
44%
80%
20%
Familienstand
Verzweifelt
02 %
Sag ich nicht
18 %
Weiss ich nicht
09 %
In einer Beziehung
28 %
Single
40 %
78%
22%
50%
50%
55%
45%
68%
32%
83%
40%

Area Grouping

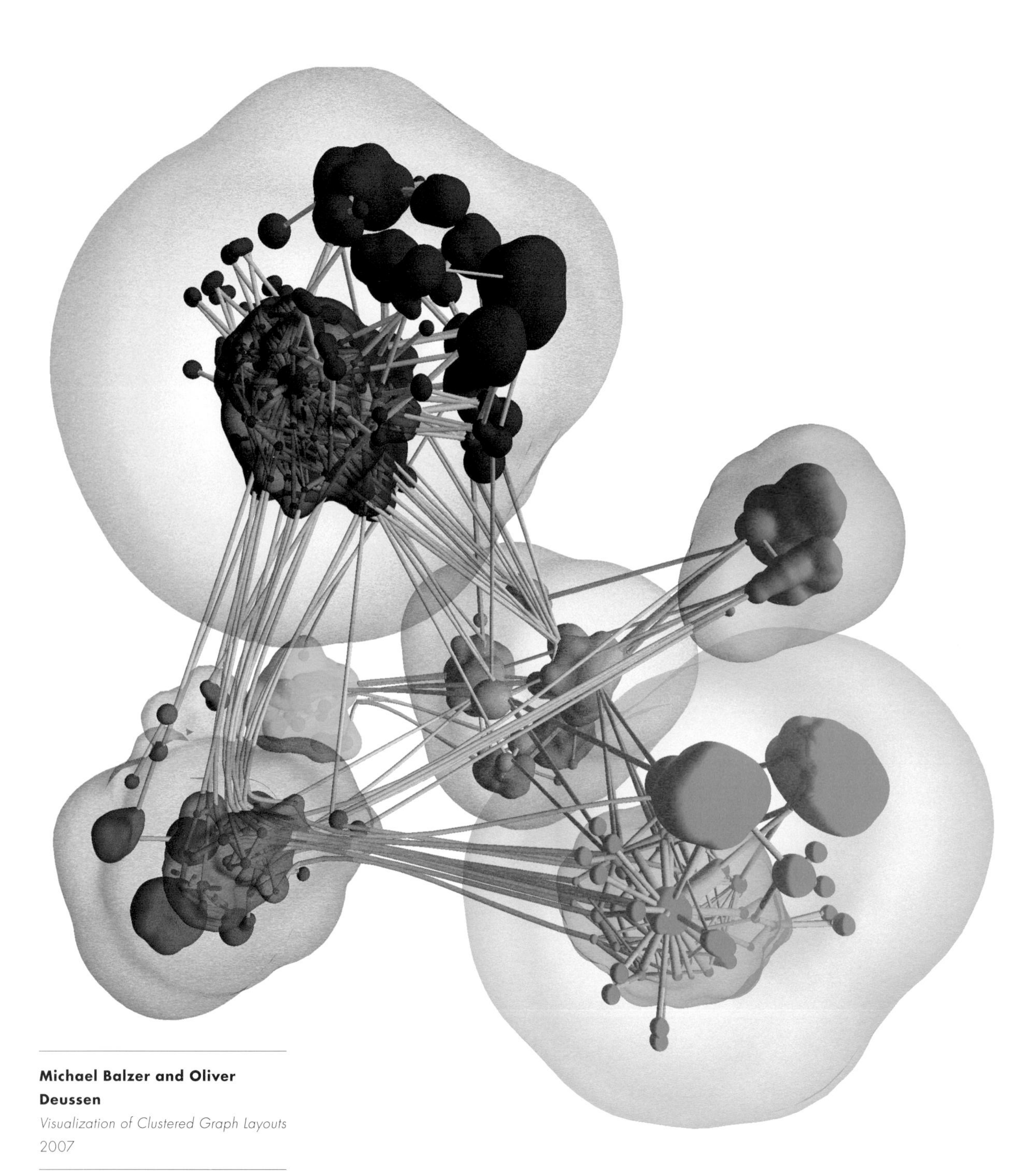

Michael Balzer and Oliver Deussen
Visualization of Clustered Graph Layouts
2007

An example of a visualization technique that generates an interactive representation of large and complex networks in two or three dimensions by representing clusters of nodes as single objects

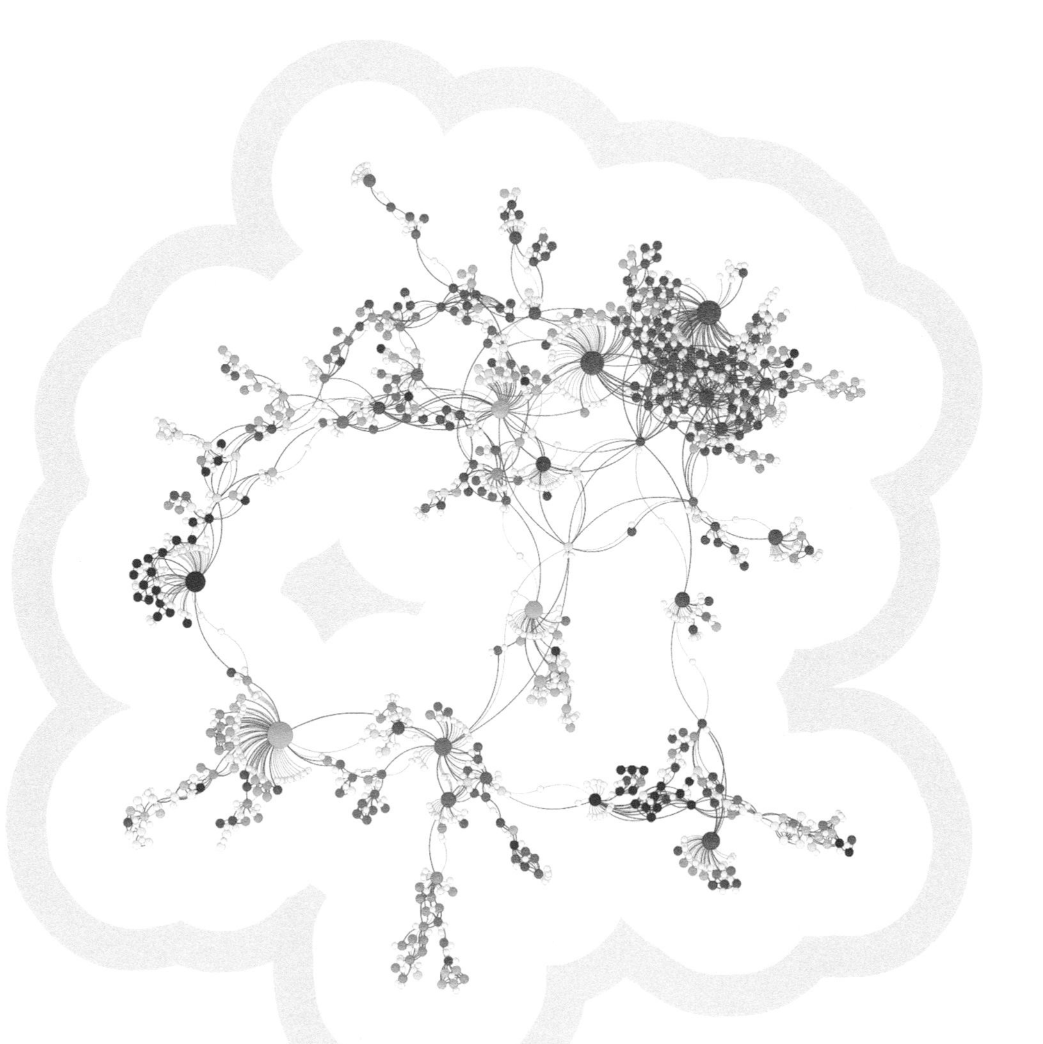

Linkfluence

Diseasome

2009

An interactive map of the human-disease network, made of 903 genes and 516 diseases, and divided into twenty-two different categories, such as cardiovascular, dermatological, metabolic, muscular, nutritional. White nodes represent genes; color nodes, diseases. Edges show correlations between genes and diseases, as well as between diseases that share common genes.

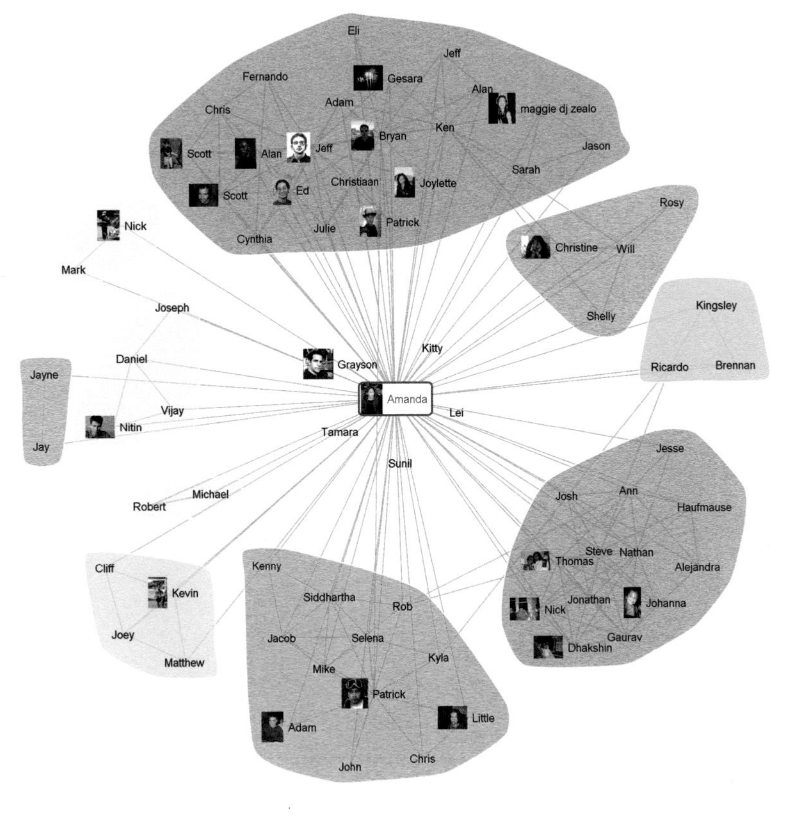

Jeffrey Heer

Automated community analysis

2004

A community-analysis visualization, which clusters individuals in groups bounded by different colors. It is generated using Vizster, an interactive visualization tool for the exploration of the community structure of social-networking services, such as Friendster, Tribe.net, and Orkut.

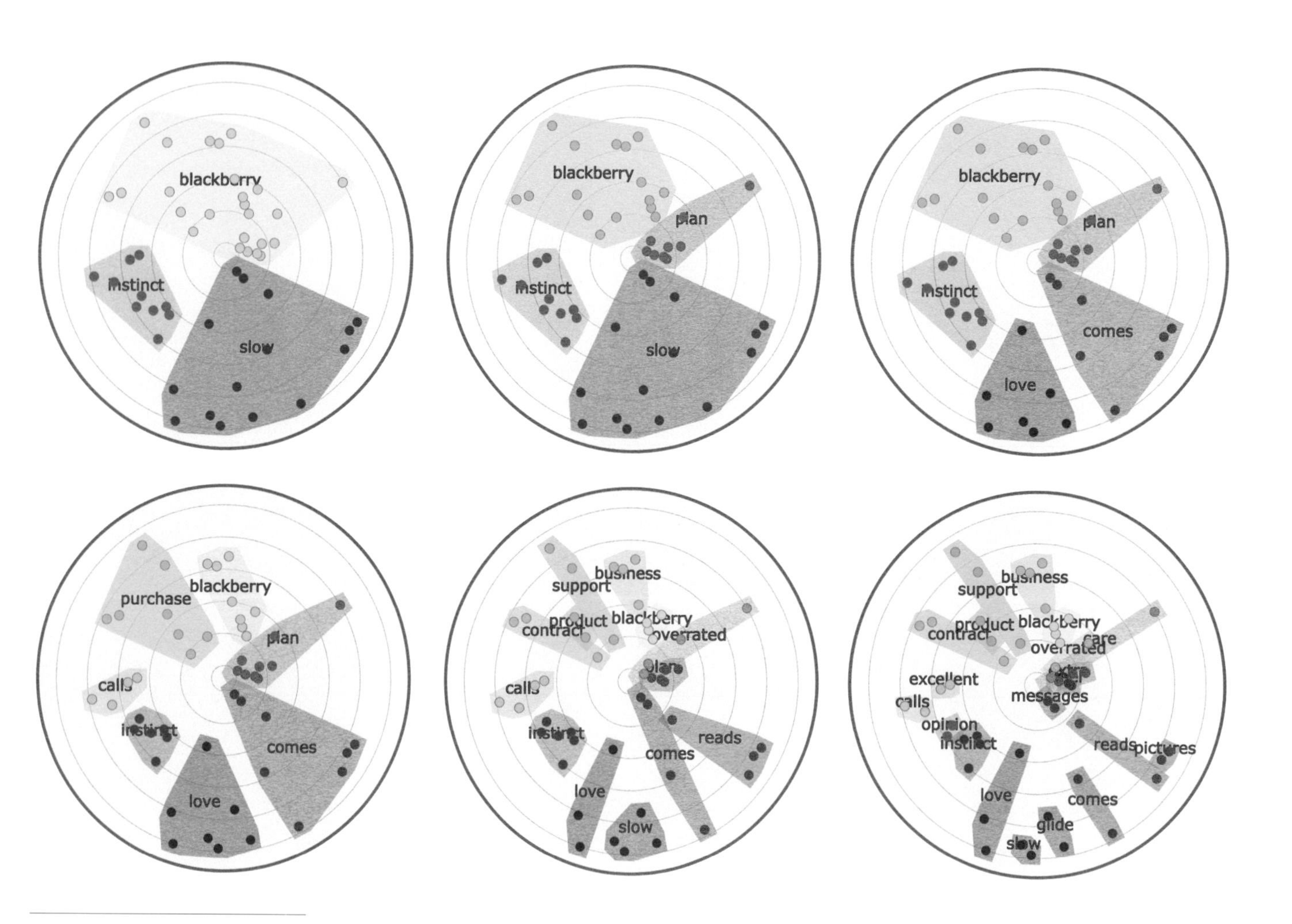

Ernesto Mislej
Cocovas
2006

A view into different visualization modes of Cocovas, an analysis tool for investigating the relevance and similarity of online-search results. Its radial diagram (shown here in a sequence) clusters similar results in various categories highlighted by different colors. Users can adjust the different levels of clustering, from broad to smaller, more specific categories.

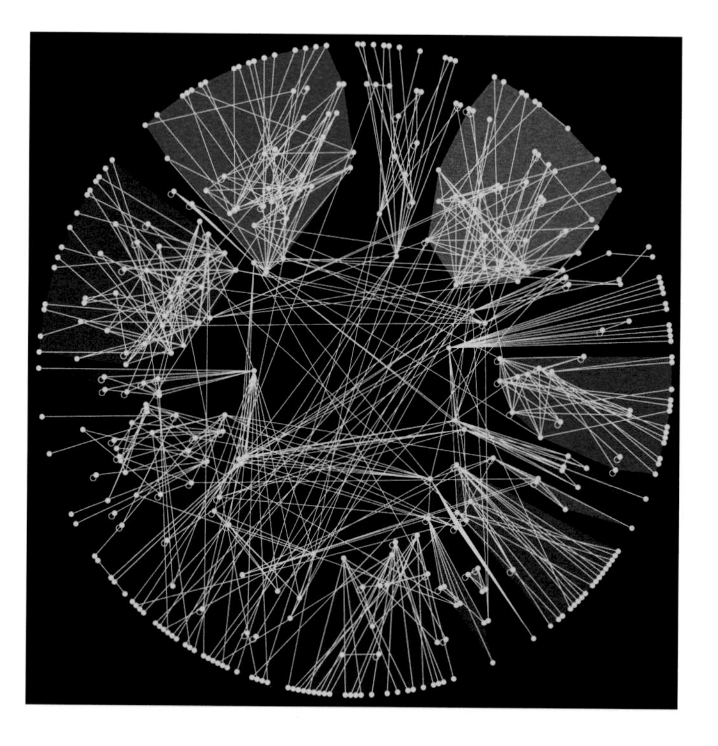

Eytan Adar
GUESS
2005

An example of a visualization created from an open-source program written in Jython (an implementation of the Python programming language written in Java) to visualize networks and perform operations on them. It allows for the dynamic representation of graphs on infinite planes and with various degrees of zooming.

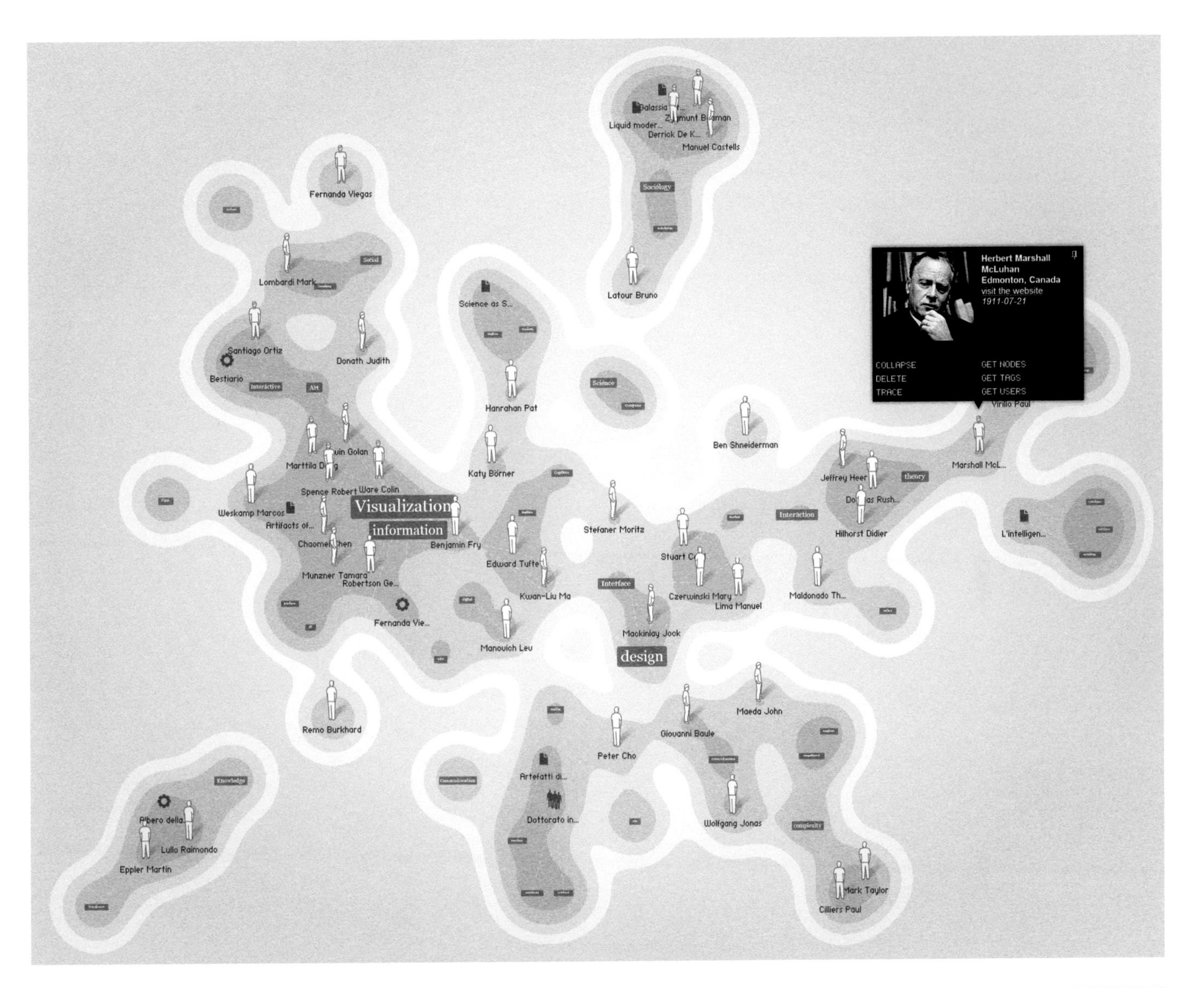

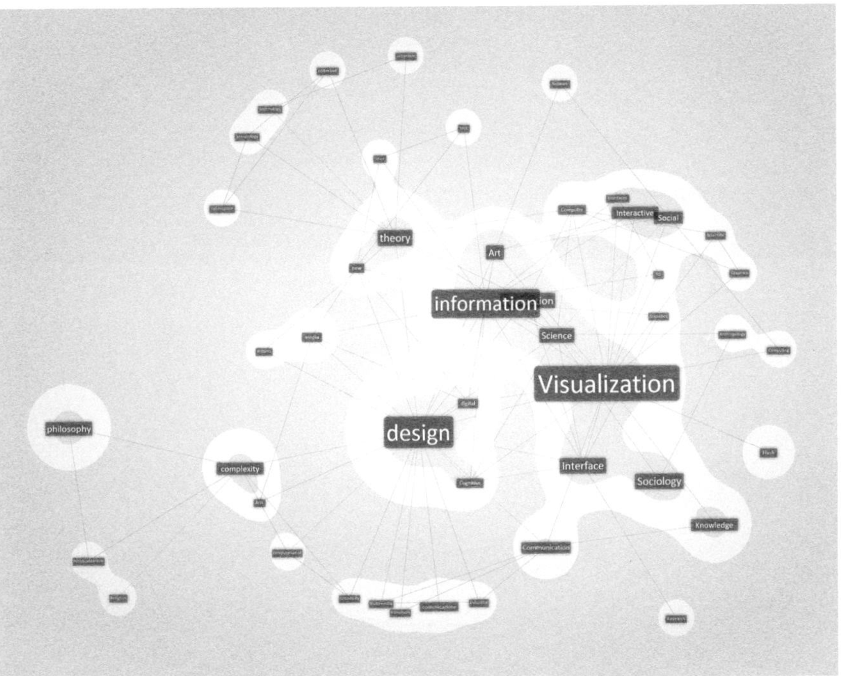

Marco Quaggiotto

Knowledge Cartography

2008

Screenshots taken from *ATLAS*, an application developed to explore the possibilities of applying cartographic techniques to mapping knowledge. *ATLAS* allows users to list their bio-bibliographic references and to map them according to four main rendering modes: semantic, socio-relational, geographic, and temporal.

Centralized Burst

Matt Rubinstein, Yarun Luon, Sameer Halai, and John Suciu
Mapping WoW Arena Teams
2007

A visualization of group formations in the arenas—a virtual environment for a player-versus-player battle—of the "massively multiplayer online role-playing game" World of Warcraft (WoW). It represents 16,534 players, 5,758 teams, and 4,065 guilds.

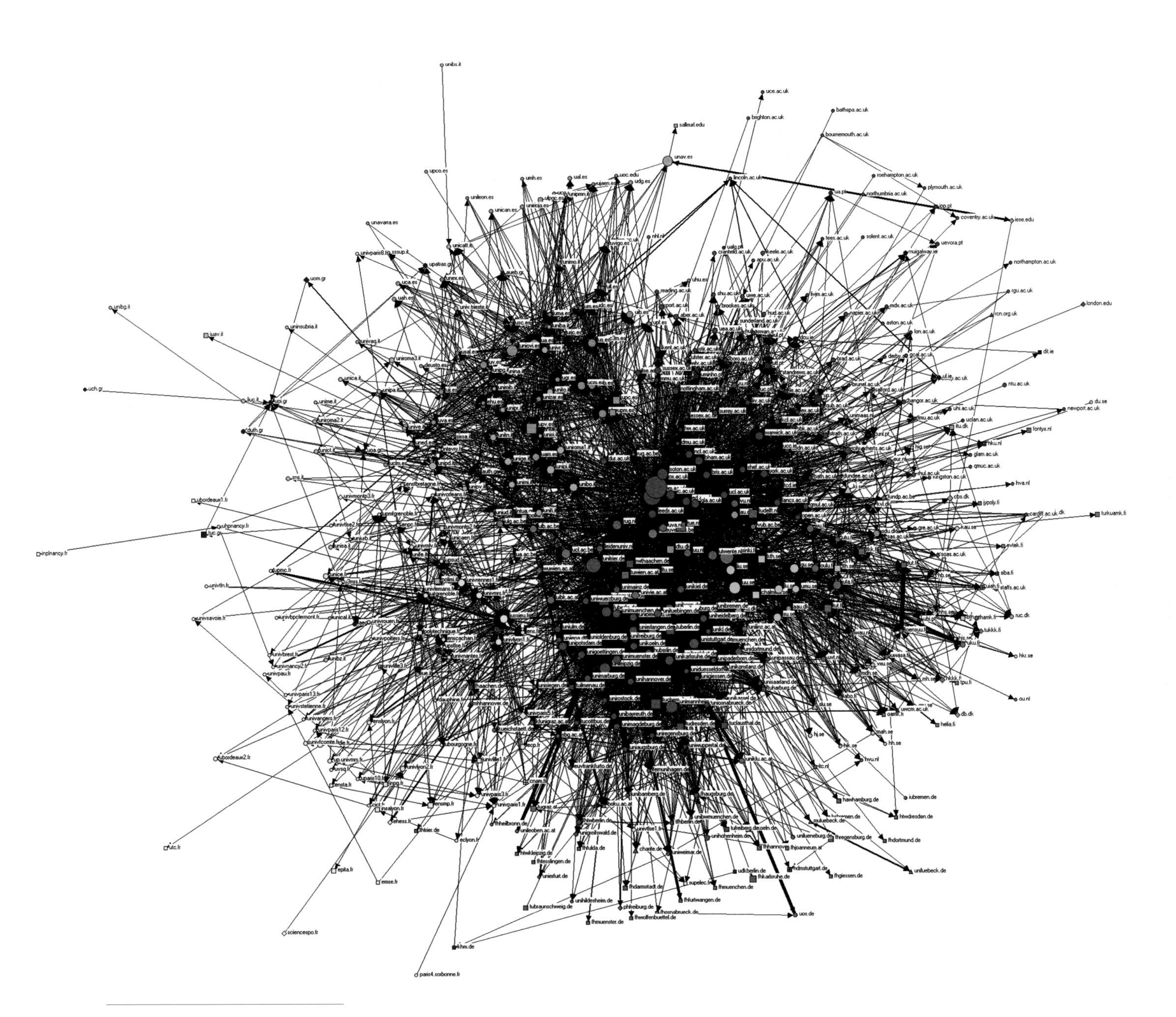

Jose Luis Ortega and Isidro Aguillo

European Academic Network

2004

A map of online academic relationships, based on exchanged hyperlinks, between 535 universities—in fourteen European countries—that belong to the European Higher Education Area (EHEA)

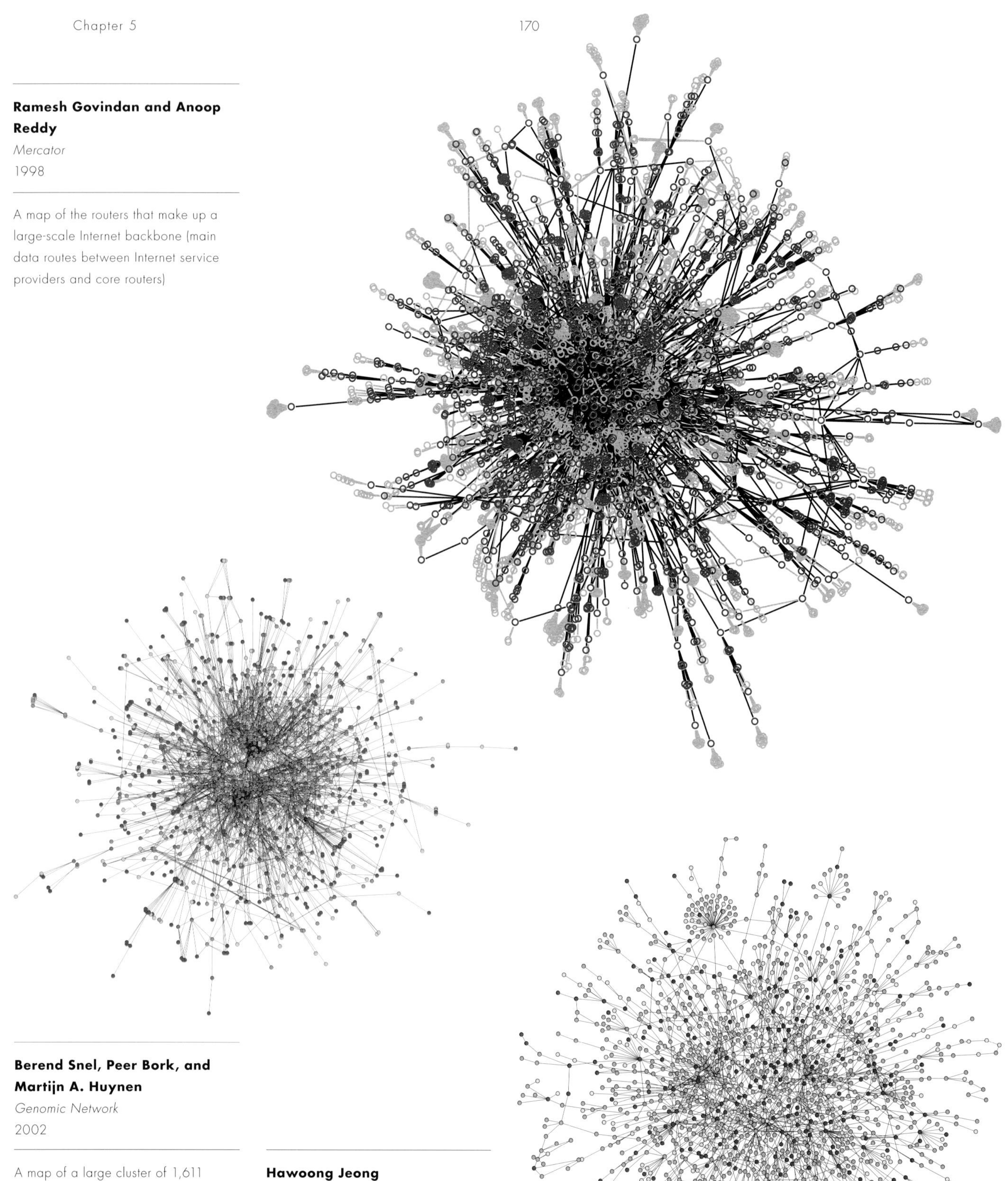

Ramesh Govindan and Anoop Reddy
Mercator
1998

A map of the routers that make up a large-scale Internet backbone (main data routes between Internet service providers and core routers)

Berend Snel, Peer Bork, and Martijn A. Huynen
Genomic Network
2002

A map of a large cluster of 1,611 groups of orthologous genes—genes that occur in two or more different species that originate from the same common ancestor. From B. Snel, P. Bork, M. A. Huynen, "The Identification of Functional Modules from the Genomic Association of Genes," *Proceedings of the National Academy of Sciences (PNAS) USA* 99 (April 30, 2002): 5890–95.

Hawoong Jeong
Protein-Protein Network
2001

A map of protein-to-protein interactions of the yeast *Saccharomyces cerevisiae*. From H. Jeong, S. P. Mason, A. L. Barabási, and Z. N. Oltvai, "Lethality and Centrality in Protein Networks," *Nature*, no. 411 (May 3, 2001): 41–42.

Peter Uetz
Protein-Protein Interaction Modeling
2003

An interaction map of the yeast proteome (entire set of proteins) published by the scientific-research community. It contains 1,548 proteins that make up 2,358 interactions.

Nicholas Christakis and James Fowler
The Spread of Obesity in a Large Social Network
2007

A social-network map of 2,200 people accessed according to their body-mass index in the year 2000. Each circle (node) represents one person in the data set. Circles with red borders denote women, and circles with blue borders indicate men. The size of each circle is proportional to the person's body-mass index. The color of the circle shows the person's obesity status: yellow is an obese person; green, a nonobese person. The color of the ties between the nodes indicates the personal relationship between them: purple represents a friendship or marital tie; orange, a familial tie. From N. A. Christakis and J. H. Fowler, "The Spread of Obesity in a Large Social Network Over 32 Years," *New England Journal of Medicine* 357, no. 4 (July 2007): 370–79.

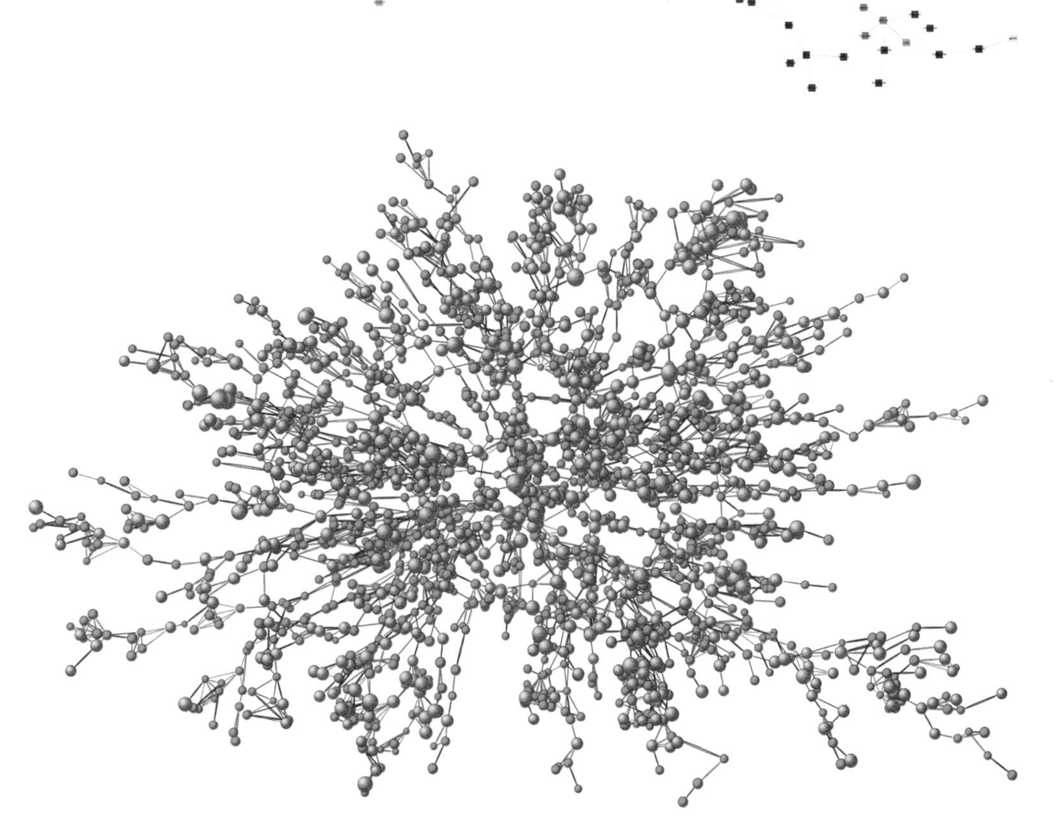

Centralized Ring

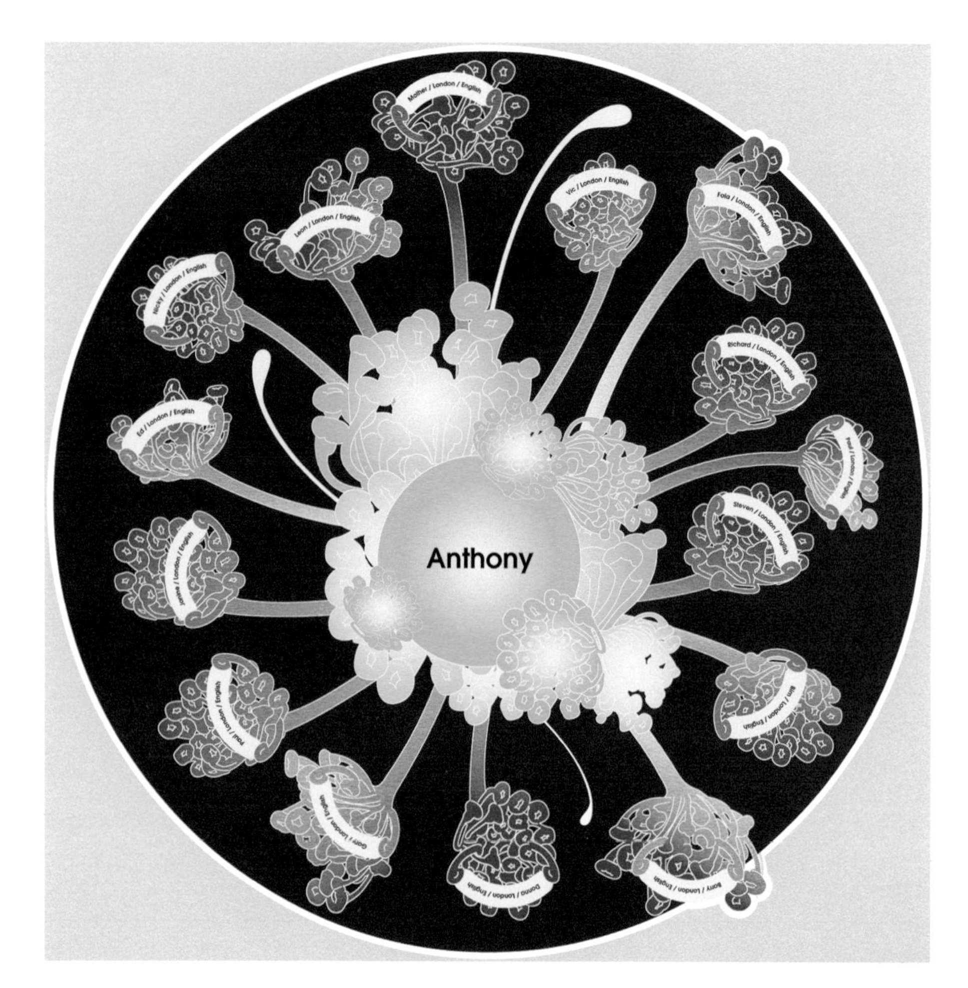

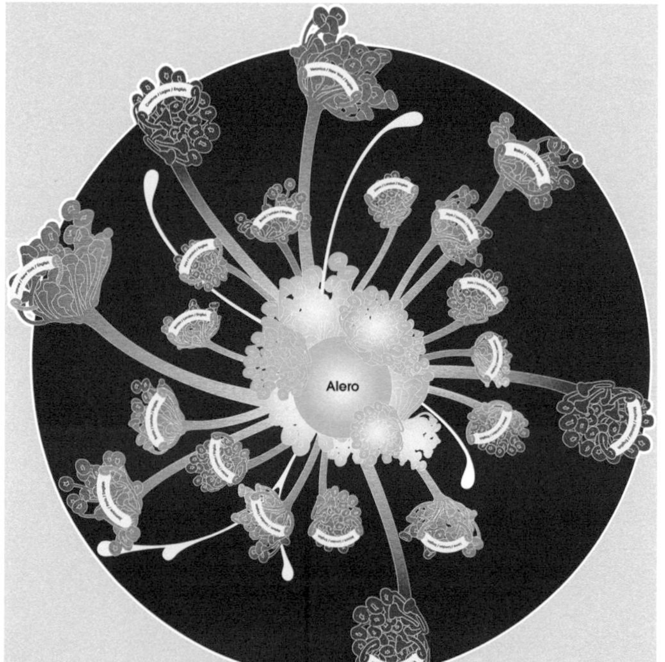

Ai Zaidi

motiroti's Priceless

2006

A part of a series of personal networks ("constellations") of the London Underground staff, exhibited in South Kensington station. Commissioned by Serpentine Gallery in collaboration with the Exhibition Road Cultural Group and Platform for Art.

Moritz Stefaner
CIA World Factbook Visualization
2004

An interactive map of geographic boundaries and linguistic ties (B, has a border with; P, is part of; S, is spoken in) between countries in the CIA world factbook database. (See also chapter 3, page 87, figure 11.)

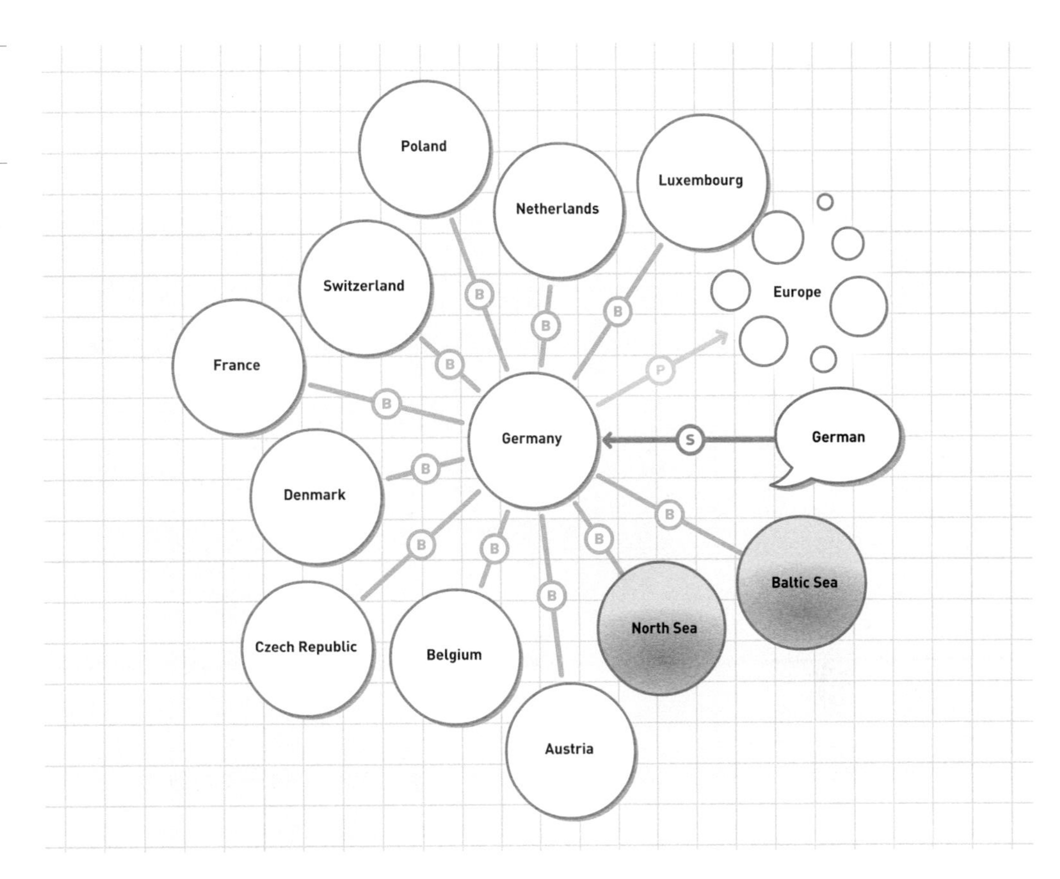

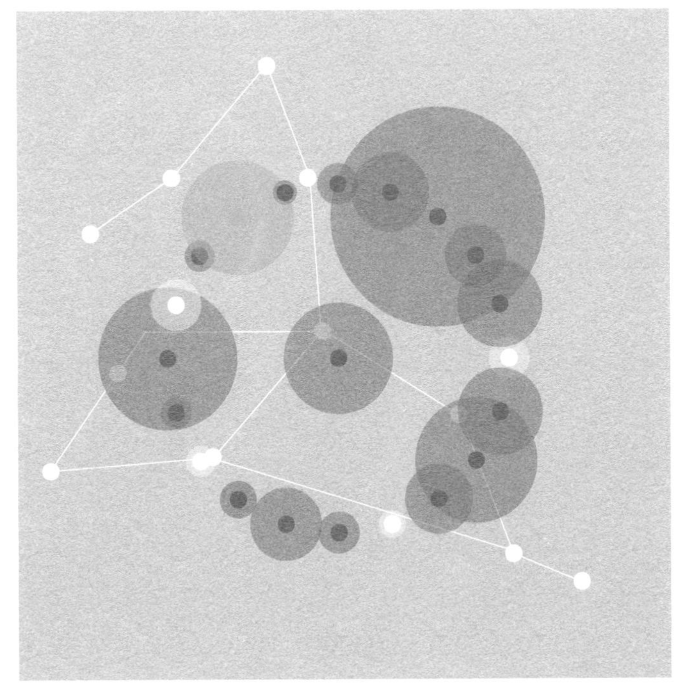

Sebastien Pierre and Olivier Zitvogel
Revealicious—SpaceNav
2005

A dynamic visualization of an individual's Del.icio.us tag (in the center) surrounded by related tags, based on bookmarking behavior

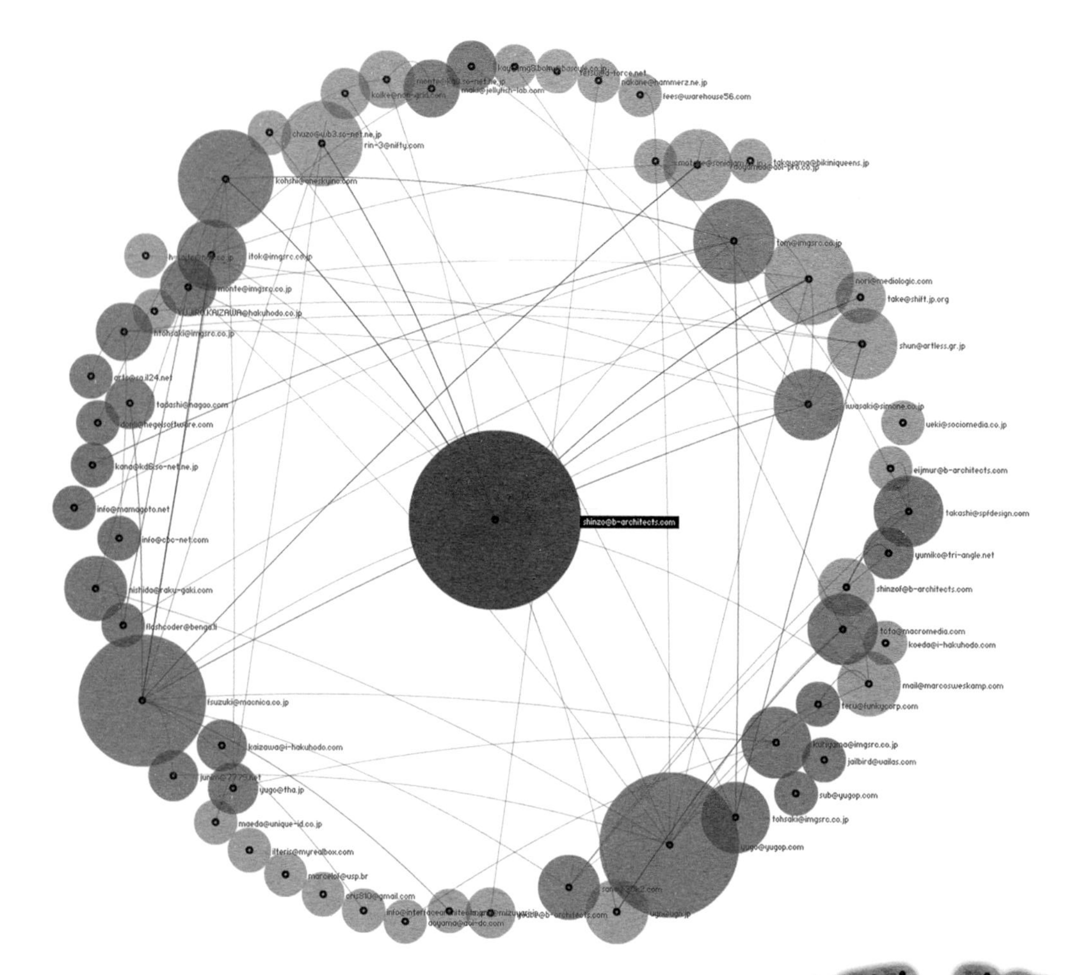

Marcos Weskamp

Social Circles

2003

A chart of social interactions (based on exchanged emails) within a mailing list. Nodes represent members of the mailing list, while its size illustrates the frequency of posts by that member. Links between members depict communication in reply to a thread.

Jute Networks

PublicMaps

2008

A sociogram of individual donations in Asheville, North Carolina, to Barack Obama's 2008 presidential campaign. (See also chapter 4, page 113.)

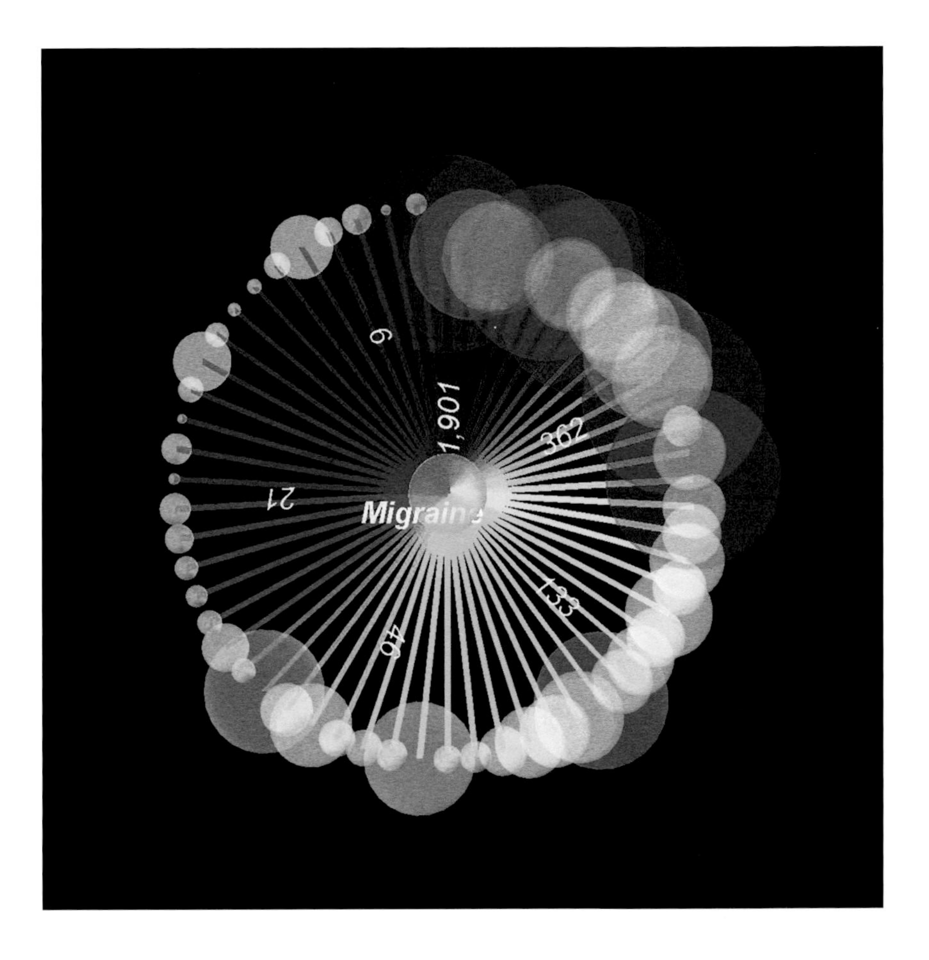

Andrey Rzhetsky et al.
Mapping Complex Diseases
2007

A map of genetic overlap between migraine and approximately sixty other diseases. Each of the outer circles represents a different disease, and the size of the circle corresponds to the size of patient samples, ranging from 46 to 136,000 people, of those suffering from the disease.

Christoph Gerstle and Florian Moritz
StudiAnalyse
2007

A visualization of the German online social network StudiVZ

Circled Globe

Aaron Koblin, Kristian Kloeckl, Andrea Vaccari, and Franscesco Calabrese
New York Talk Exchange
2008

A map of real-time exchange of information, from long-distance-telephone and IP data, flowing between New York and other cities around the world

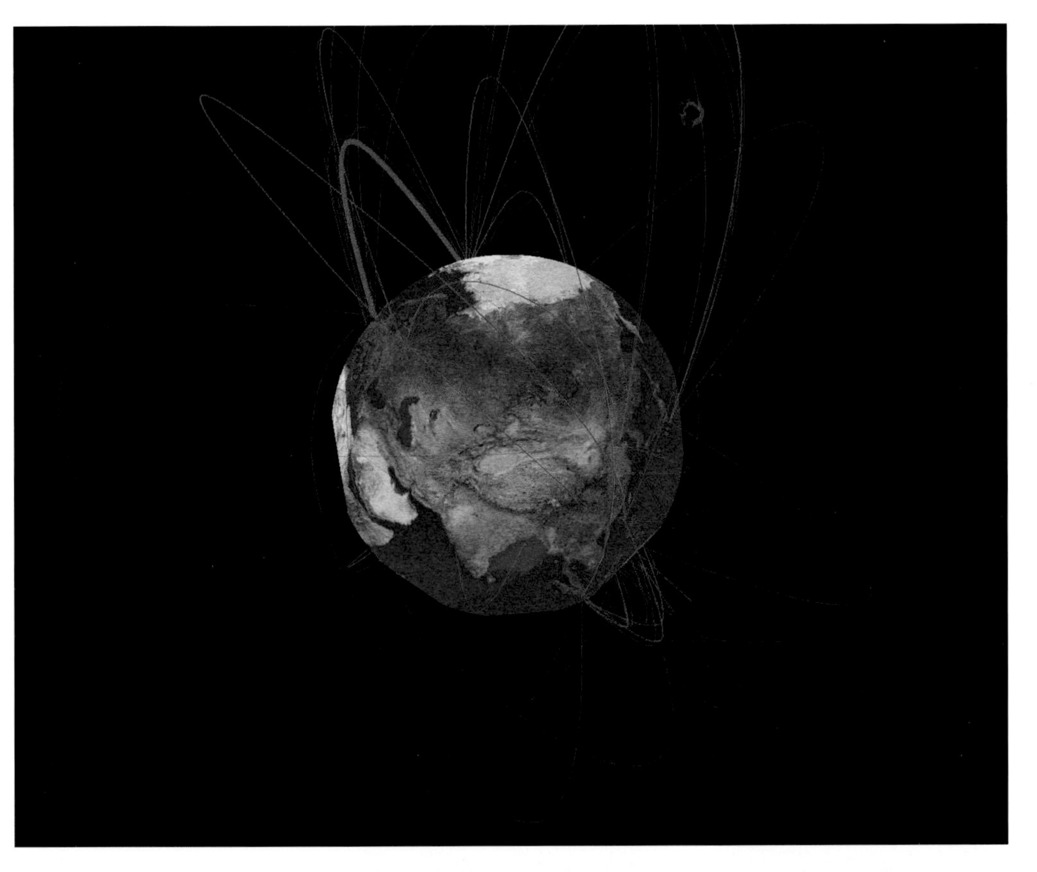

Bestiario
City Distances
2008

A three-dimensional scheme showing the relationship intensity between cities based on Google searches. The number of mentions of any two cities on the same web page was divided by the physical distance separating them. The width of the resulting arc connecting two cities indicates the relationship strength between them.

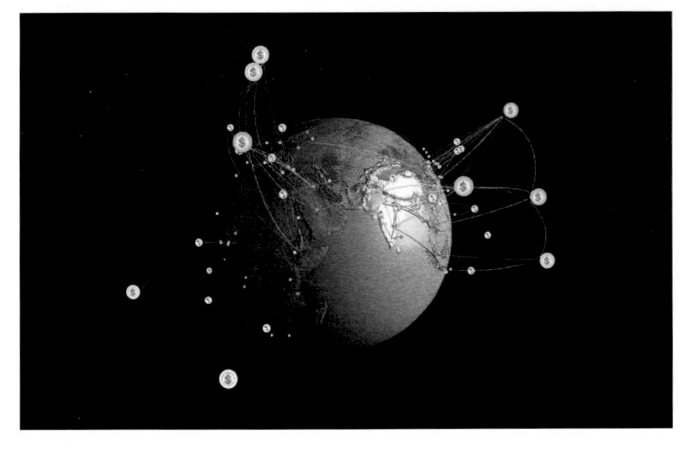

Eye-Sys
Global Cereal Supply Chain
2005

A visualization part of a video demonstration showcasing the production and consumption of breakfast cereal around the globe. It also acts as an general analogy for real-world supply-and-demand scenarios.

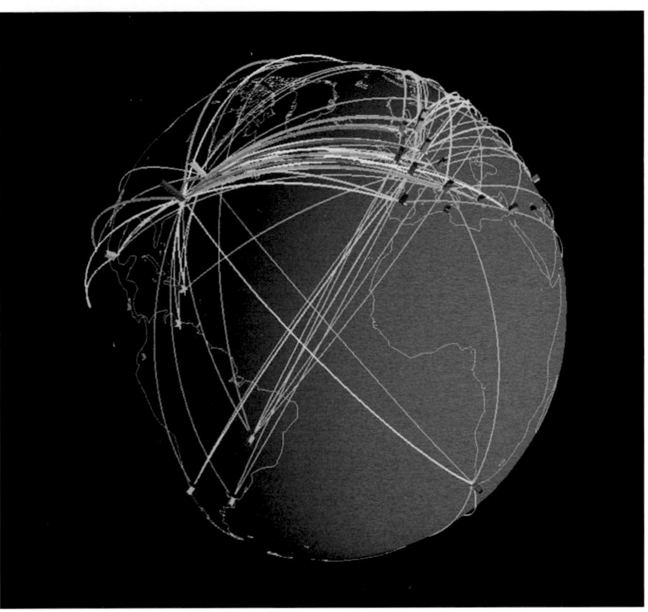

Stephen G. Eick
3D Geographic Network Display
1996

A visualization showing the internet traffic between fifty countries over a two-hour period, as measured by the National Science Foundation Network (NSFNET) backbone in 1993. The color and thickness of the lines correspond to the time and intensity of the traffic, respectively. From Stephen G. Eick. "Aspects of Network Visualization," *Computer Graphics and Applications* 16, no. 2 (March 1996): 69–72.

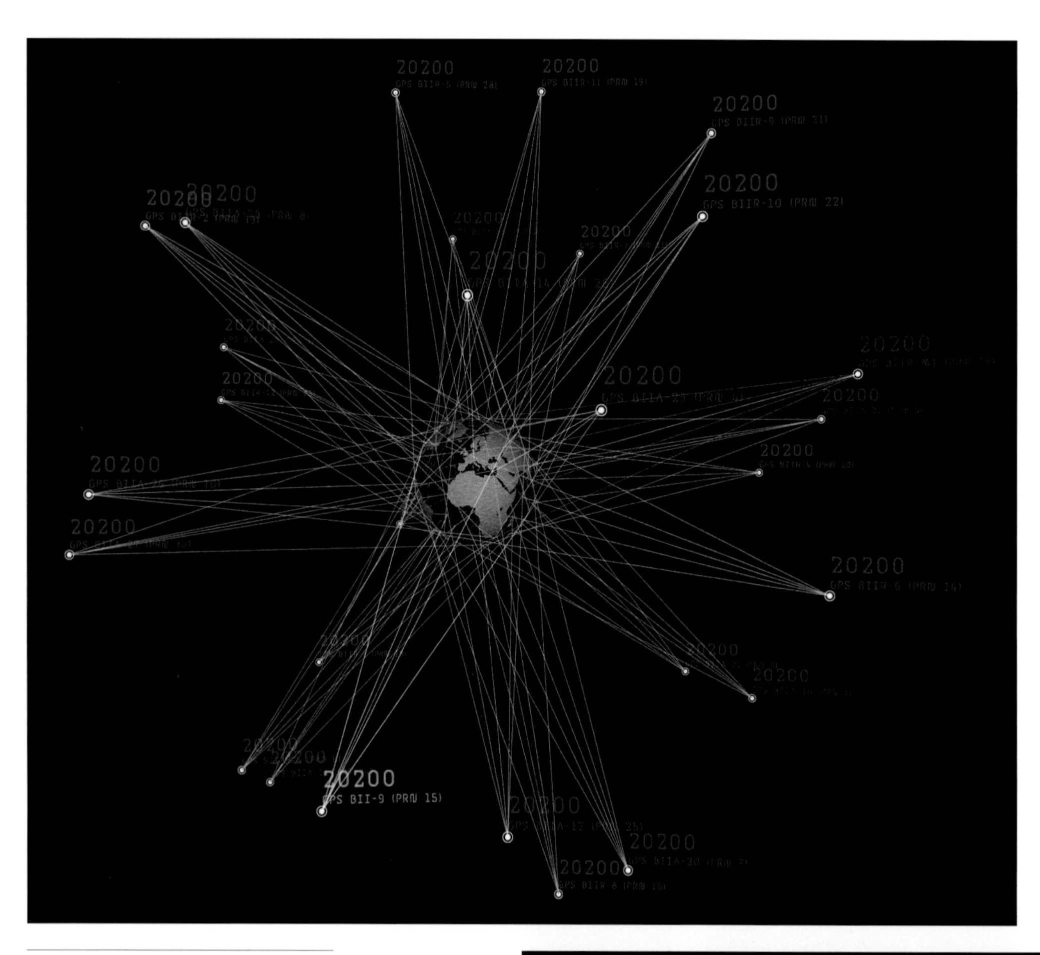

Shiftcontrol, Hosoya Schaefer Architects, and Büro Destruct
Mobiglobe
2006

An interactive visualization of geostationary GPS satellites

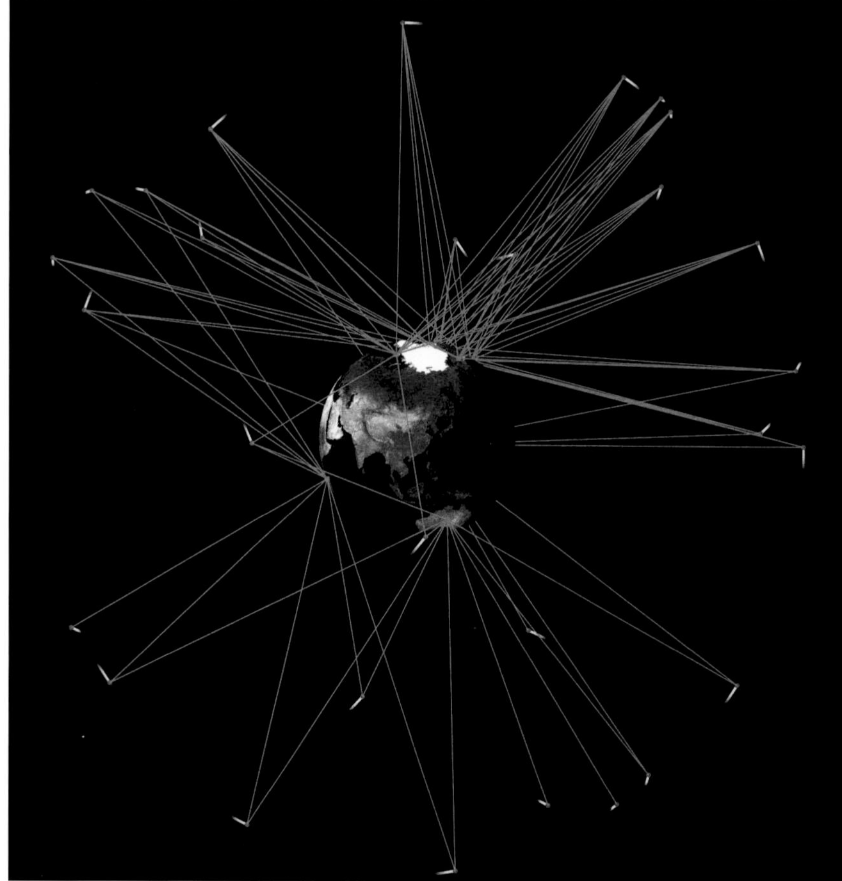

(right)
Advanced Analytic
FreeFall
2006

A visualization showing the simulated flight path of more than 650 satellites

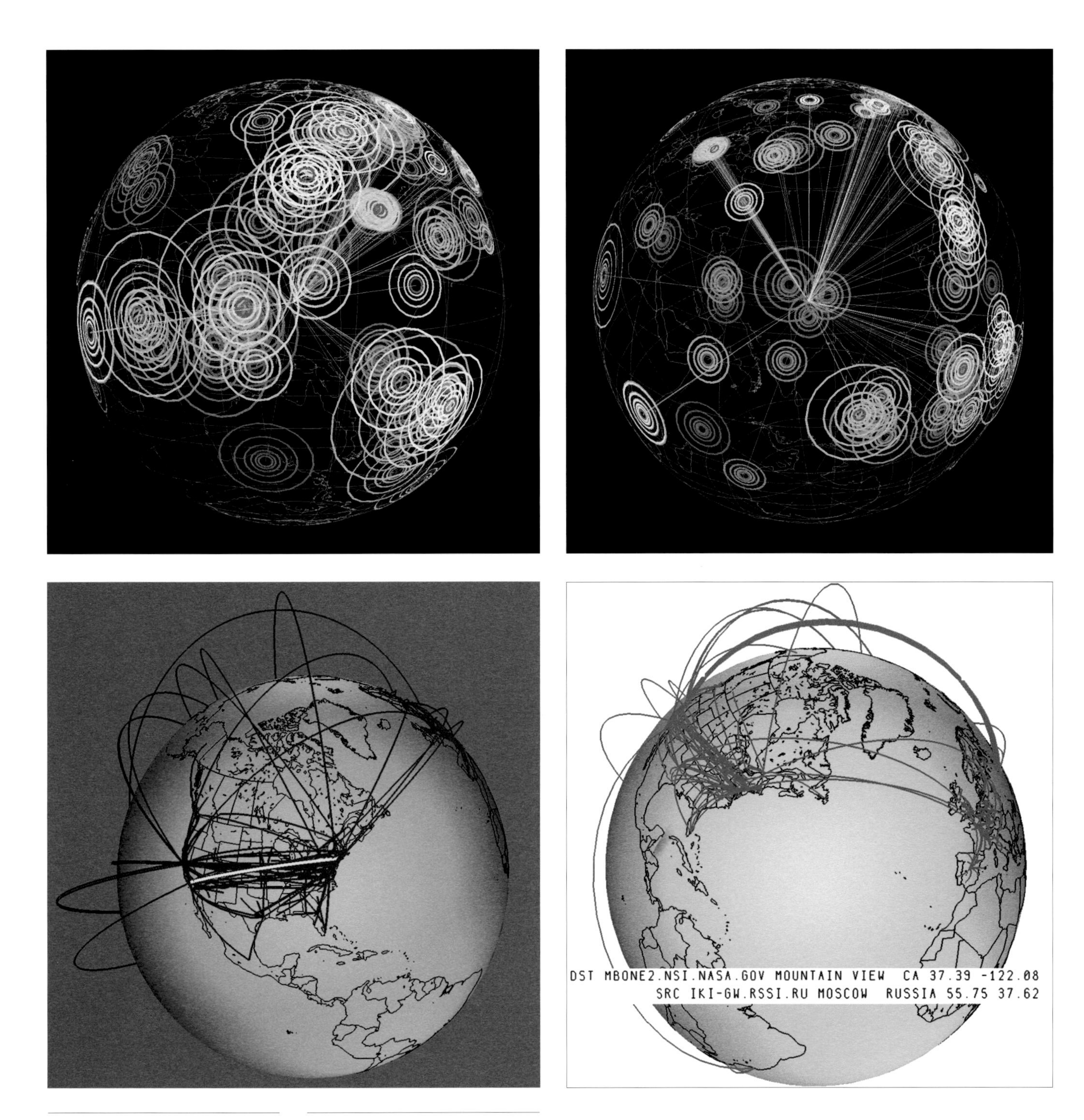

(top row)
Richard Wolton
EarthQuake 3D
2004

An interactive three-dimensional visualization of the occurrence and magnitude of the twenty most recent earthquakes

(bottom row)
Tamara Munzner, Eric Hoffman, K. Claffy, and Bill Fenner
MBone topology
1996

A geographic representation of the Internet Multicast Backbone (MBone) topology in 1996. Developed in the early 1990s, the MBone is a virtual network built on top of the internet to deliver packets of multimedia data. From Tamara Munzner, Eric Hoffman, K. Claffy, and Bill Fenner, "Visualizing the Global Topology of the MBone," 85–92 (paper presented at Proceedings of the IEEE Symposium on Information Visualization [InfoVis], San Francisco, CA, October 28–29, 1996).

Circular Ties

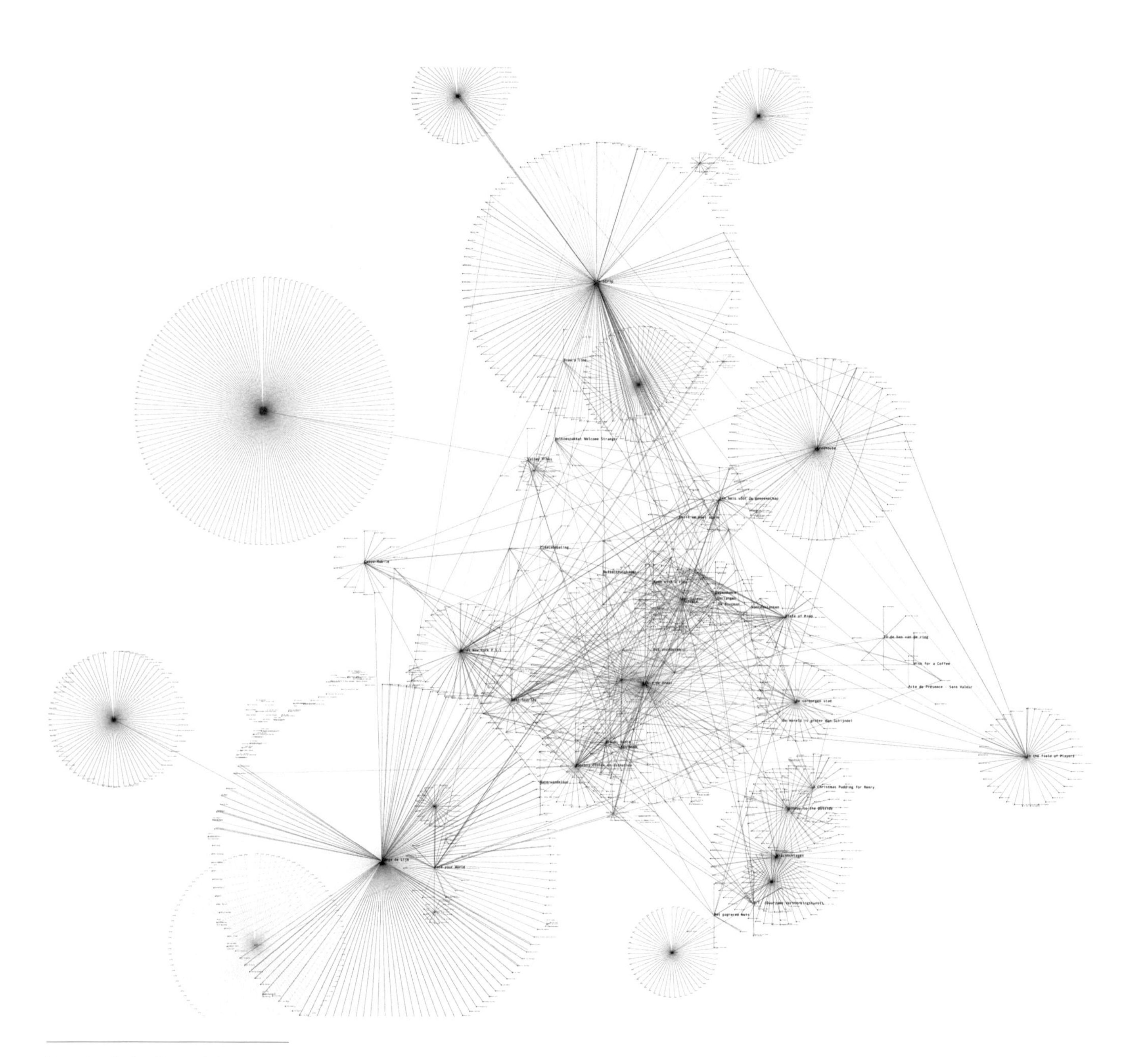

Maurits de Bruijn and Jeanne van Heeswijk
Typologies and Capacities
2006

A social-network map of collaborators involved in all of Dutch artist Jeanne van Heeswijk's projects. Each project, represented by a prominent hub, is surrounded by the names of its direct collaborators and linked to other projects with which they were also involved.

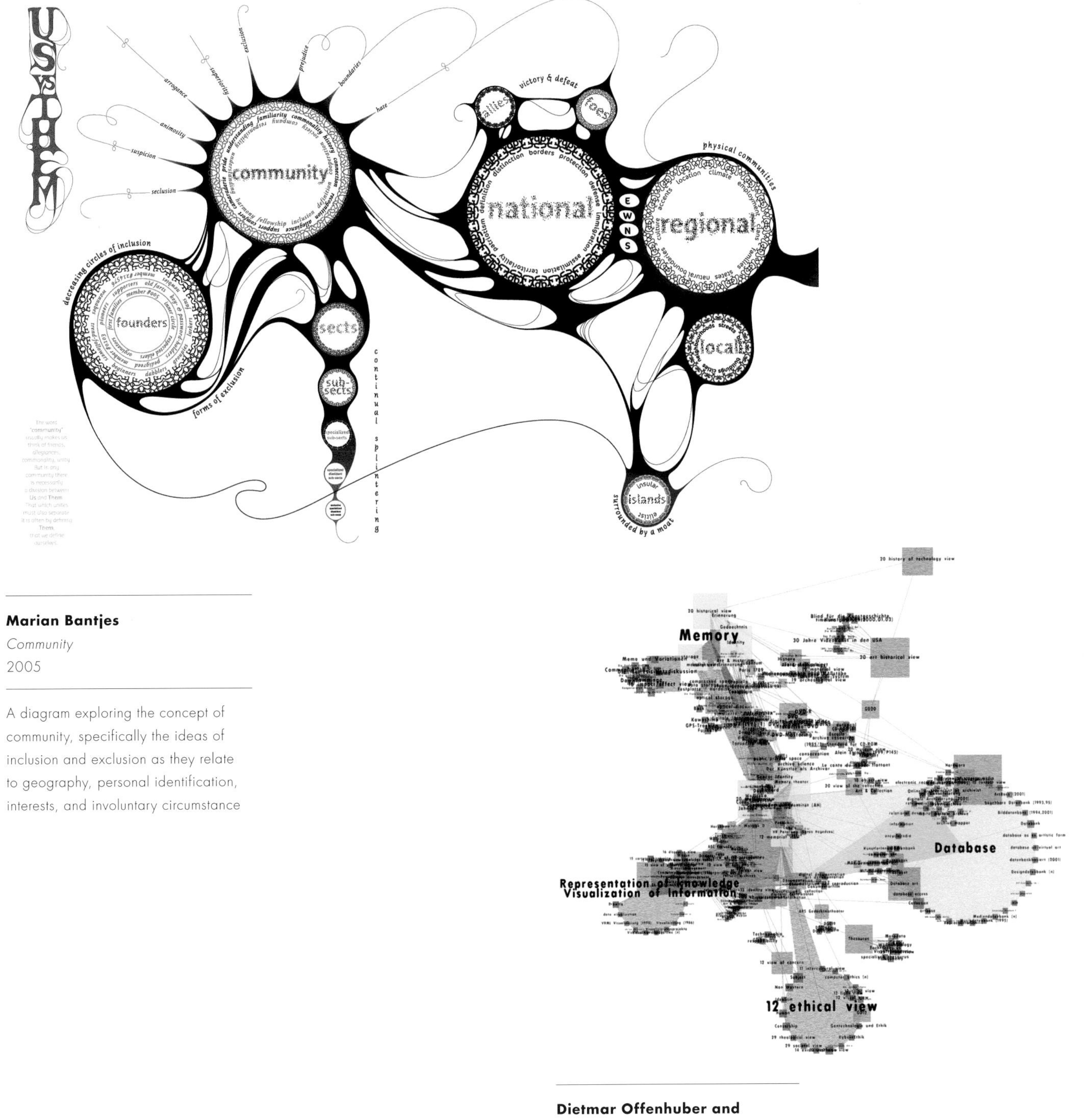

Marian Bantjes

Community

2005

A diagram exploring the concept of community, specifically the ideas of inclusion and exclusion as they relate to geography, personal identification, interests, and involuntary circumstance

Dietmar Offenhuber and Gerhard Dirmoser

SemaSpace

2006

A map of the Ars Electronica social network, showing all of the projects and people involved in the art festival between 1996 and 2003

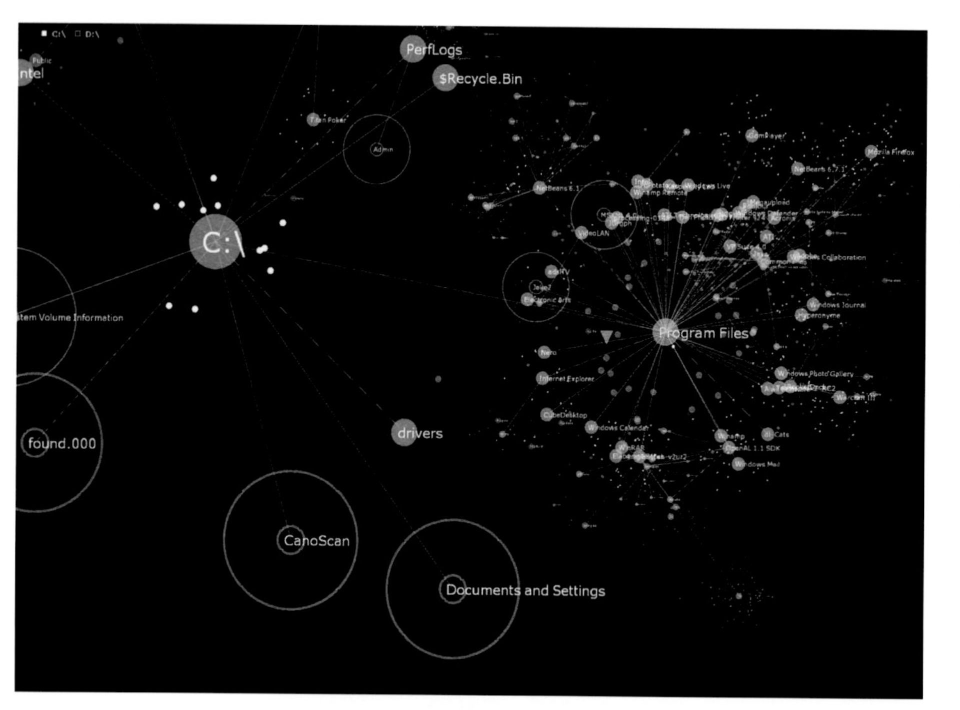

Pascal Chirol
Hyperonyme
2007

A graphic representation of the internal-folder and pyramidal-file structure of a computer's hard drive

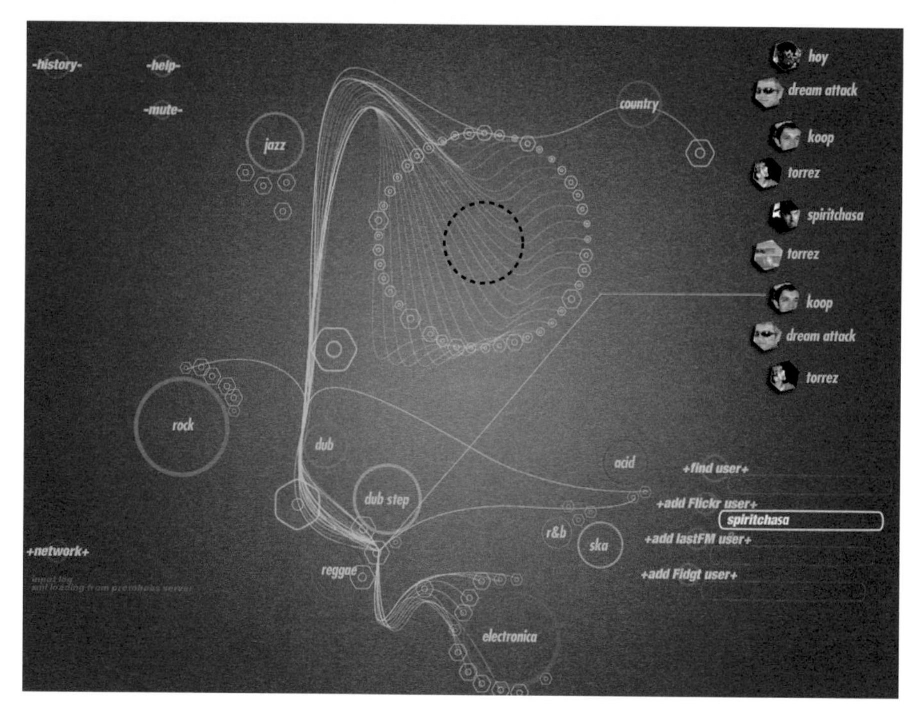

Eduardo Sciammarella
Fidg't Visualizer
2007

A dynamic visualization of multiple online social networks of a given user, including Flickr and Last.fm

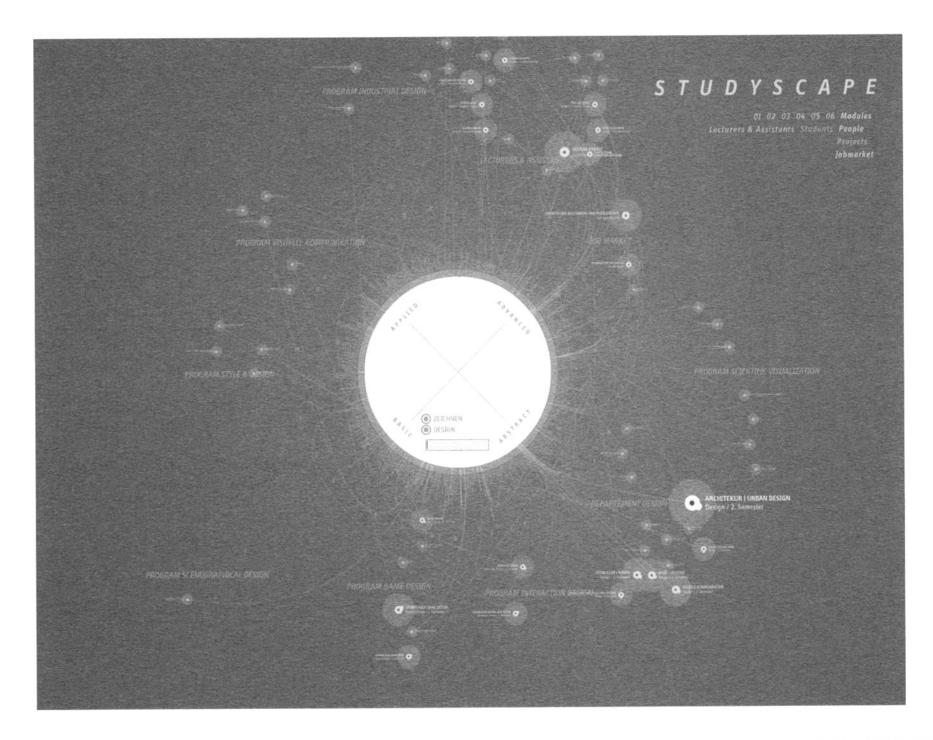

Patrick Vuarnoz
Studyscape
2005

A visual output of a search engine that matches search terms to related academic courses, research projects, job opportunities, and members of a university community (lecturers, assistants, and students) (above); detail (right)

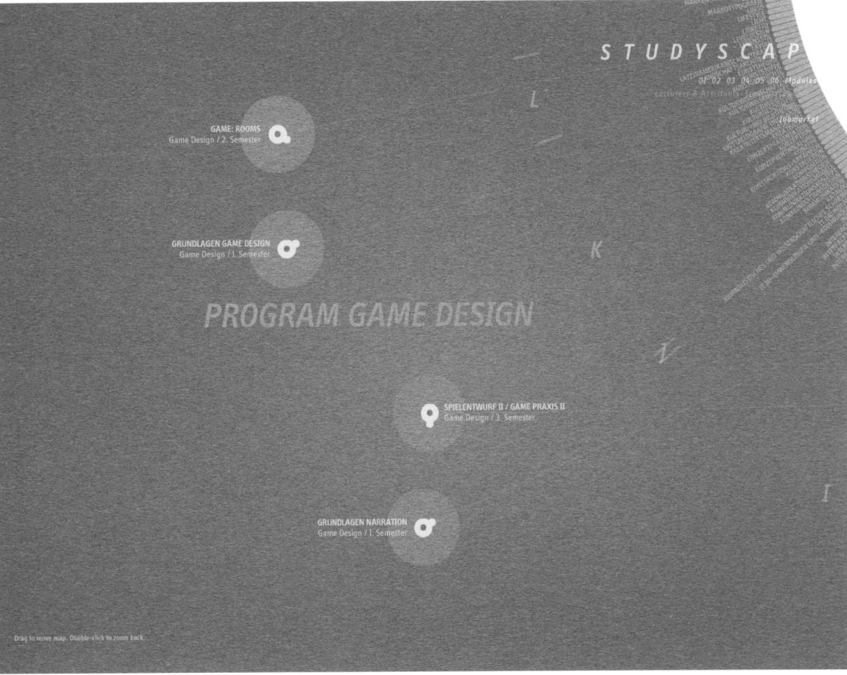

Elliptical Implosion

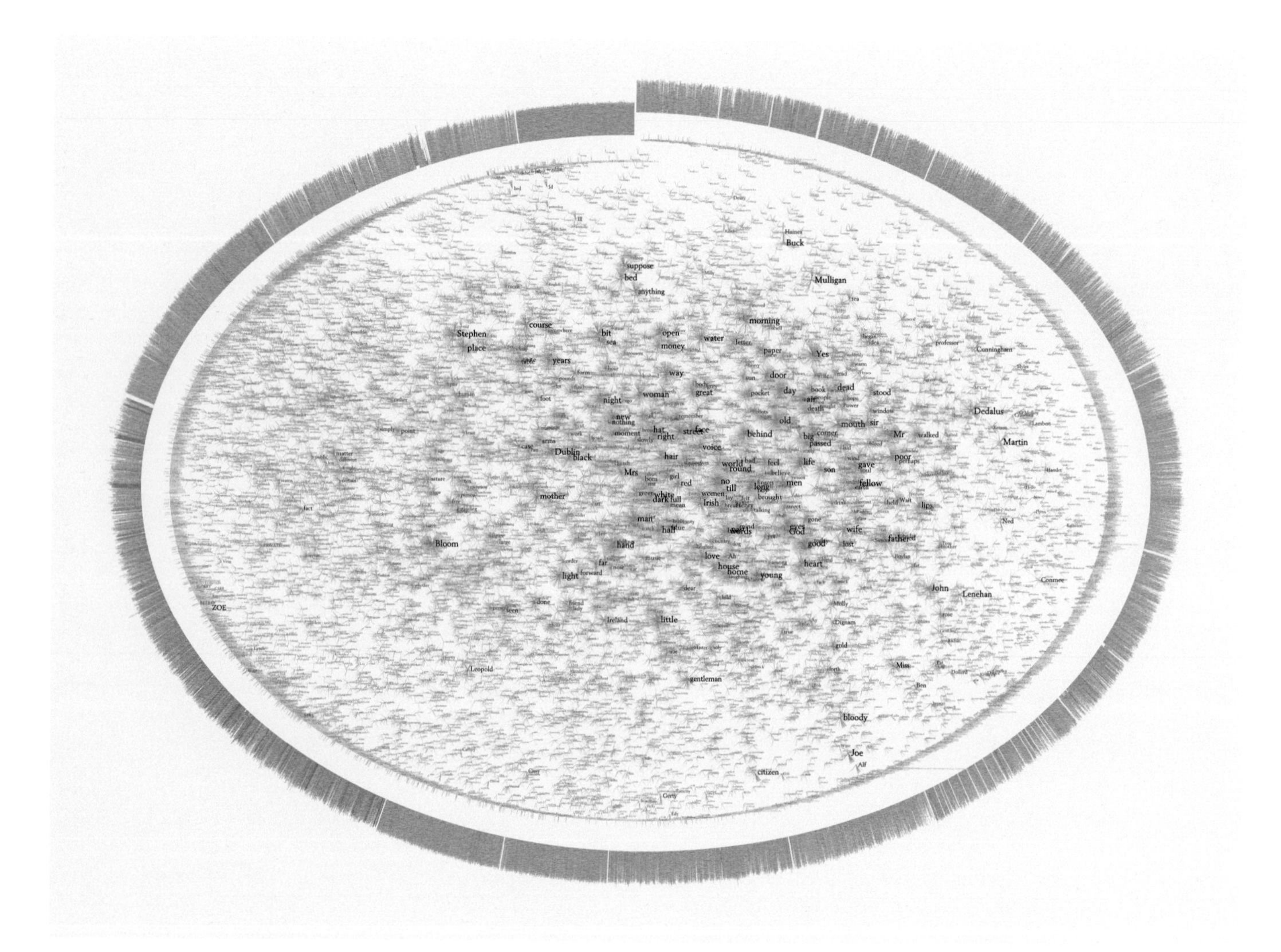

W. Bradford Paley
TextArc: Ulysses
2009

This interactive visualization, generated using TextArc, details James Joyce's use of language throughout *Ulysses* (1922)

(overleaf)
W. Bradford Paley
TextArc: Alice in Wonderland
2009

An interactive map highlighting word frequency and associations between terms in Lewis Carroll's novel *Alice in Wonderland* (1865). (See also chapter 3, page 123.)

(opposite, bottom)
Muckety
MucketyMaps
2008

A network diagram disclosing connections between people, corporations, and other organizations appearing in the news

(above)

Greg Judelman and Maria Lantin

flowerGarden

2005

A visualization generated from a web-based tool for mapping real-time social networking. It was implemented at the three-day Bodies in Play: Shaping and Mapping Mobile Applications summit in Banff, Alberta, Canada, May 2005. The fifty participants were invited to input information about who they spoke to and what they discussed during the duration of the event. Each participant is visually represented by a flower, with a petal growing on the flower in real time as a new conversation is entered. The flowers of individuals who have conversed with one another are connected by green vines, and the proximity of two flowers directly corresponds to the number of conversations between them. The topics discussed appear in the background according to how often they are debated.

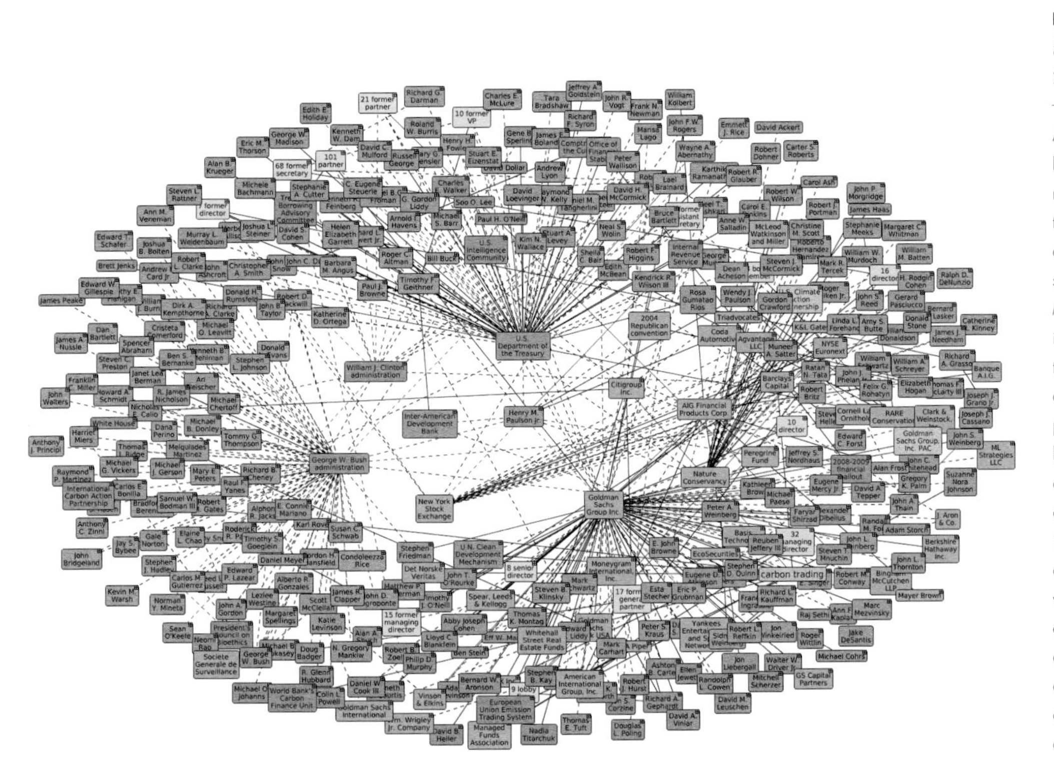

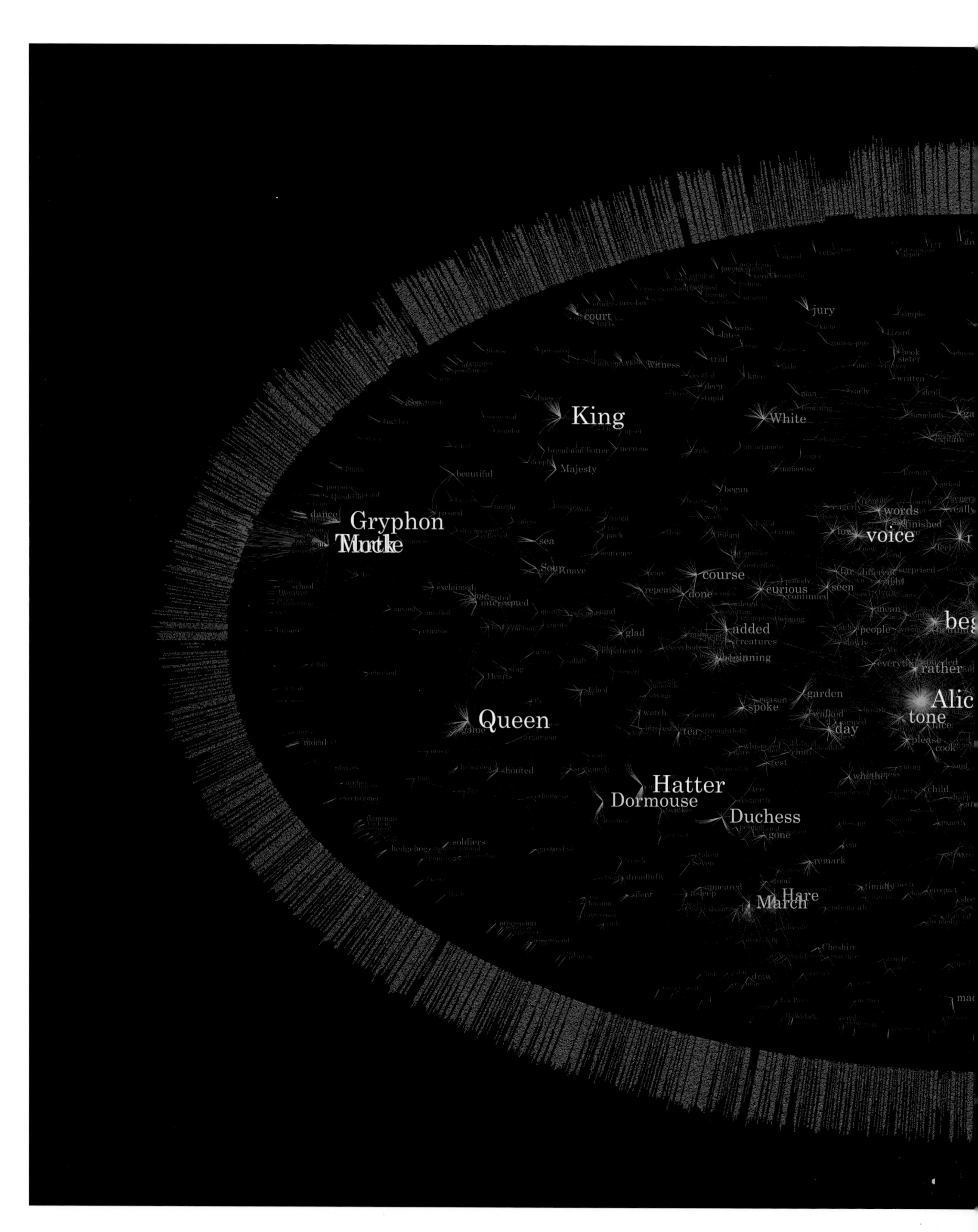
King
Gryphon
Queen
Hatter
Dormouse
Duchess
March Hare
voice
White
jury
court
course
curious
added
Majesty
words
rather
tone
garden
day
spoke
seen
Cheshire
soldiers
shouted
remark

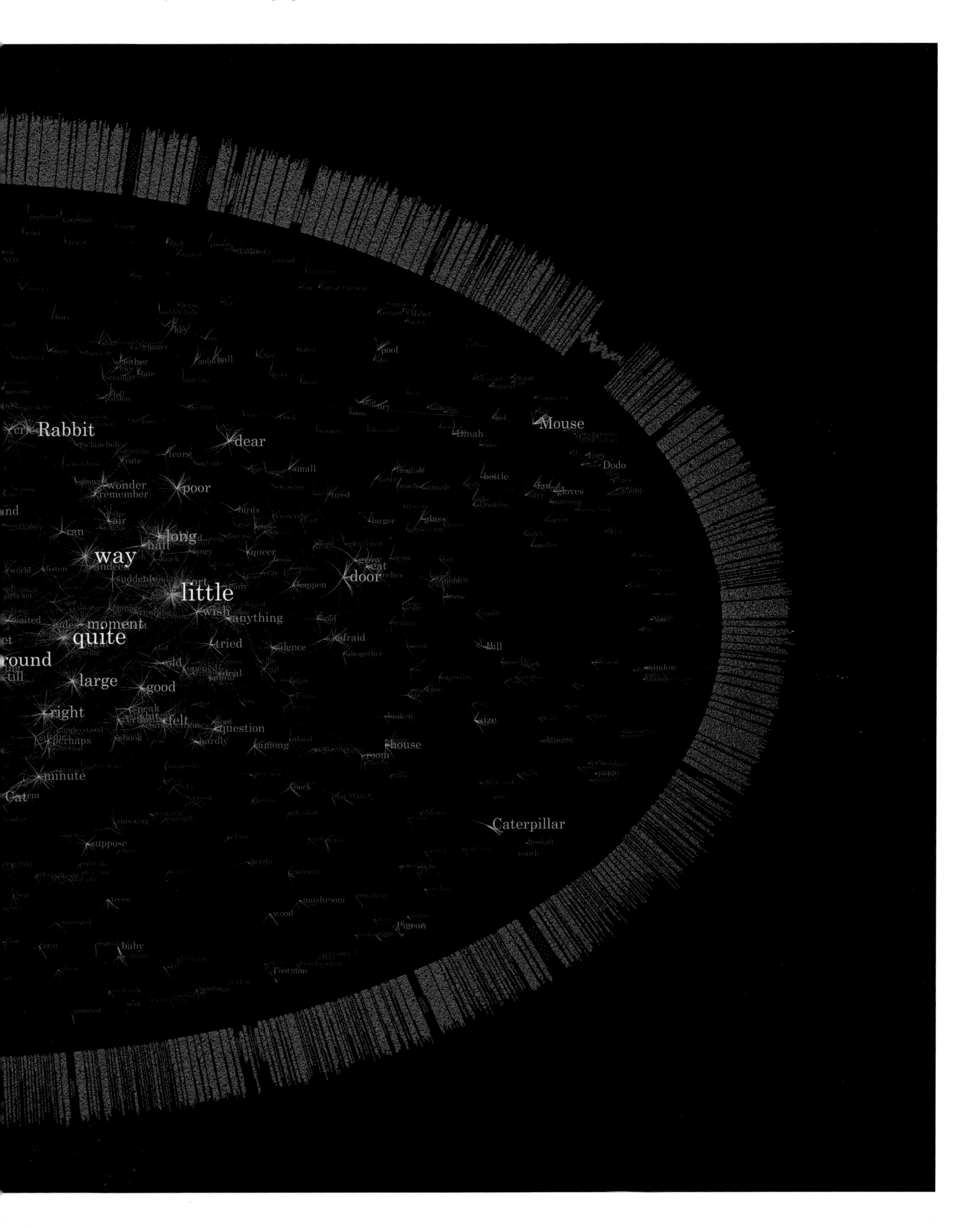
Rabbit
dear
wonder
remember
poor
small
tears
ran
long
hall
way
indeed
listen
world
little
door
wish
anything
moment
quite
tried
silence
afraid
round
till
large
good
right
felt
question
perhaps
hardly
among
house
room
minute
Cat
Caterpillar
suppose
mushroom
wood
Pigeon
baby
Footman
Mouse
Dinah
Dodo
Lory
bottle
gloves
pool
Bill
size
window
altogether
birds
queer
happen
hookah
youth
trees
sneezing

Flow Chart

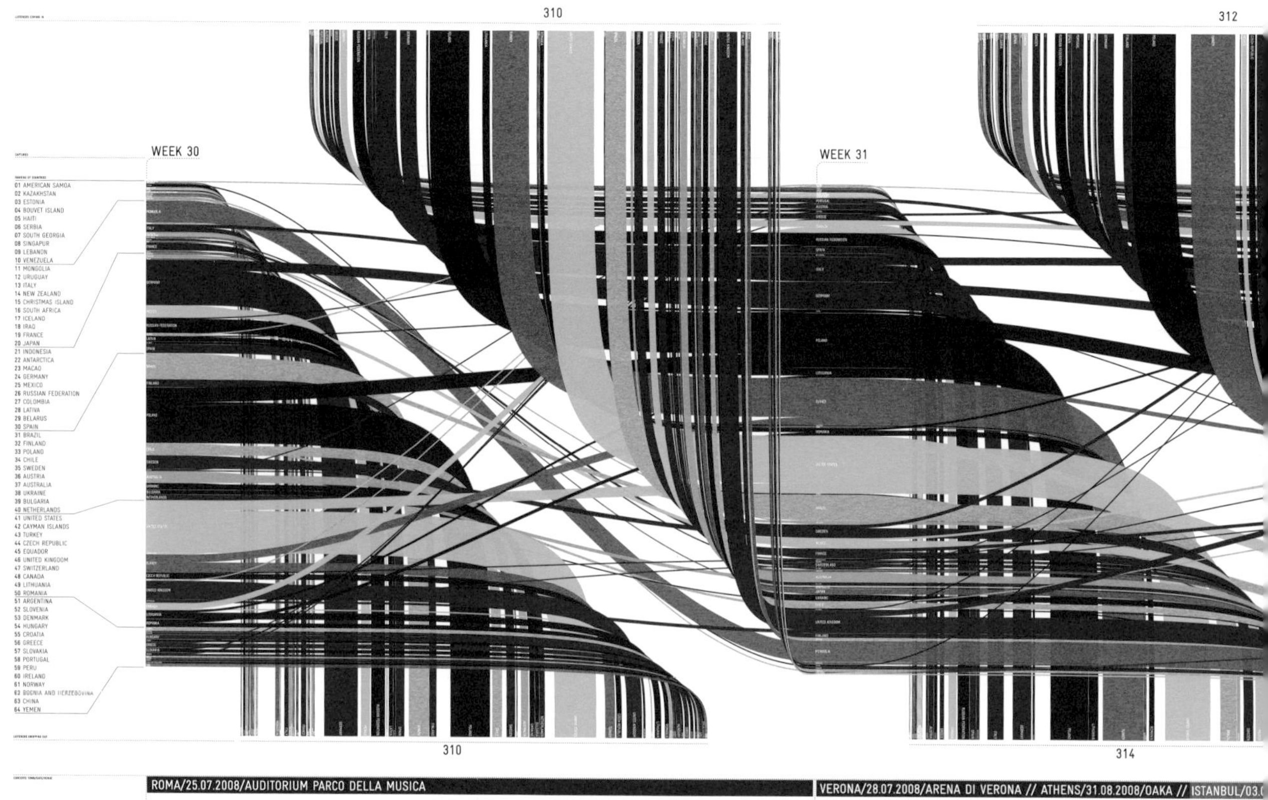

Christopher Adjei and Nils Holland-Cunz
Monitoring and Visualizing Last.fm
2008

A visualization depicting the fluctuating numbers of Last.fm Björk fans worldwide during four one-week periods. Each color represents a continent; each stripe, a different country (above). Detail (opposite).

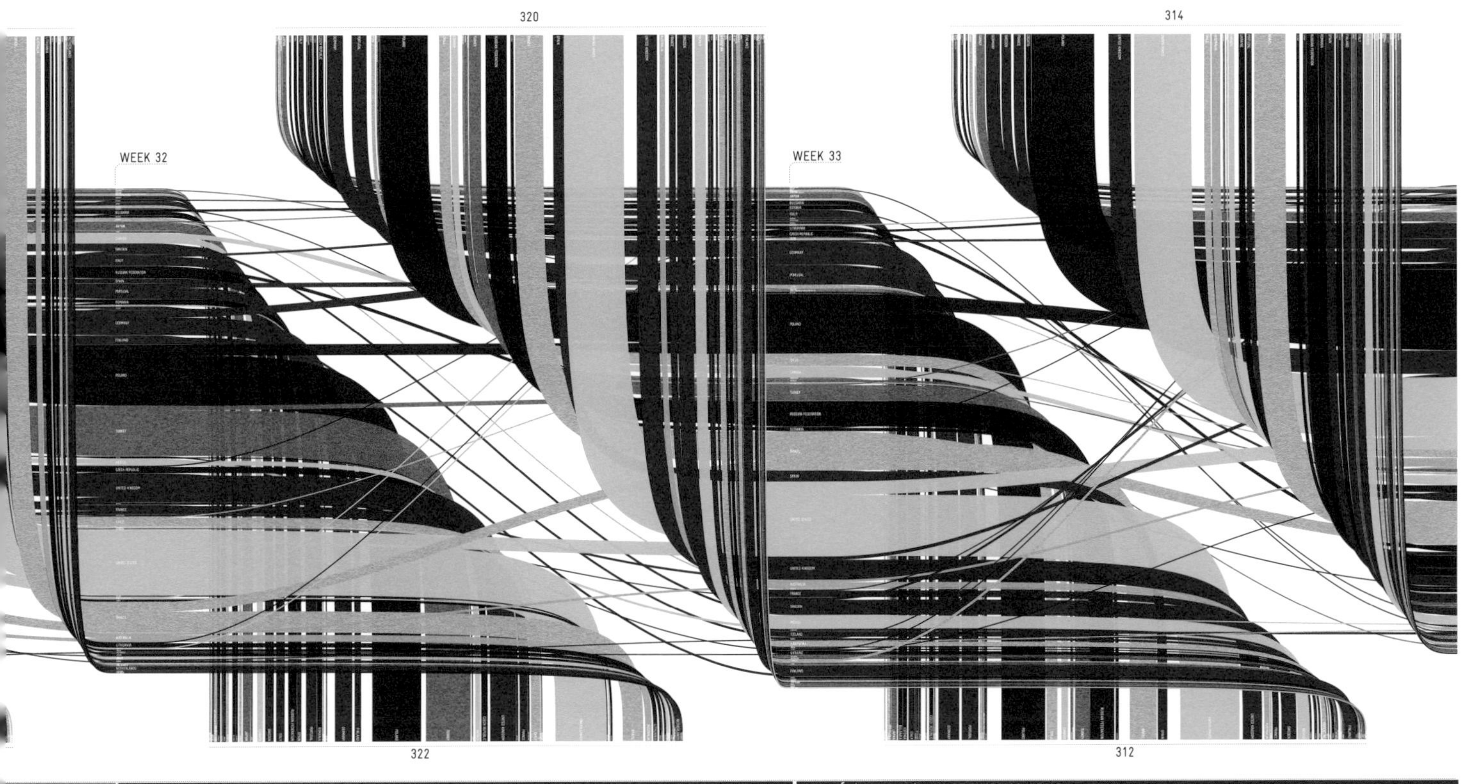
320
314
WEEK 32
WEEK 33
322
312
RENA
ZAMBUJEIRA DO MAR/07.08.2008/HERDADE DA CASA BRANCA // ZARAGOZA/10.08.2008/ANFITEATRO 43
ALMERÍA/15.08.2008/PLAYA DE GUARDIAS VIEJAS
WEEK32/04.08.2008–10.08.2008
WEEK33/11.08.2008–17.08.2008

TURKEY
IRAN, ISLAMIC REPUBLIC OF
BULGARIA
UNITED STATES
CHILE
FINLAND
BRAZIL
POLAND
BELGIUM
ITALY
AUSTRALIA
UNITED KINGDOM
CANADA
MEXICO
SWEDEN
RUSSIAN FEDERATION

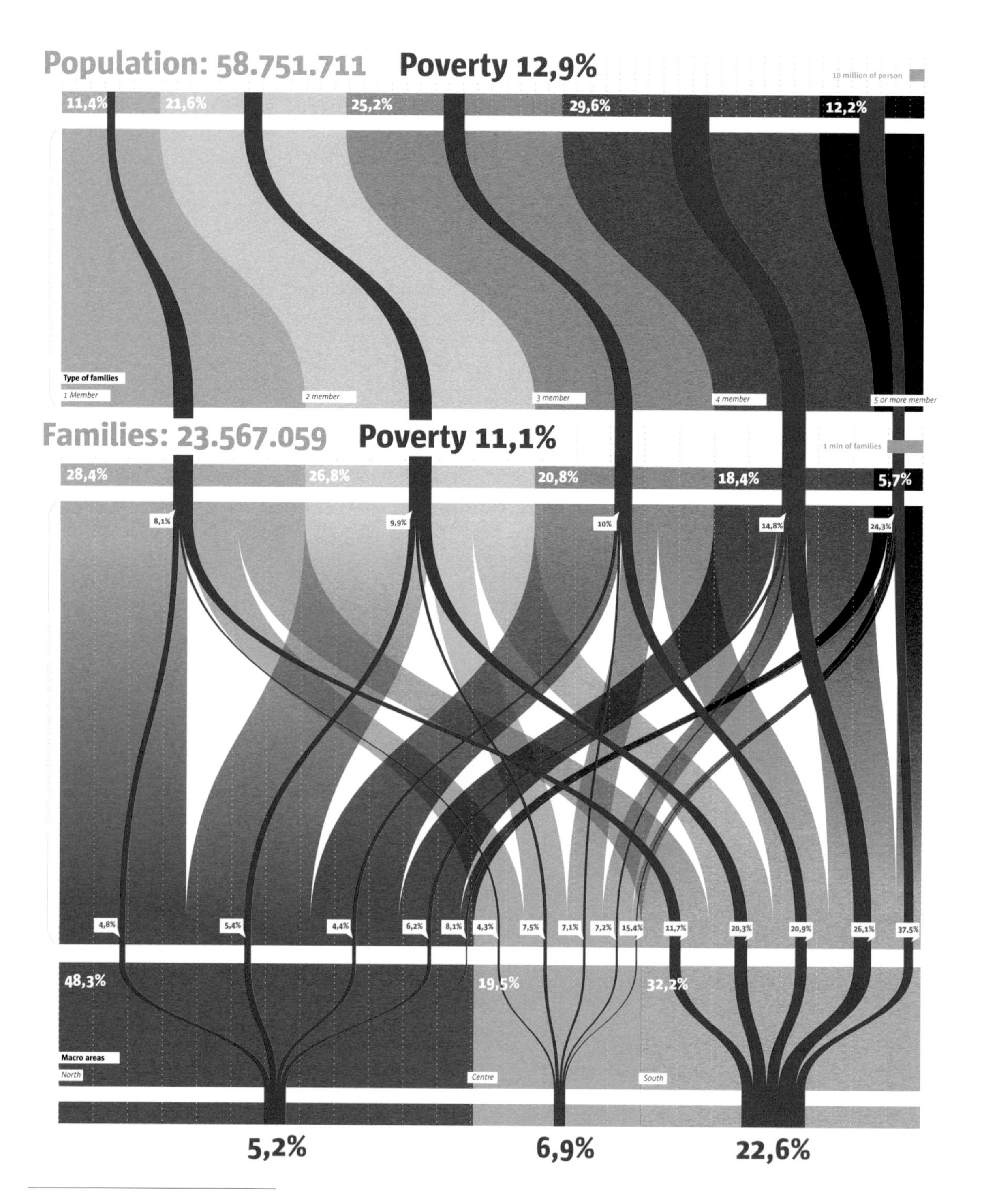

Density Design: Mario Porpora
The Poverty Red Thread
2008

A map of the poverty line in Italy organized according to family typologies (number of family members), and further categorized by location (the north, center, or south of Italy)

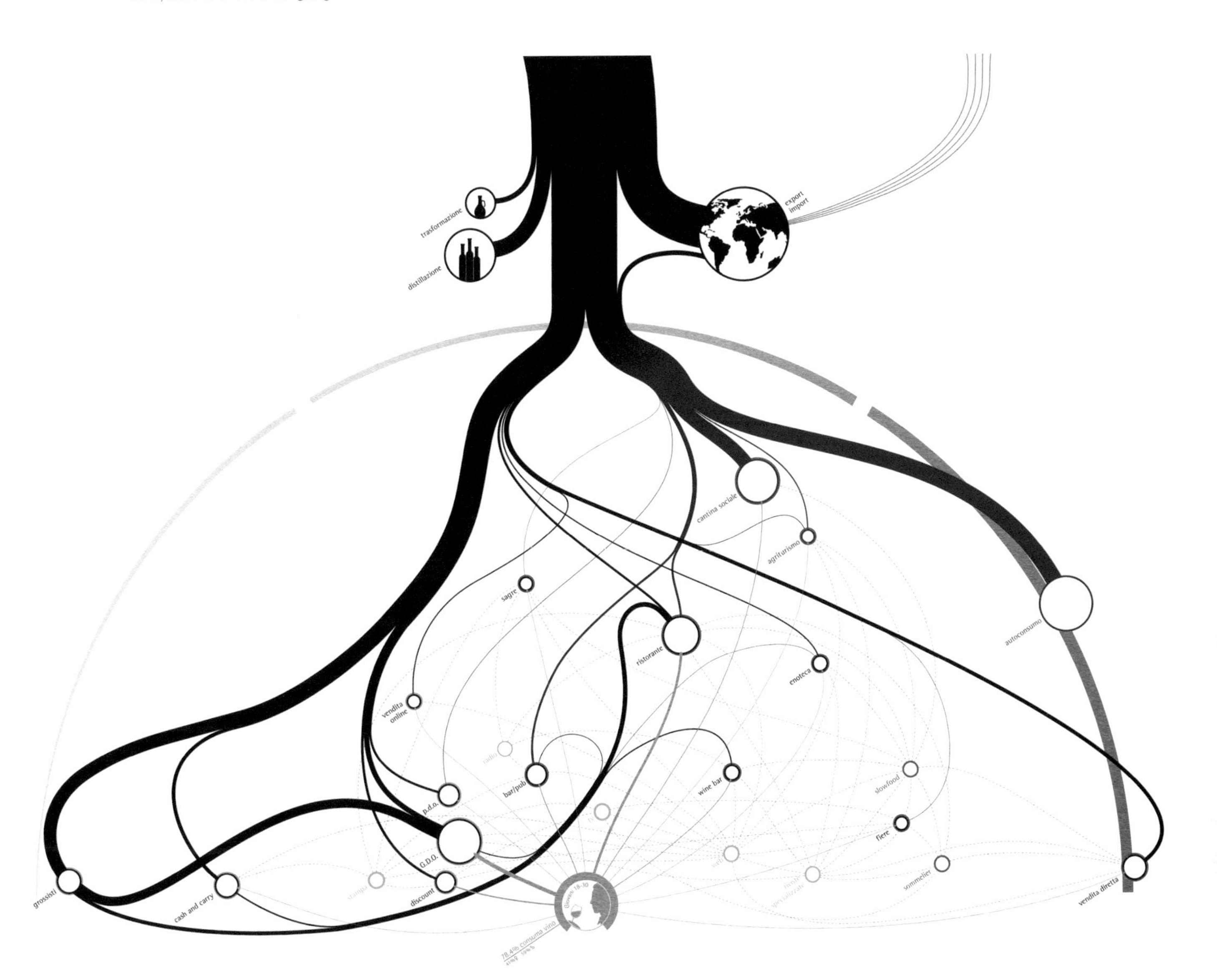

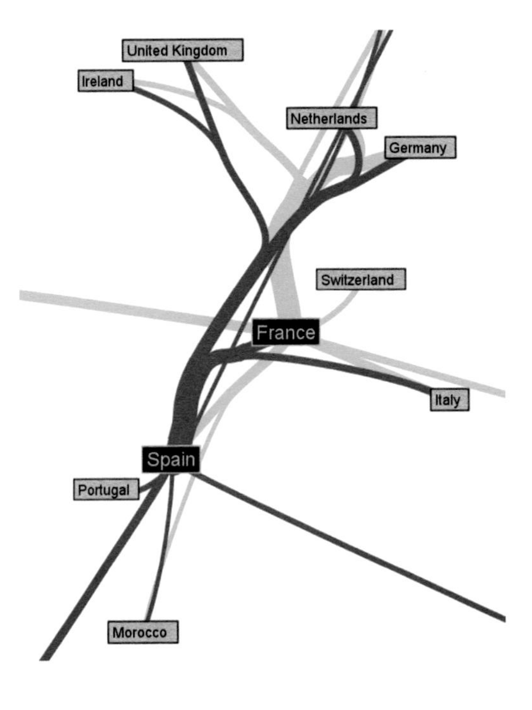

(top)

Density Design: Bergamini Andrea, Bertola Guido, Discacciati Pietro, Mangiaracina Francesca, and Merenda Stefania

The Italian Wine System

2006

A part of a larger map on the distribution of Italian wines, from production to consumption, both domestically and abroad

(right)

Doantam Phan, Ling Xiao, Ron Yeh, Pat Hanrahan, and Terry Winograd

Flow Map Layout

2005

A pioneering computer-generated flow map—traditionally drawn by hand—that shows the movement of objects from one location to another (e.g., the number of people in a migration, the amount of traded goods, or energy exchanged)

Organic Rhizome

Robert King

Locus

2004

Examples generated from an experimental tool for visualizing instant messaging conversations

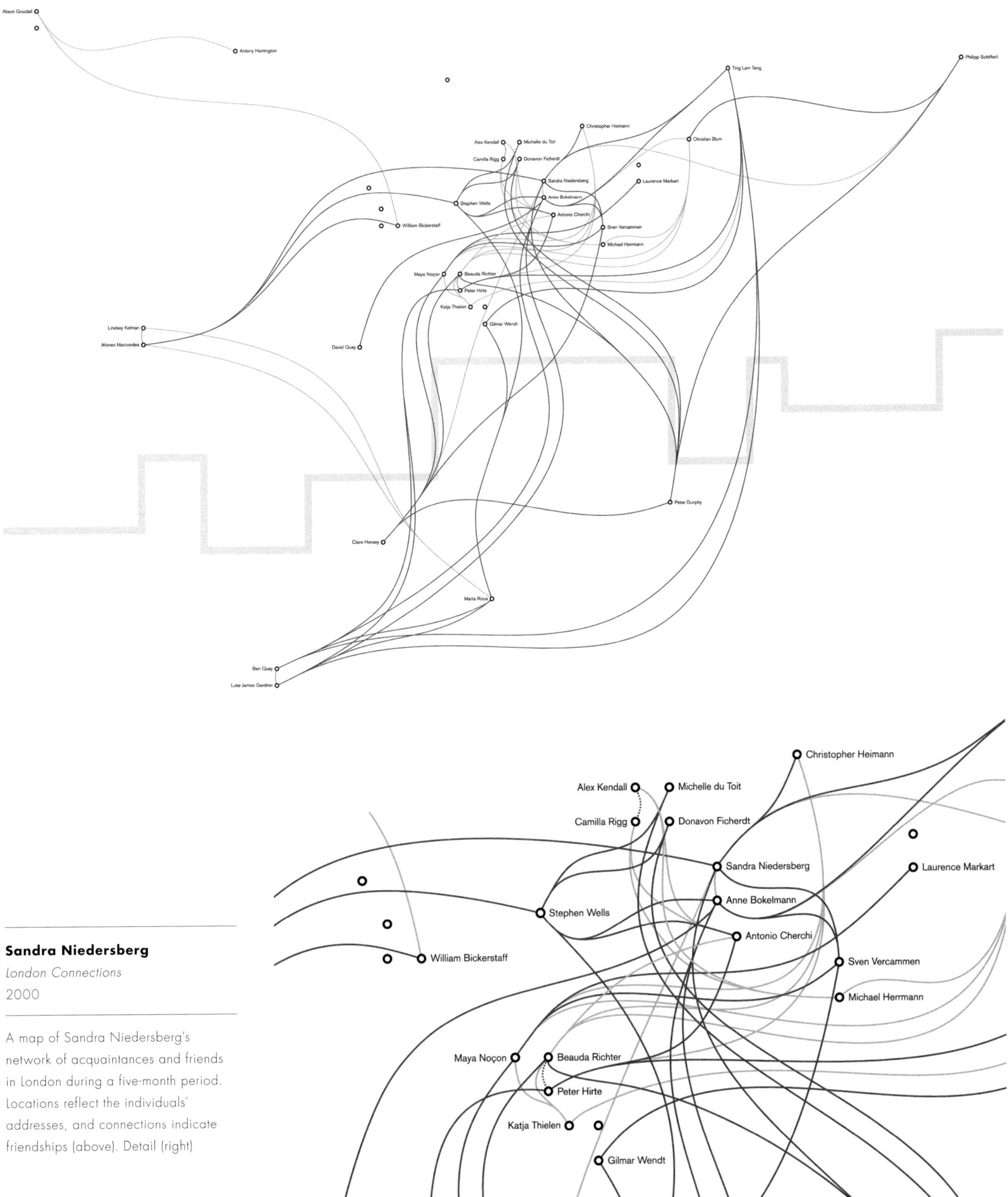

Sandra Niedersberg
London Connections
2000

A map of Sandra Niedersberg's network of acquaintances and friends in London during a five-month period. Locations reflect the individuals' addresses, and connections indicate friendships (above). Detail (right)

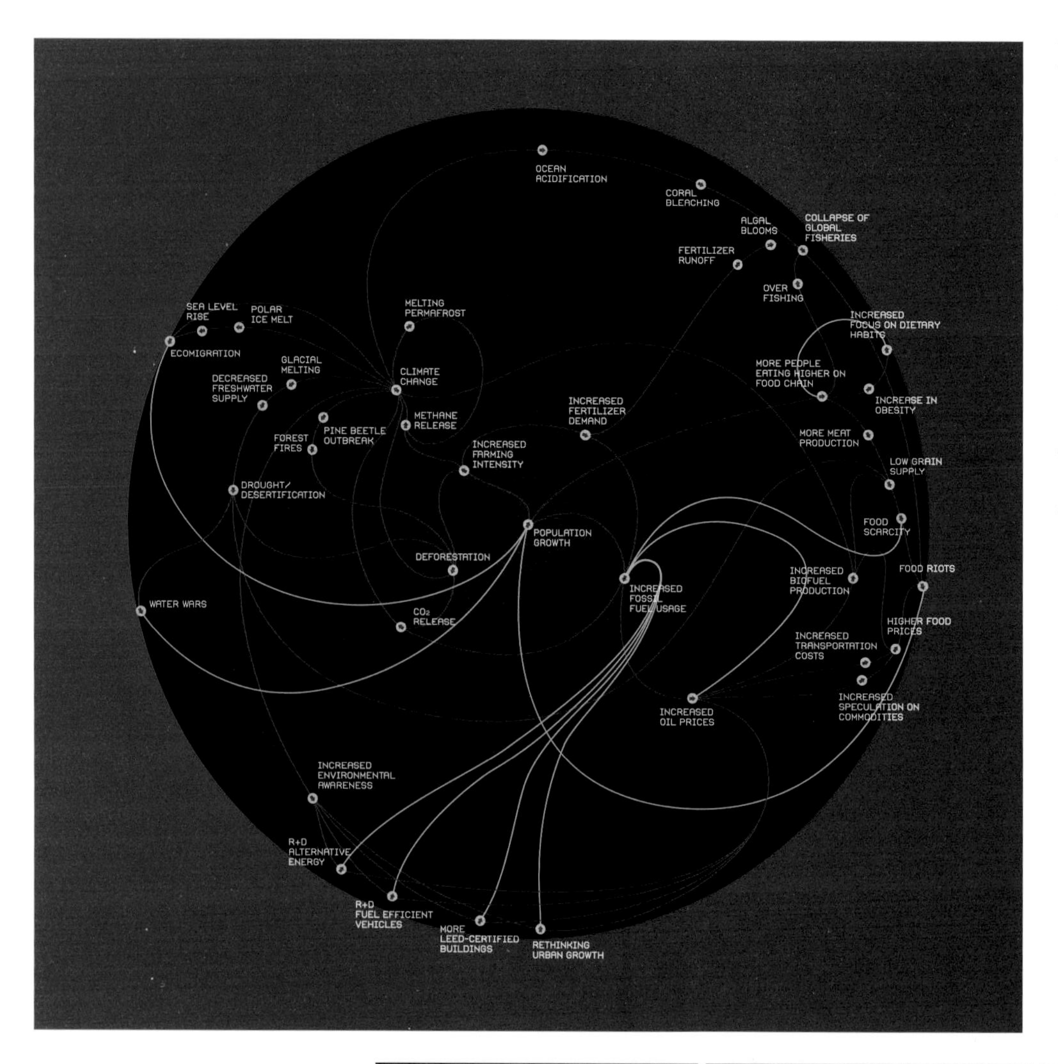

Tyler Lang and Elsa Chaves

Connecting Distant Dots

2009

A map of connections between the various social and environmental forces that affect our planet. The inner concentric circles are the main causes of environmental damage (population growth, farming, deforestation), and rippling outward are the effects. Red lines depict augmenting forces (CO_2 release, methane release, drought), and blue lines show inhibitory forces (ecomigration, food riots, alternative energy). This is a reprint of an illustration that originally appeared in *Seed* 2, no. 20 (2009).

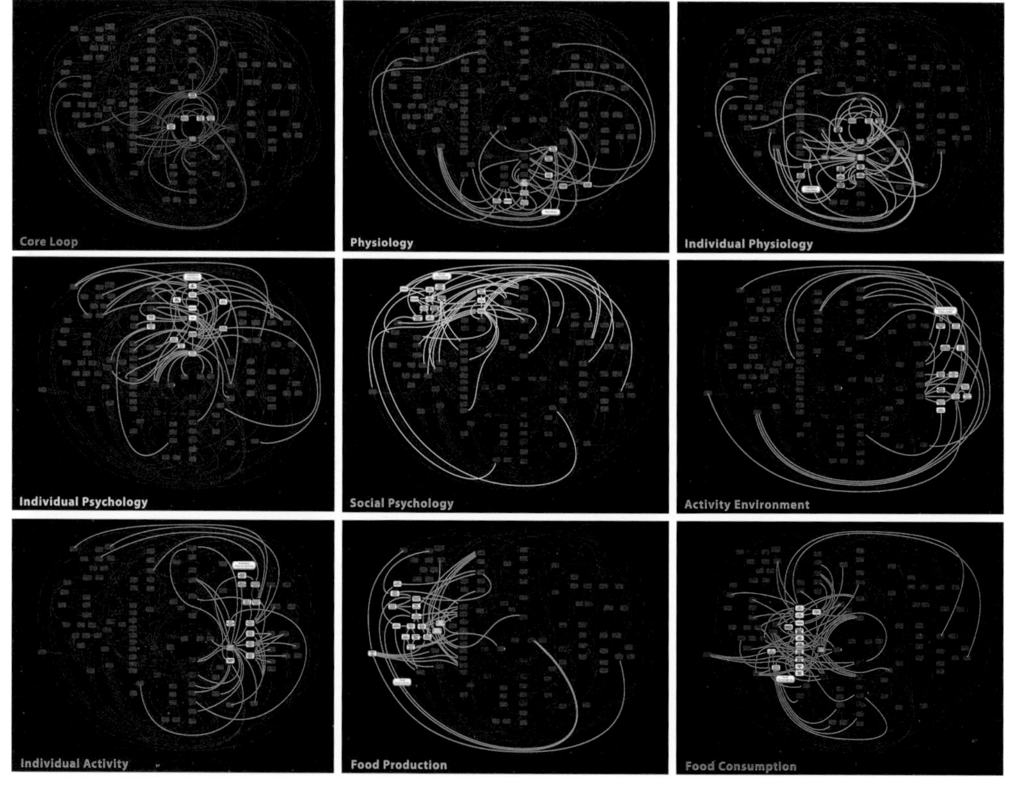

Philippe Vandenbroeck, Jo Goossens, and Marshall Clemens

Obesity System Influence Diagram

2007

A systemic look at nine factors contributing to the obesity epidemic, such as social psychology and food consumption

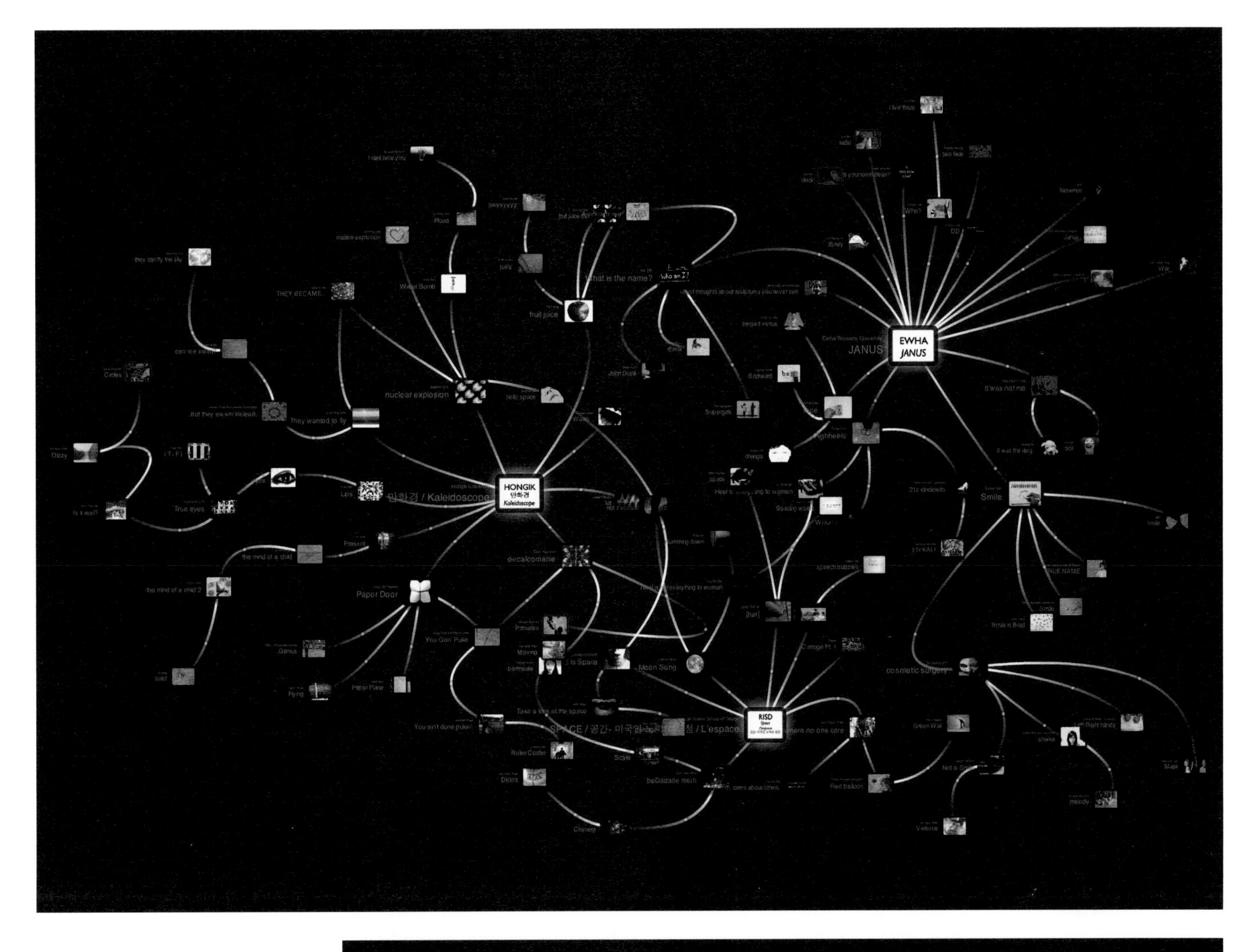

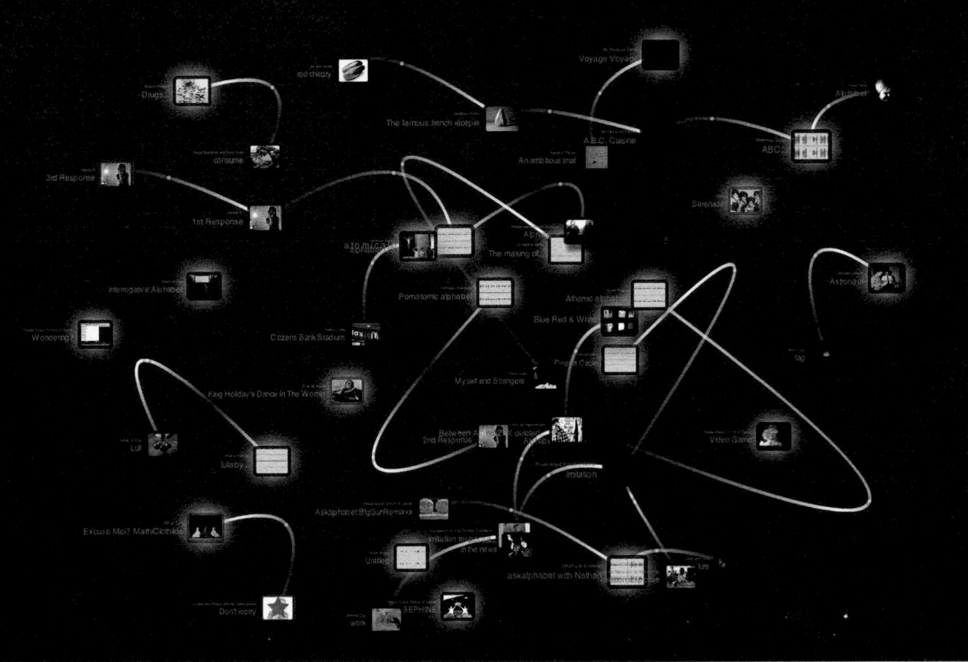

Daniel Peltz, Dennis Hlynsky, and Chuan Khoo
RISD.tv Call & Response
2007

Screenshots from a collaborative visualization framework for video production at Rhode Island School of Design (RISD), which highlights various tagging relationships between videos submitted to the database

Radial Convergence

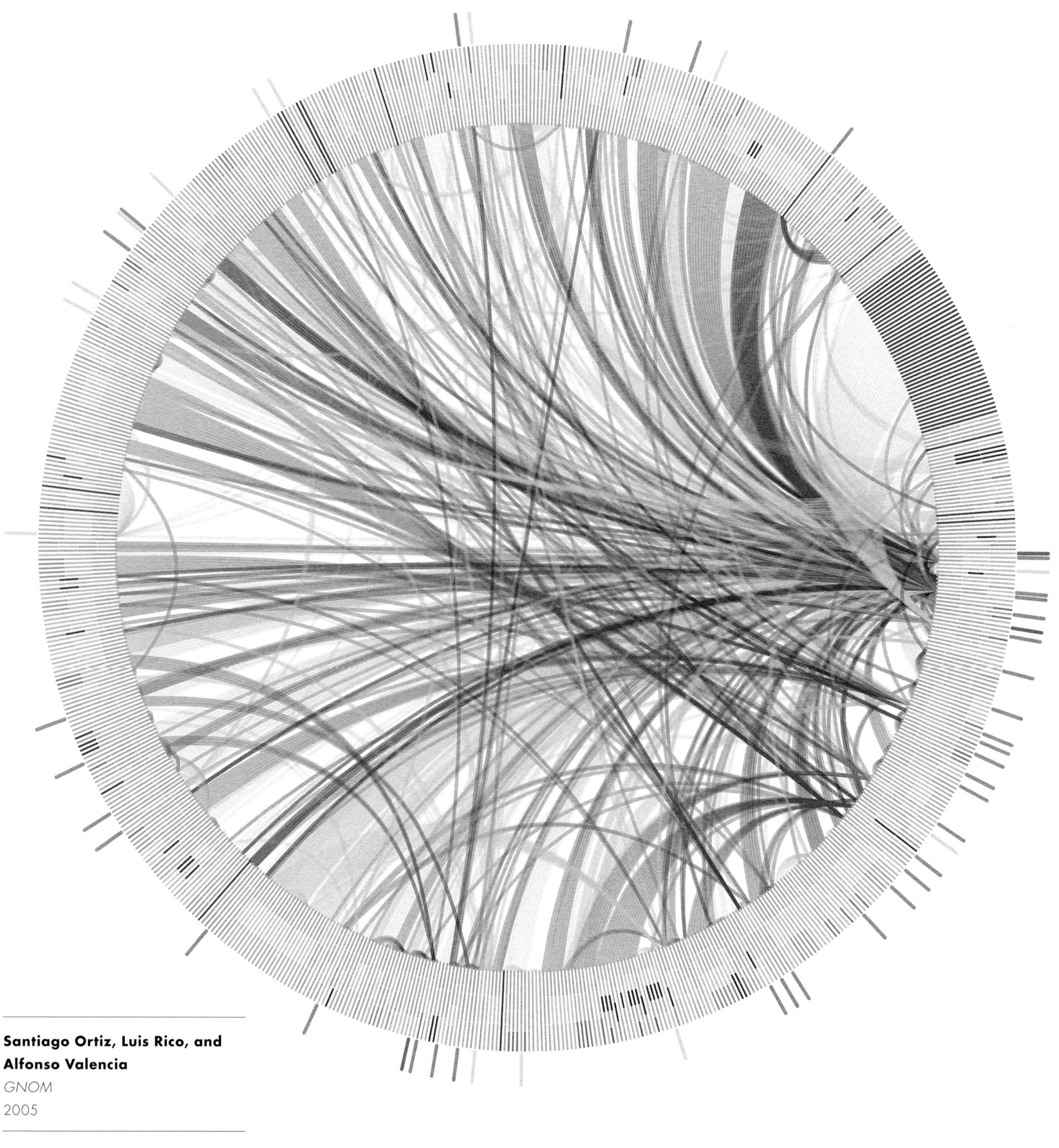

Santiago Ortiz, Luis Rico, and Alfonso Valencia
GNOM
2005

A visualization of the interaction network of genes of the bacteria *Escherichia coli*. Each gene is represented by a segment of color on the outside ring. The color of the lines within the circle express the nature of the relationship between genes.

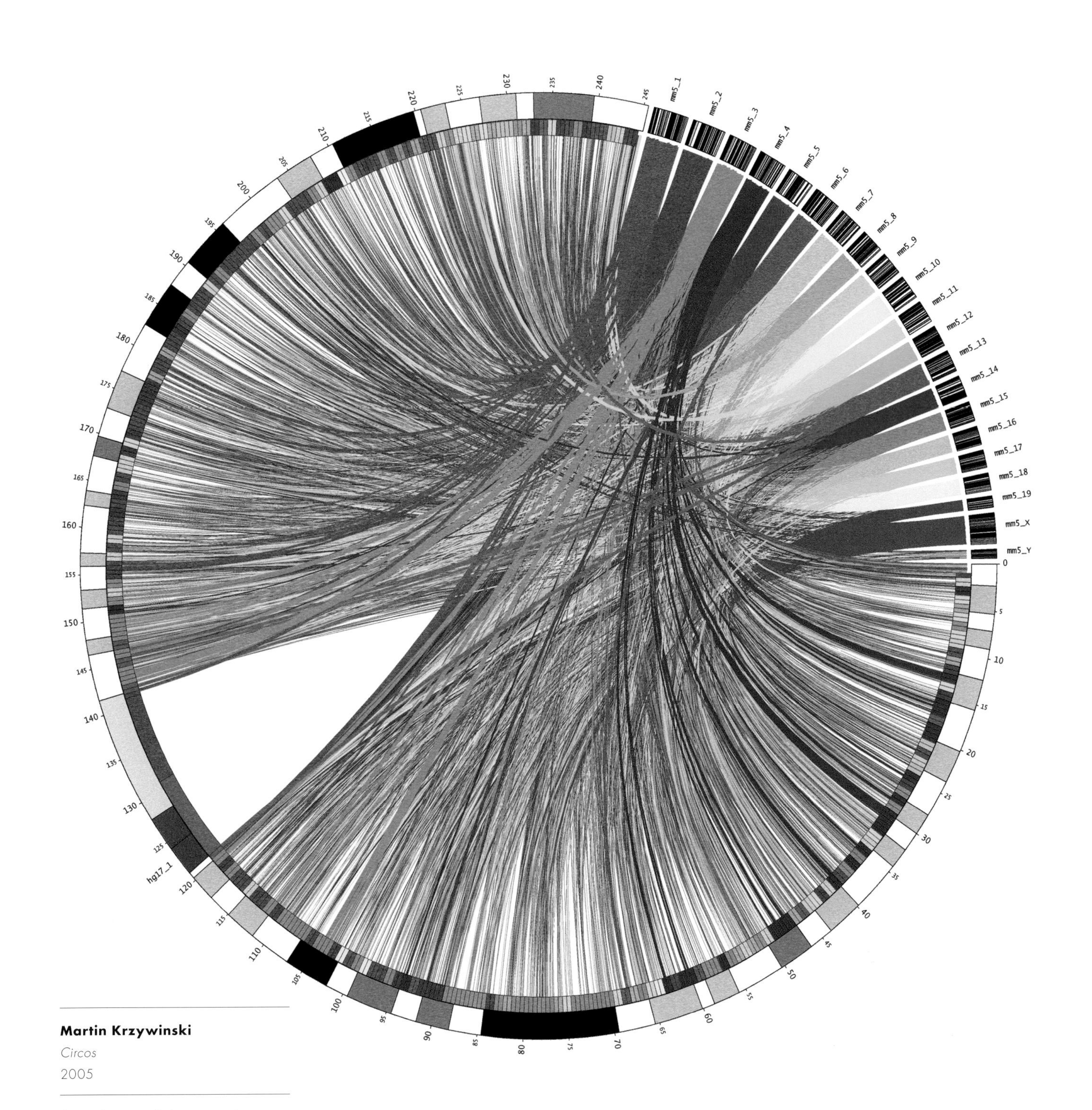

Martin Krzywinski

Circos

2005

A visualization of chromosomal relationships within one genome

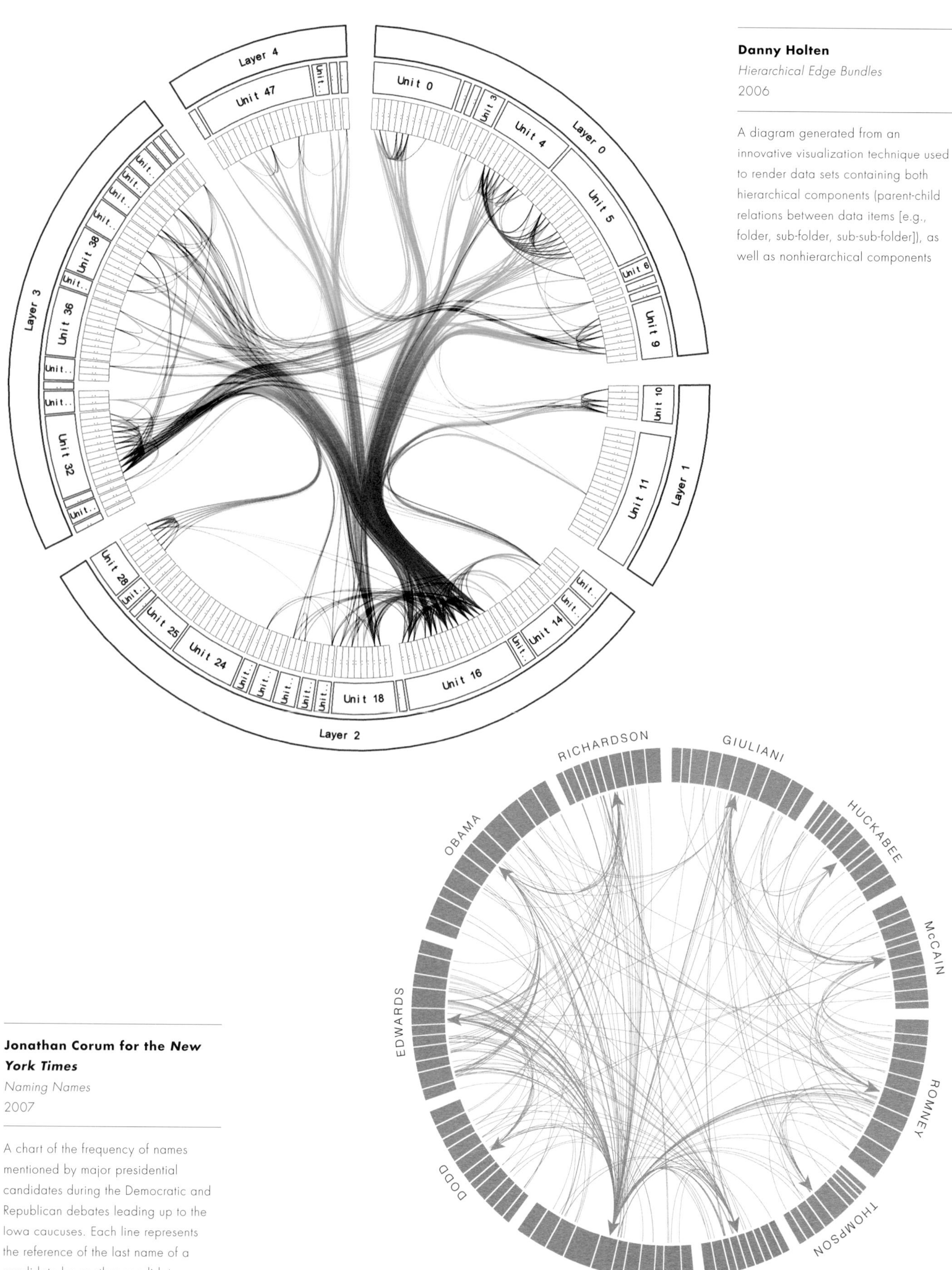

Danny Holten
Hierarchical Edge Bundles
2006

A diagram generated from an innovative visualization technique used to render data sets containing both hierarchical components (parent-child relations between data items [e.g., folder, sub-folder, sub-sub-folder]), as well as nonhierarchical components

Jonathan Corum for the *New York Times*
Naming Names
2007

A chart of the frequency of names mentioned by major presidential candidates during the Democratic and Republican debates leading up to the Iowa caucuses. Each line represents the reference of the last name of a candidate by another candidate.

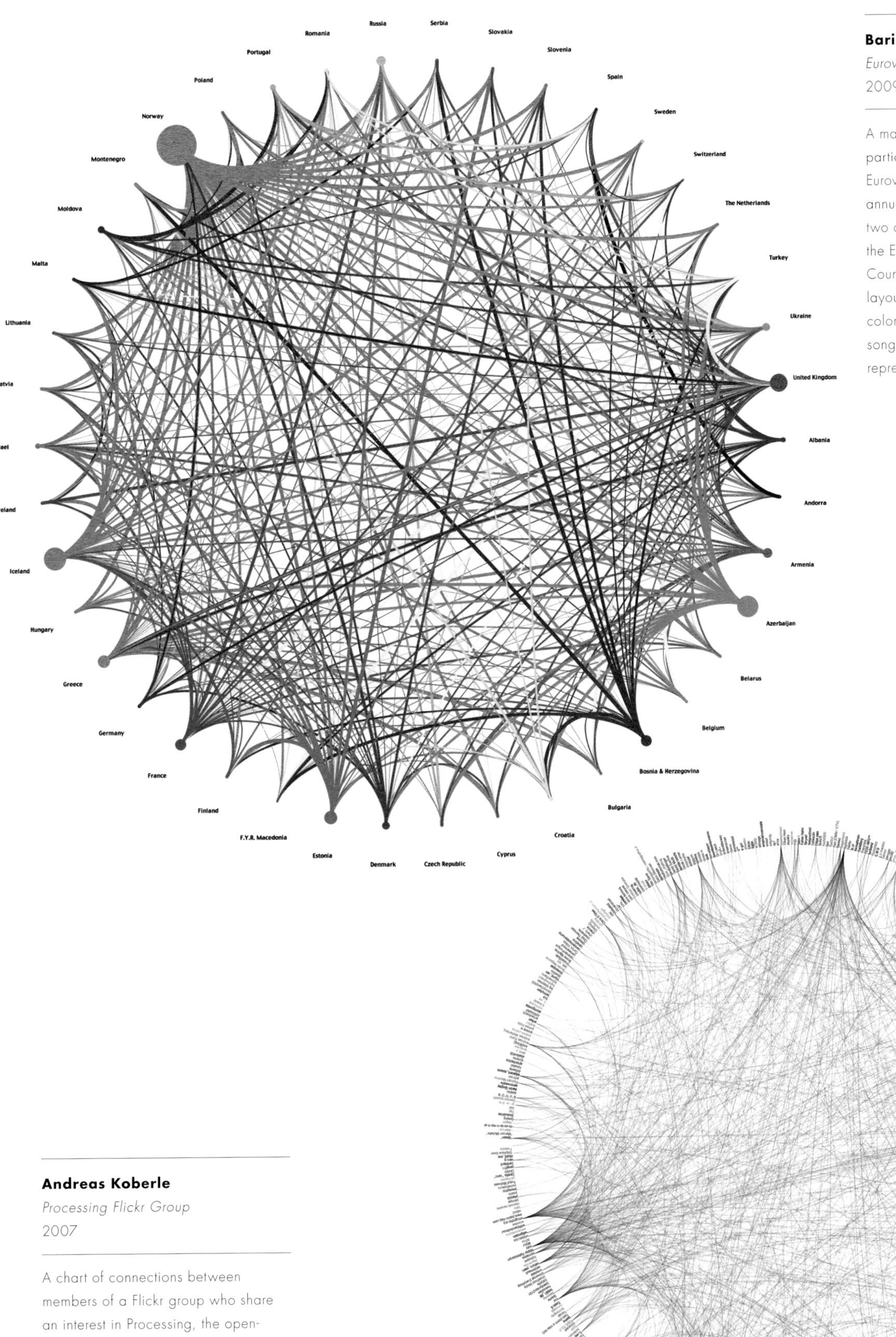

Baris Gumustas
Eurovision 2009 Results
2009

A map of voting patterns of participating countries in the Moscow Eurovision 2009 song contest, an annual competition held among forty-two countries that are members of the European Broadcasting Union. Countries are arranged in a radial layout and represented by a unique color. Links signify a vote for a country's song, and the weight of the link represents the value of the votes.

Andreas Koberle
Processing Flickr Group
2007

A chart of connections between members of a Flickr group who share an interest in Processing, the open-source programming language and development environment

Radial Implosion

Richard Rogers

Climate Change: U.S. Groups in International Context

2004

A map of interlinks among prominent climate-change websites

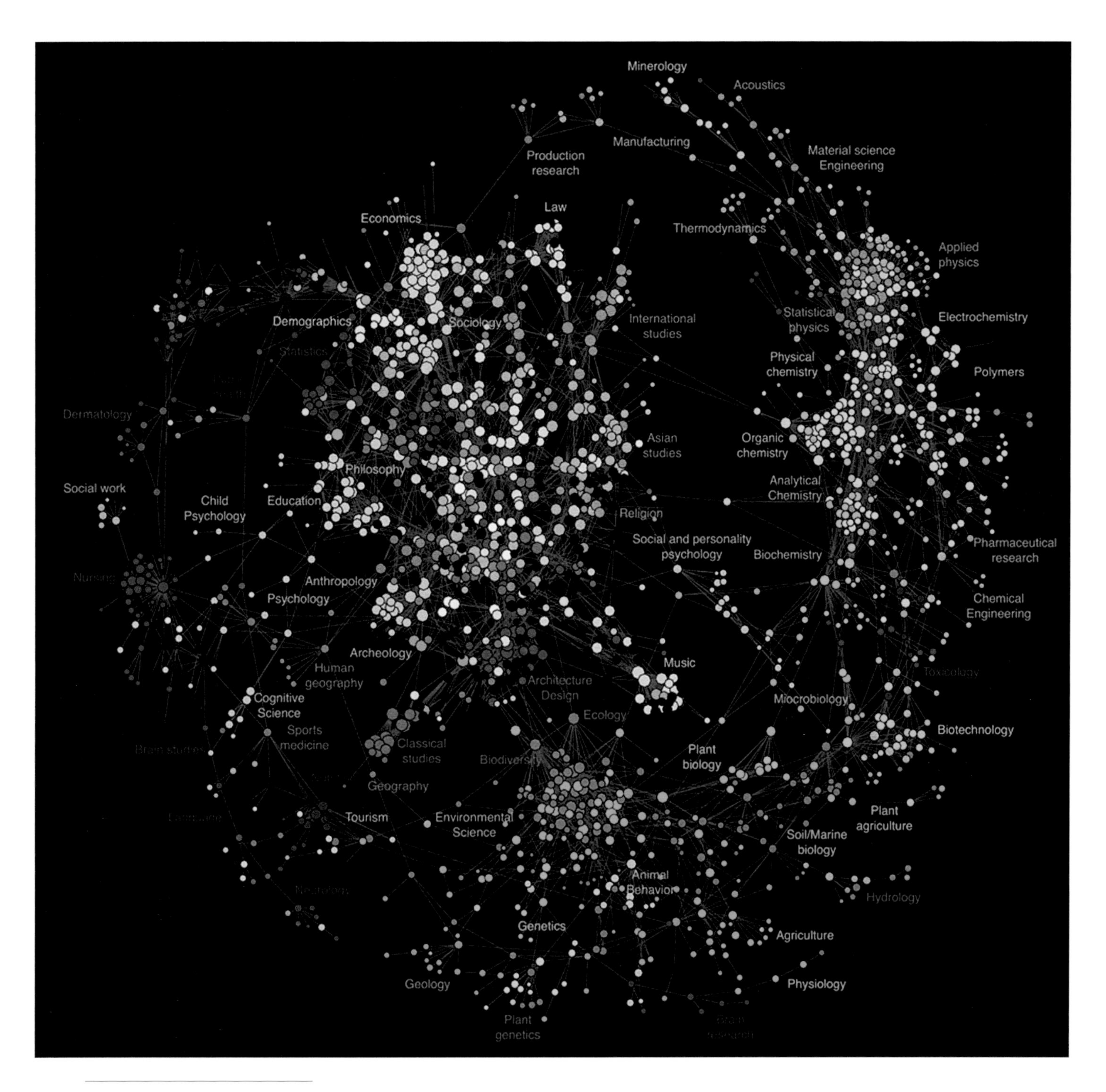

J. Bollen, H. Van de Sompel, A. Hagberg, L. Bettencourt, R. Chute, et al.
Map of Science
2009

A network of associations between scientific journals that includes nearly one billion user interactions recorded between 2007 and 2008 in scholarly web portals of significant publishers, aggregators, and institutional consortia. As users select journals, their click stream ties together journals. Circles represent individual journals; labels have been assigned to local clusters of journals that correspond to particular scientific disciplines. The colors correspond to the journal-classification system of the Art & Architecture Thesaurus, a controlled vocabulary of art, architecture, and material-culture terms. From J. Bollen, H. Van de Sompel, A. Hagberg, L. Bettencourt, R. Chute, et al., "Clickstream Data Yields High-Resolution Maps of Science." *PLoS ONE* 4(3): e4803.

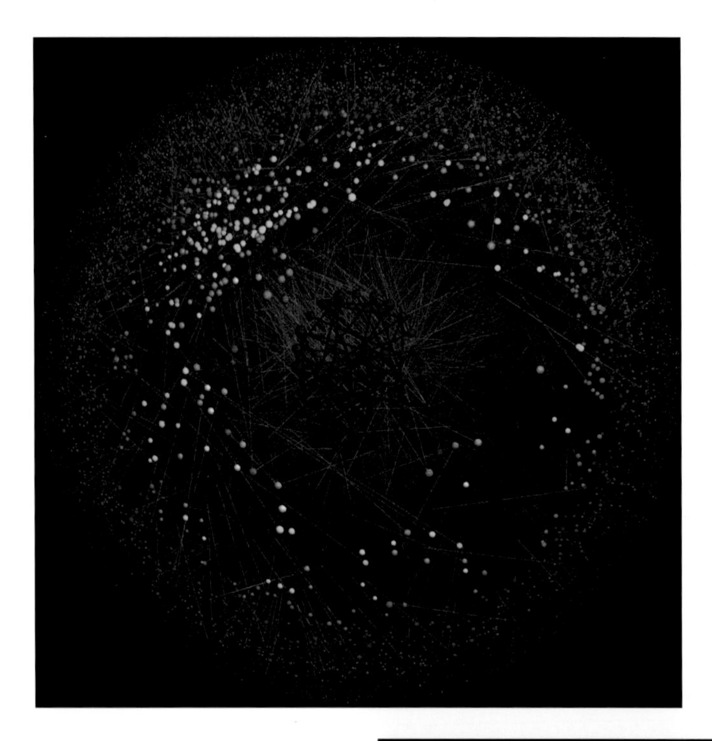

Alvarez-Hamelin, Beiró, Dall'Asta, Barrat, and Vespignani
Internet Autonomous Systems
2007.

A map of interconnectivity between autonomous systems (a collection of IP prefixes controlled by one or more network operators), based on data obtained from Cooperative Association for Internet Data Analysis (CAIDA). Colors highlight the level of connectivity of autonomous systems: red for the most connected nodes through the colors of the rainbow to violet for least connected ones. (See also chapter 4, page 120.)

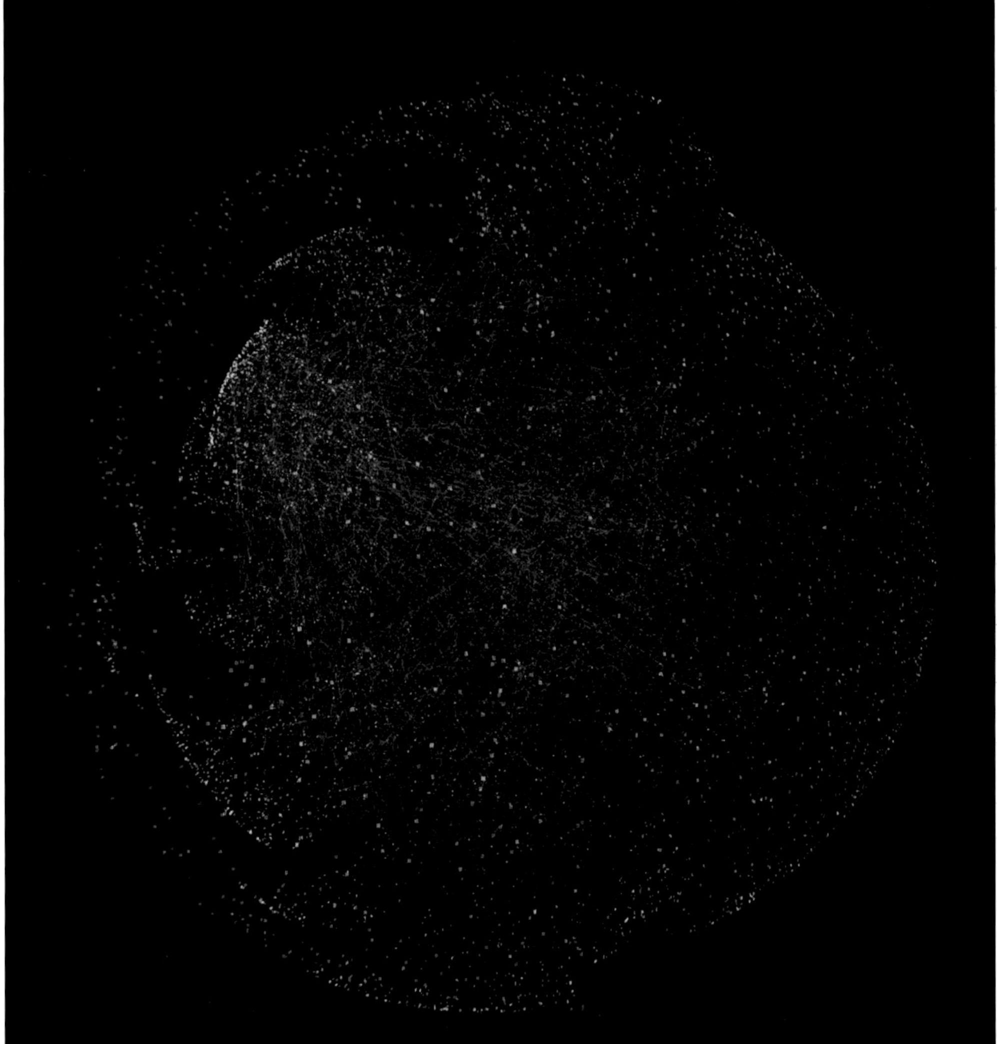

Firstborn and Digital Kitchen
Operation Smile
2007

A large-scale visualization of hundreds of smiling portraits of visitors to New York City's South Street Seaport. (See also chapter 6, page 233, figure 15.)

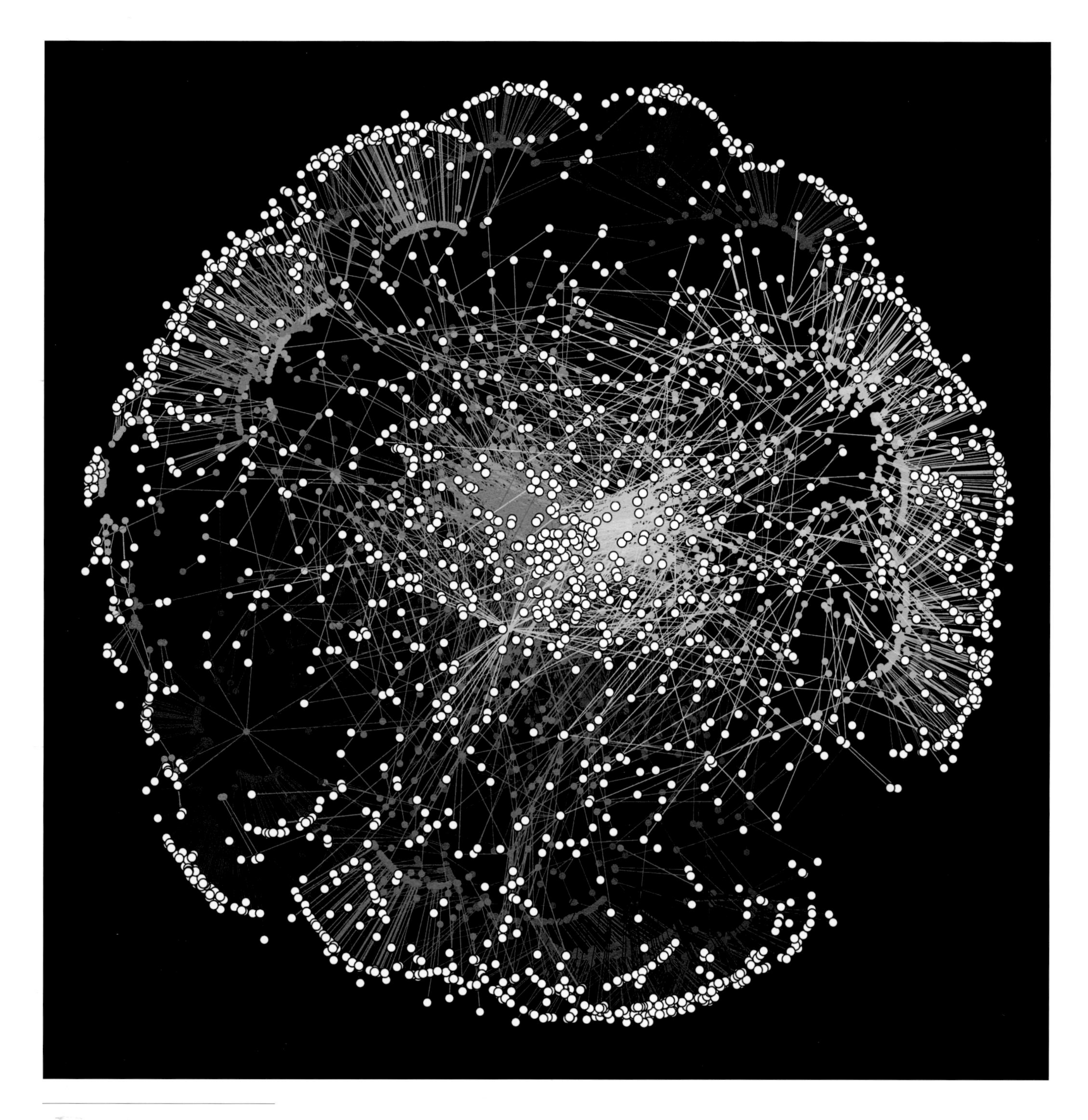

FAS.research
Collaboration Structure of Cultural Projects
2005

A map of collaborators at the annual Ars Electronica Festival, from its launch in 1979 to the 2004 event. Each white node represents one of the 2,575 artists who worked on any of the 5,176 projects. A line connects any two collaborating artists; a different color represents each festival year.

Ramifications

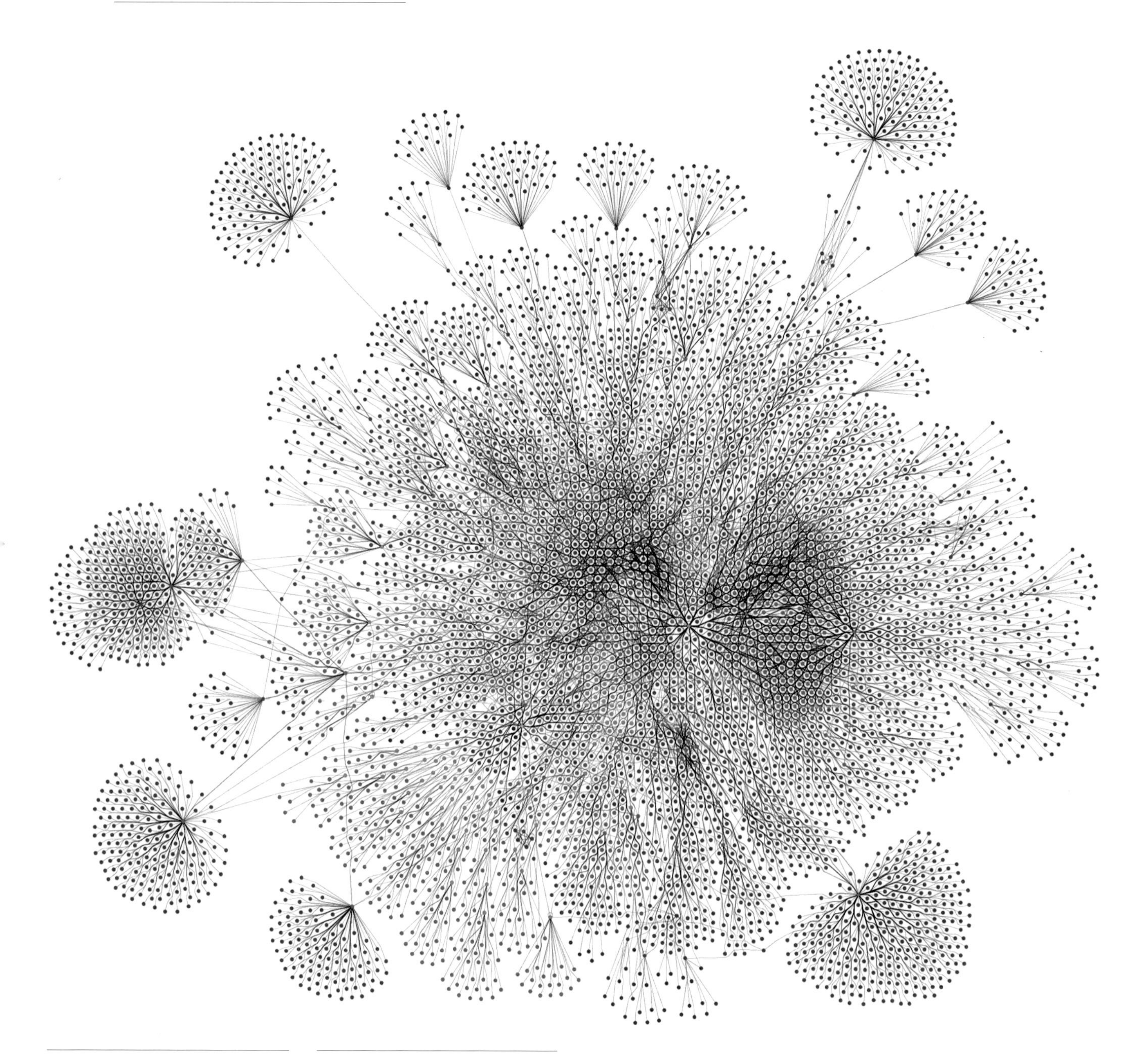

yWorks
yFiles Visualization
2005

A visualization of abstract nodes created through a Java library of layout algorithms for graphs, charts, and diagrams

(opposite)
Stefanie Posavec
Writing Without Words
2008

A chart of the structure of part one of Jack Kerouac's *On the Road* (1957). Each splitting of the branch into progressively smaller sections parallels the organization of the content from chapters to paragraphs, sentences, and words. Each color relates to one of eleven thematic categories created by Posavec for the book (e.g., travel, work and survival, sketches of regional life).

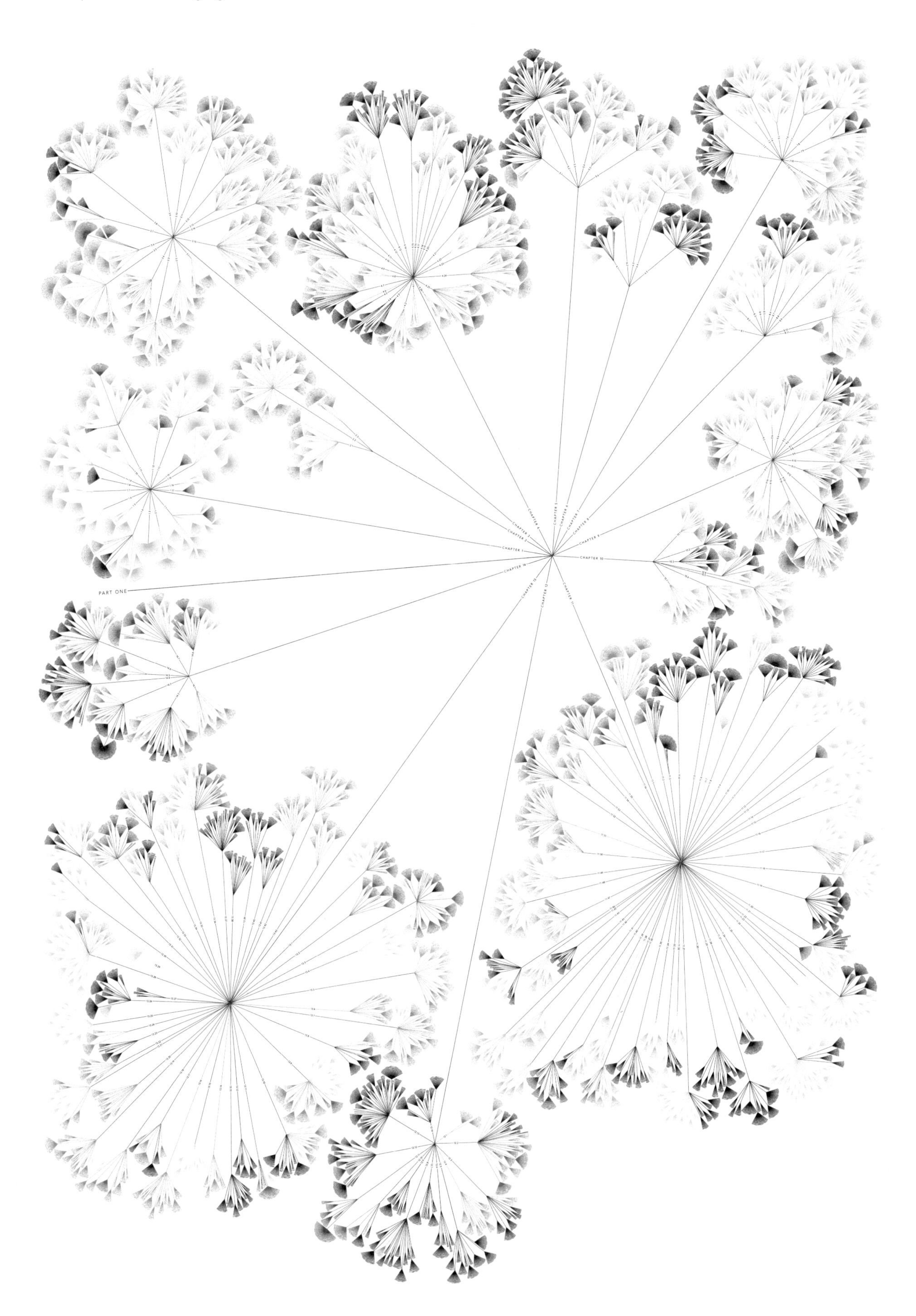
PART ONE
CHAPTER 1
CHAPTER 2
CHAPTER 3
CHAPTER 4
CHAPTER 5
CHAPTER 6
CHAPTER 7
CHAPTER 8
CHAPTER 9
CHAPTER 10
CHAPTER 11
CHAPTER 12
CHAPTER 13
CHAPTER 14

Jukka-Pekka Onnela et al.
Mobile Communication Network
2007

The structure of the mobile-communication network around a single individual. From J-P. Onnela, J. Saramäki, J. Hyvönen, G. Szabó, D. Lazer, K. Kaski, J. Kertész, and A-L. Barabási, "Structure and Tie Strengths in Mobile Communication Networks," *Proceedings of the National Academy of Sciences (PNAS) USA* 104, no. 18 (2007): 7332–36.

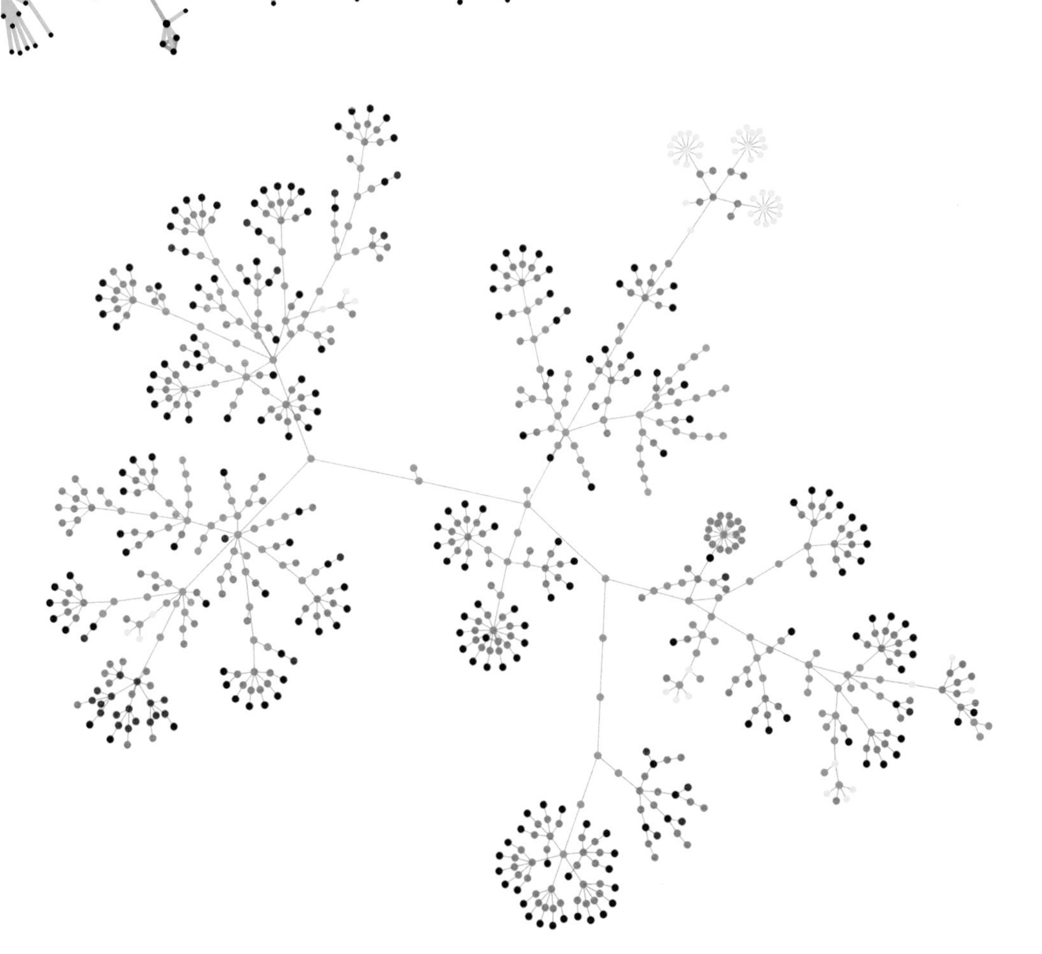

Marcel Salathé
Websites as Graphs
2006

A network diagram depicting the HTML-tag structure of MSN.com home page

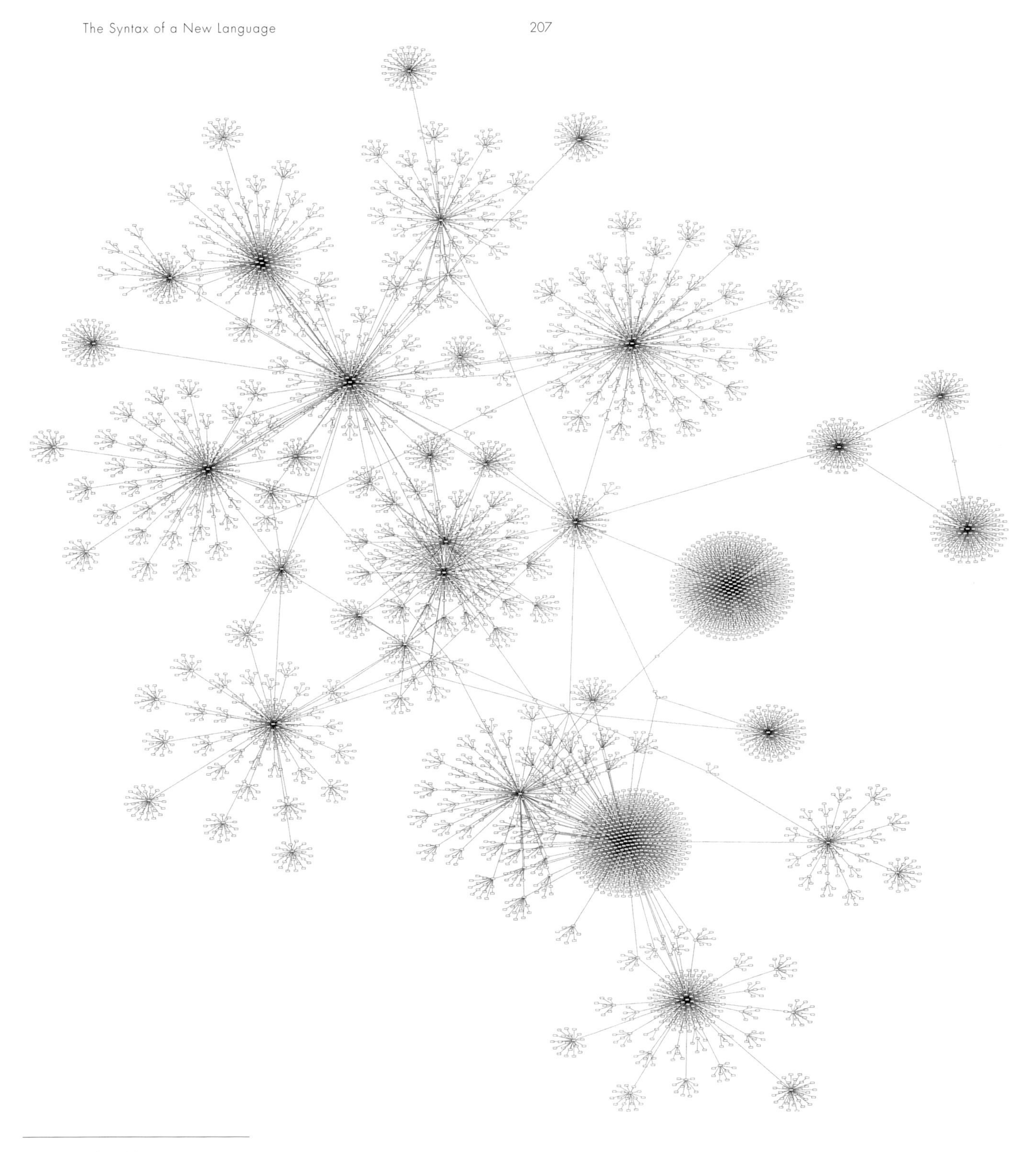

Michael Schmuhl
graphopt
2003

A screenshot of a visualization program that helps optimize computer-generated network diagrams

Scaling Circles

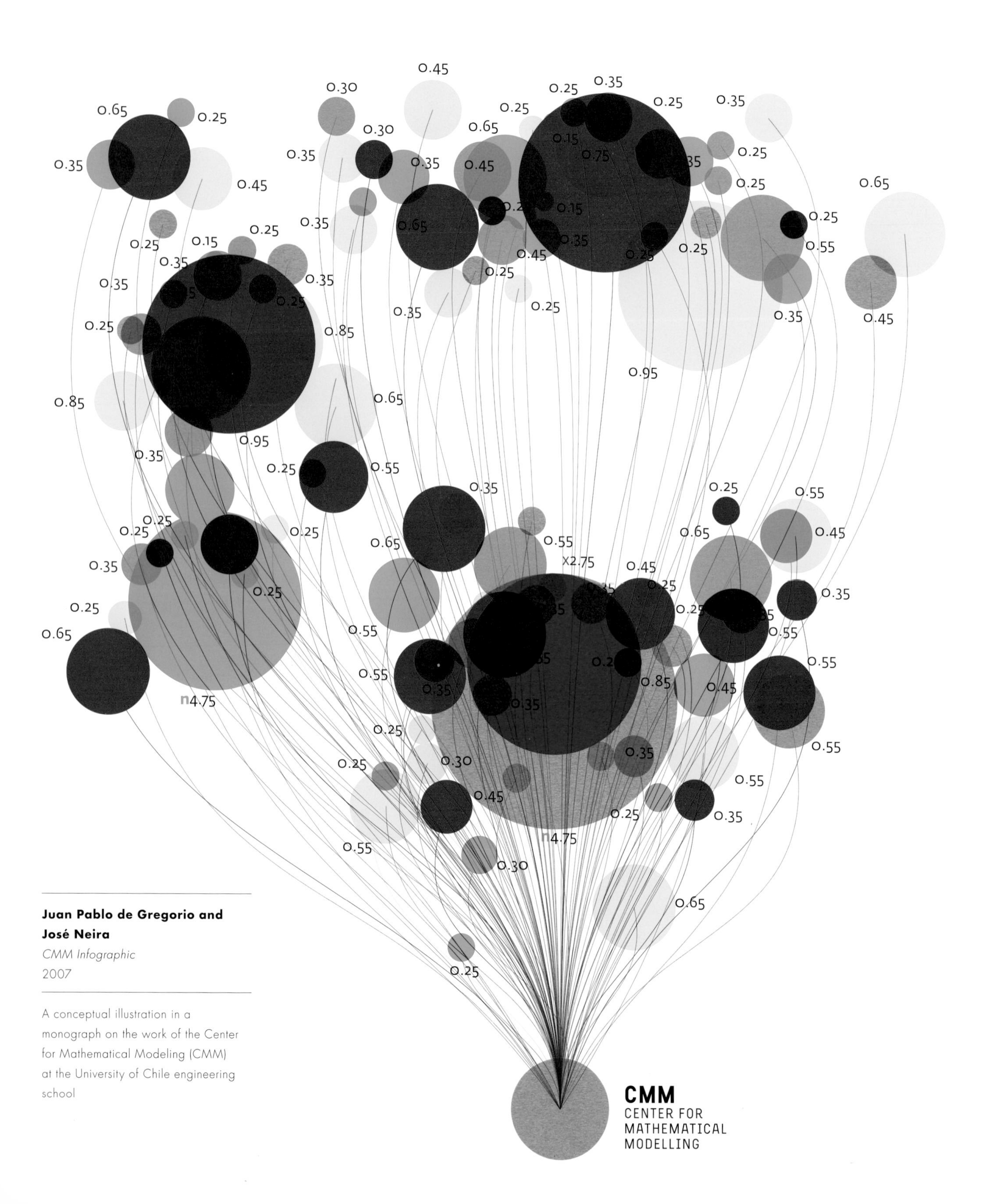

Juan Pablo de Gregorio and José Neira

CMM Infographic

2007

A conceptual illustration in a monograph on the work of the Center for Mathematical Modeling (CMM) at the University of Chile engineering school

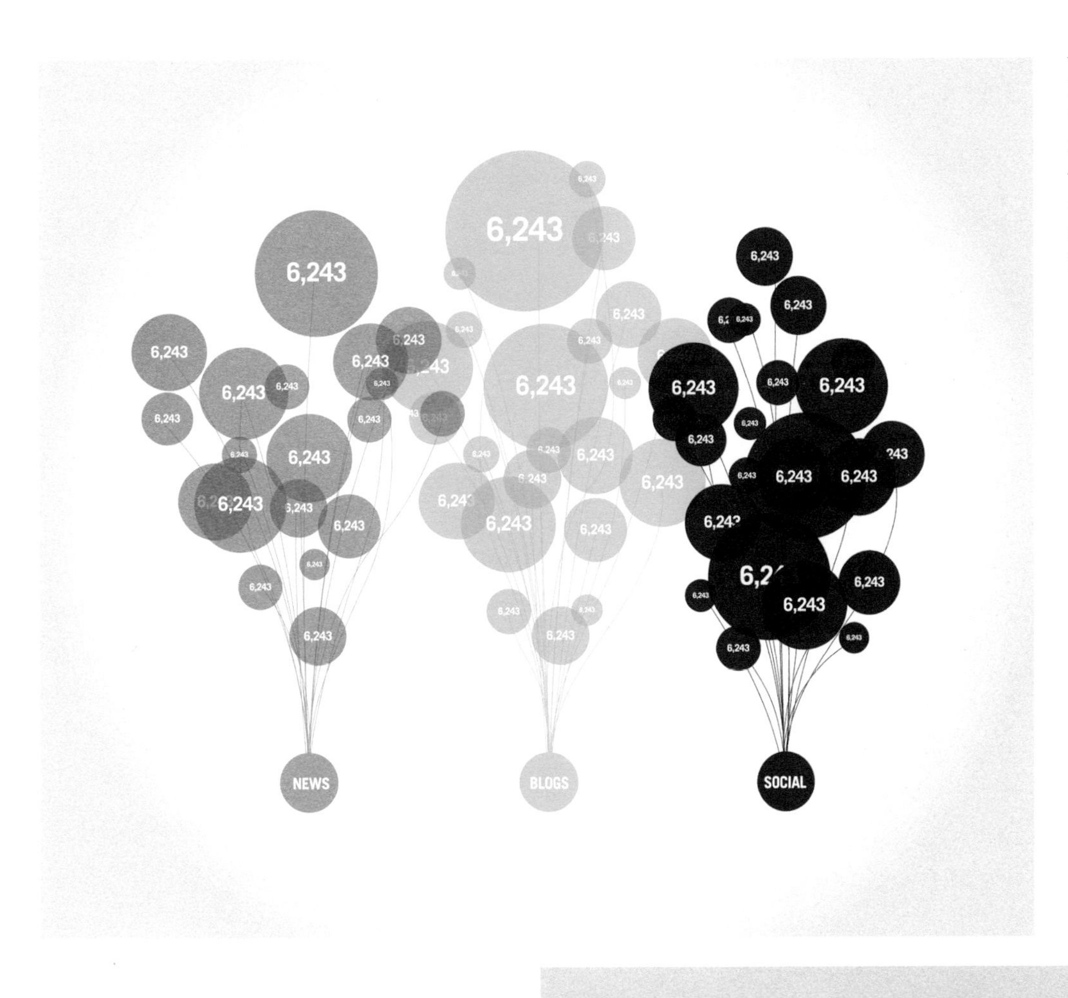

Instrument and Gridplane

Visualizing Online Media

2008

A conceptual visualization based on the idea of aggregating online media buzz from various social-media outlets

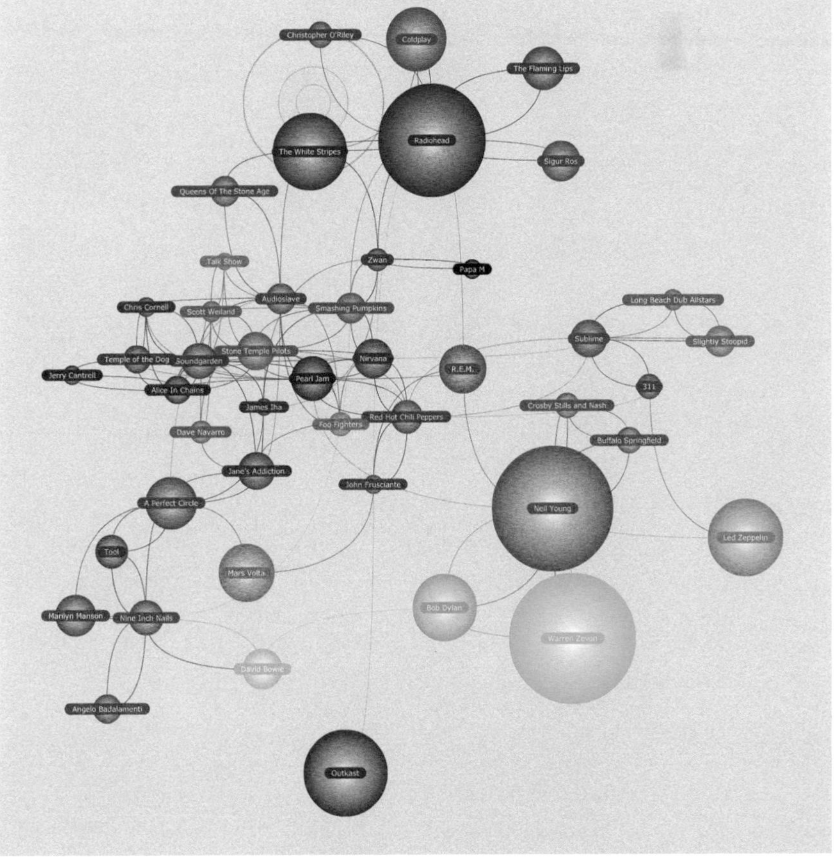

Frederic Vavrille

LivePlasma

2005

A snapshot of an interactive online visualization generated using a visual-discovery engine that maps and displays music and movie search results from Amazon's application programming interface (API). When a search term is submitted, the word appears and is immediately surrounded by related terms, the proximity based on the relatedness of the terms. In this example, the name of an artist, Pearl Jam, was selected. The bands that are similar in genre are closer to the Pearl Jam node.

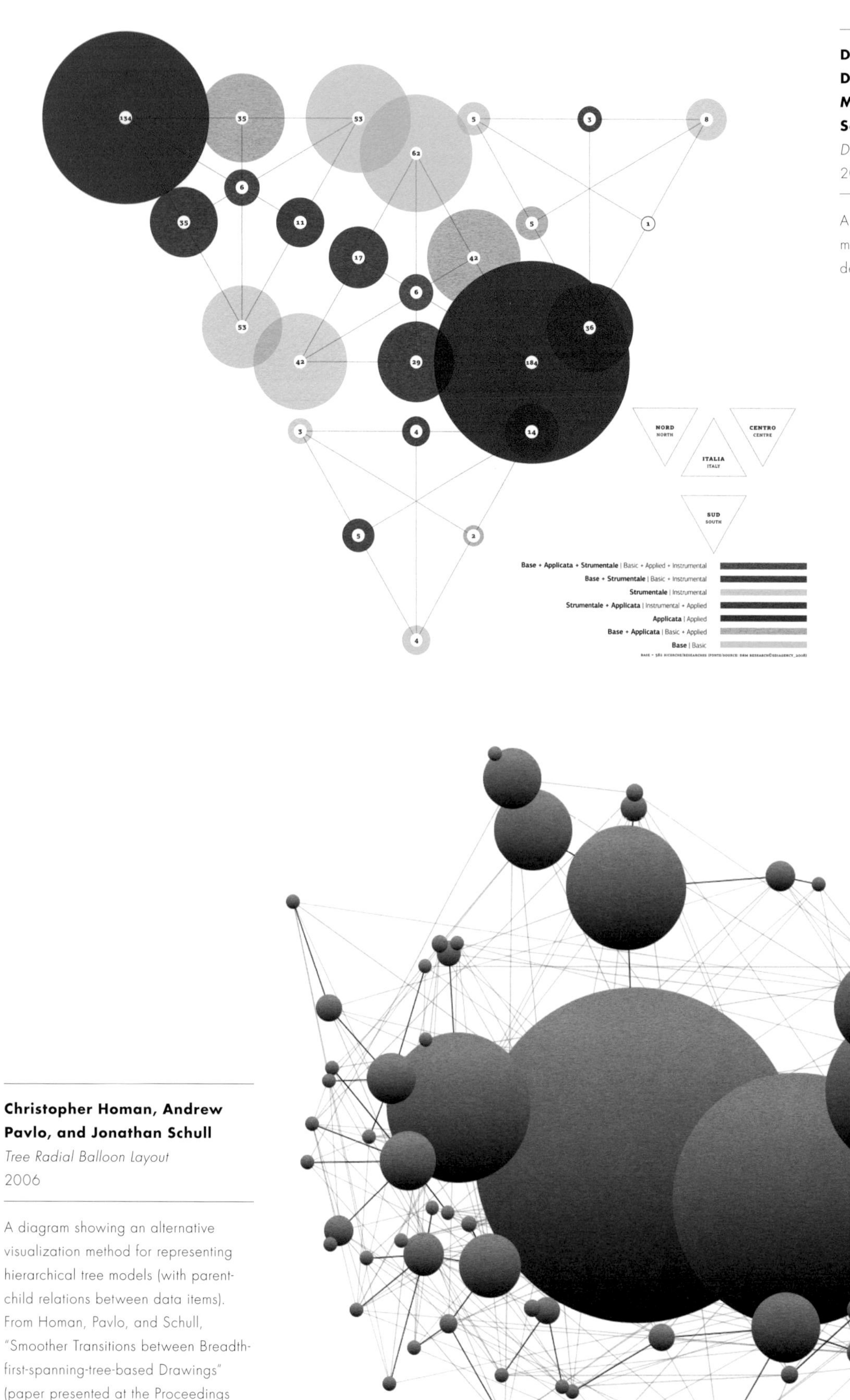

Density Design: Donato Ricci, Daniele Guido, Luca Masud, Mauro Napoli, and Gaia Scagnetti

Design Research Nature

2008

A map of the most common research methodologies used by various Italian design firms

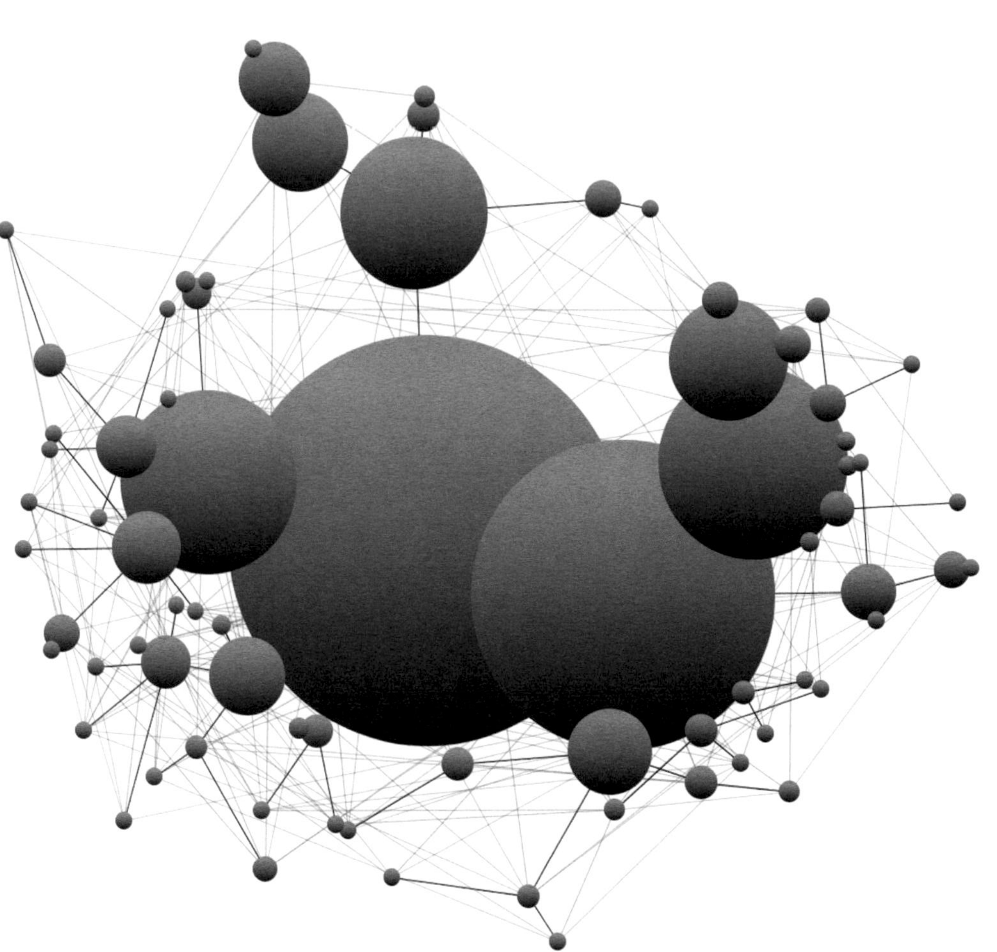

Christopher Homan, Andrew Pavlo, and Jonathan Schull

Tree Radial Balloon Layout

2006

A diagram showing an alternative visualization method for representing hierarchical tree models (with parent-child relations between data items). From Homan, Pavlo, and Schull, "Smoother Transitions between Breadth-first-spanning-tree-based Drawings" (paper presented at the Proceedings of Graph Drawing, 14th International Symposium, Karlsruhe, Germany, September 18–20, 2006).

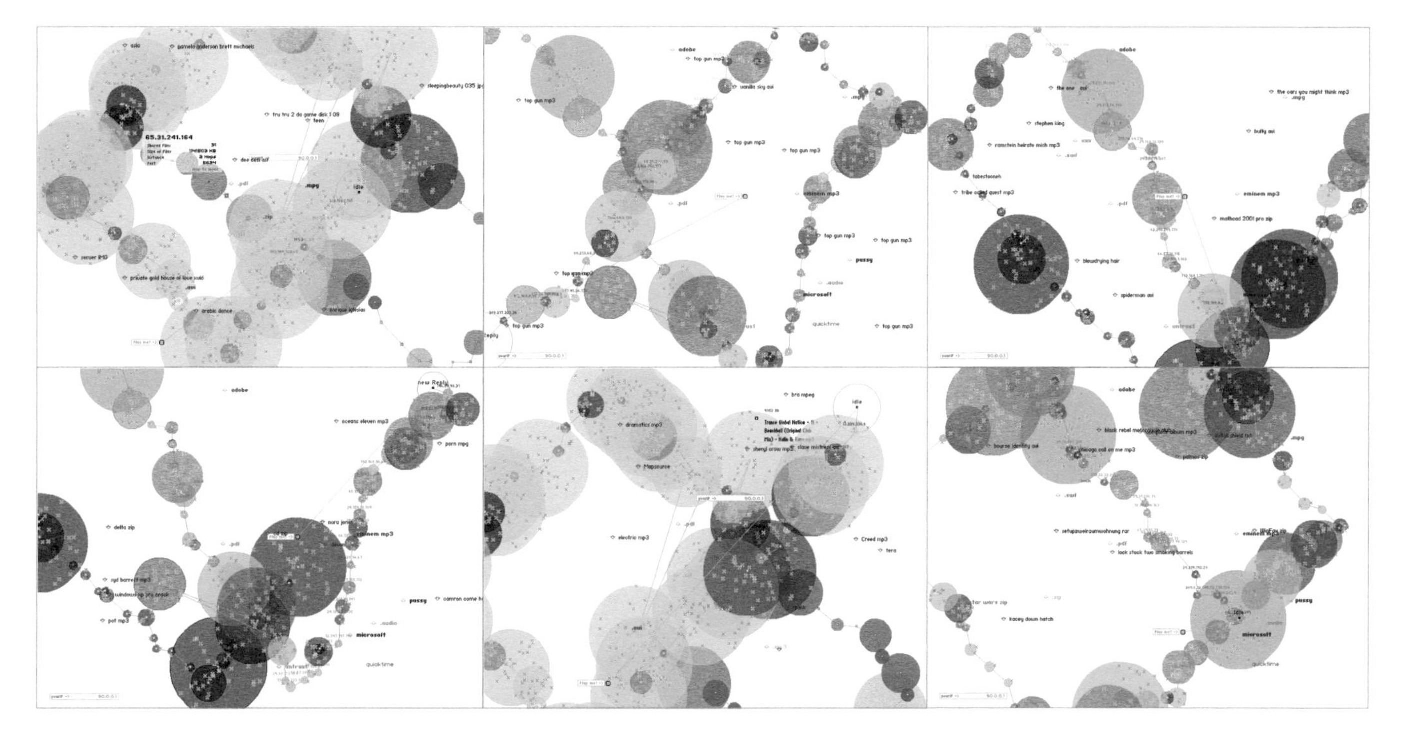

Anne Pascual and Marcus Hauer
Minitasking
2002

A sequence of screenshots from a visualization tool for browsing the peer-to-peer Gnutella network

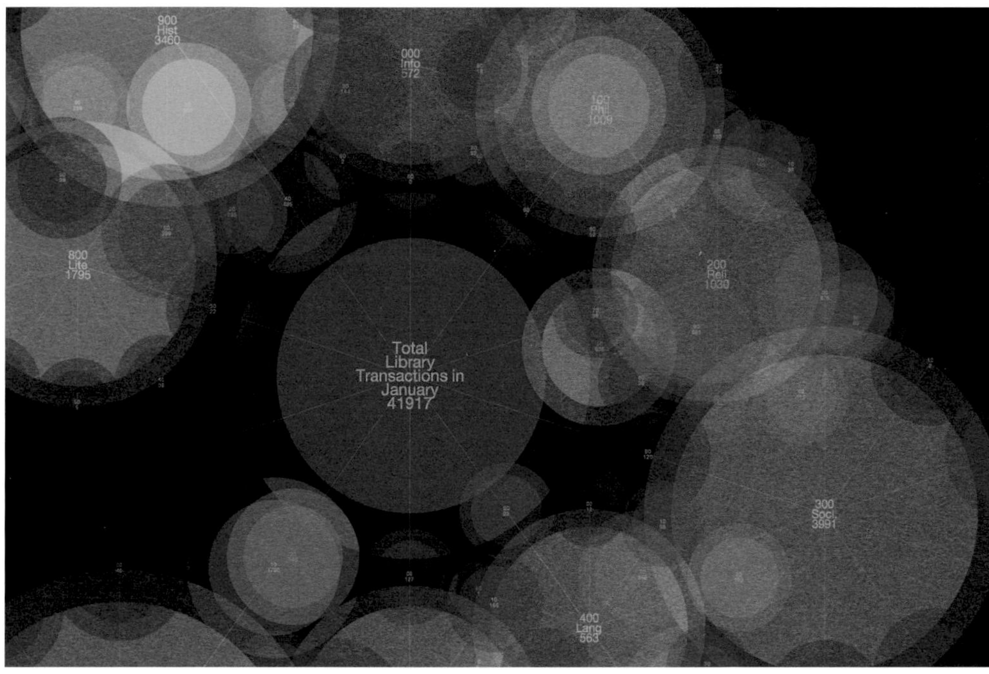

Syed Reza Ali
Dewey Circles
2009

A visualization showing the borrowing patterns of Seattle Public Library patrons using the Dewey Decimal Classification system

Segmented Radial Convergence

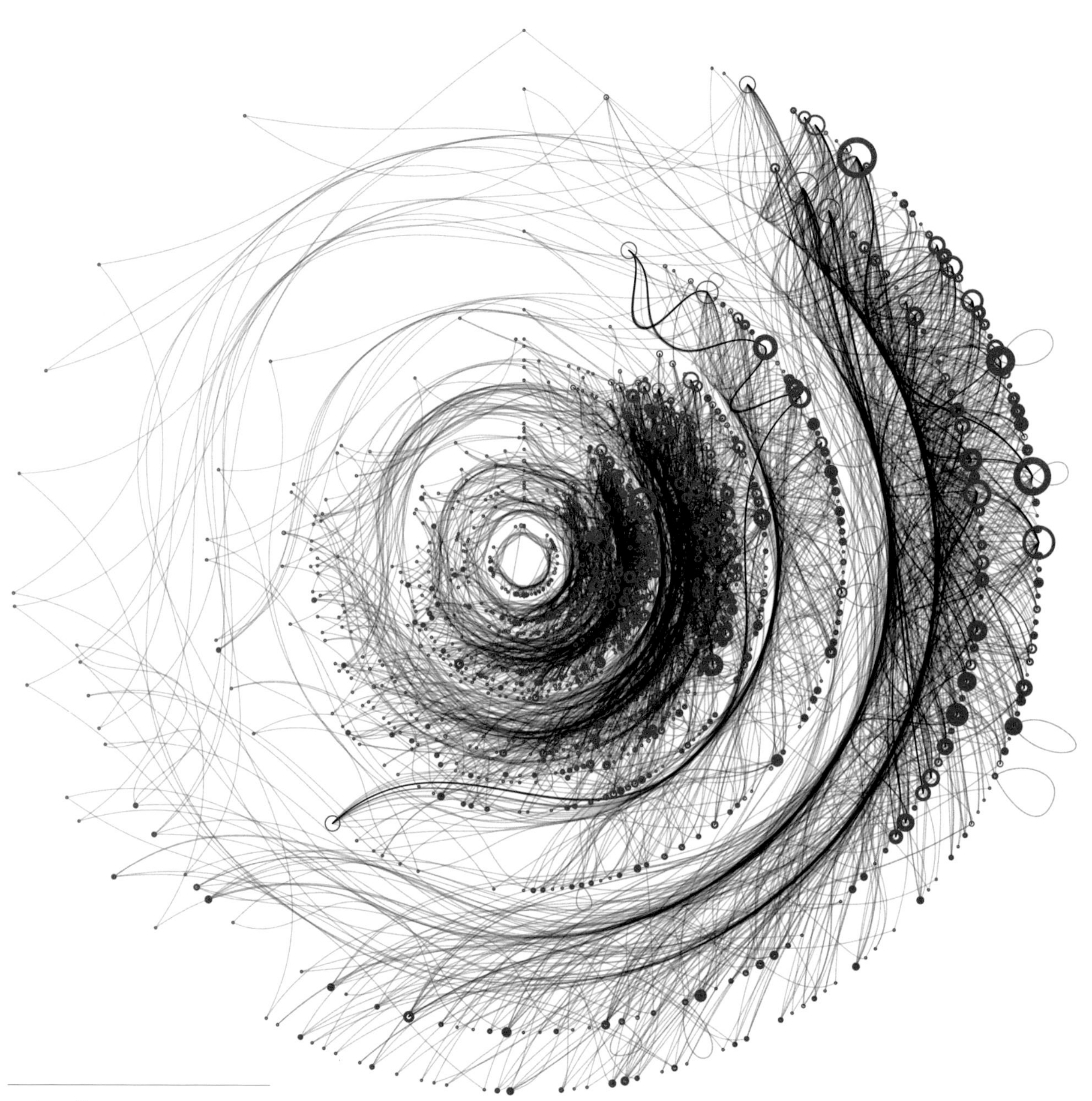

Boris Muller
Poetry on the Road
2006

A visualization of various poems featured on a poster for an international literature festival in Bremen, Germany. A number was assigned to every letter of the alphabet and added together to determine the numerical value of a word, e.g., the word *poetry* would equal 99. Using this system, entire poems were arranged on circular paths. The diameter of the circle is based on the length of the poem: short poems are in the center, while the longer ones form the outer circles. Red rings on the circular path represent a number, and its thickness is directly proportional to the amount of words that share that number. Gray lines connect the words of the poem in their original sequence, occasionally highlighting repetitive patterns in the poem.

Trina Brady

I Wish…

2007

A collection of wishes connecting 191 people (above); detail (right)

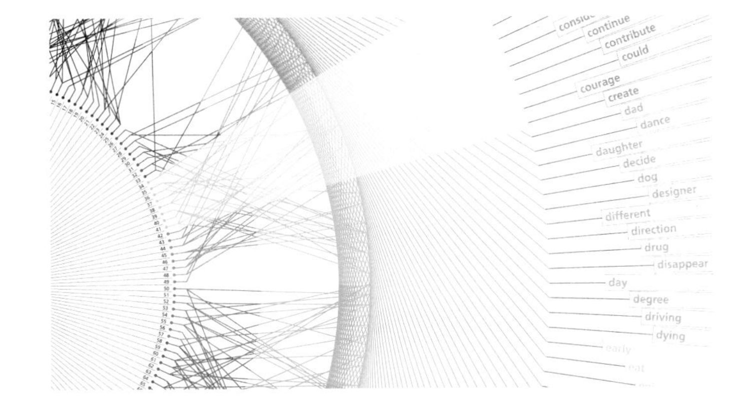

Sebastian Heycke
DOLBY
2005

An interactive visualization of the subject catalog of INCOM, an online-communication platform of the Interface Design department at the University of Applied Sciences Potsdam, Germany. INCOM's existing keywords are mapped alphabetically in a clockwise configuration on a radial structure. All of the articles in the INCOM database are represented by rectangles in the inner circle. One can hover over a keyword to see the names of related articles or over an article to discover an array of associated keywords.

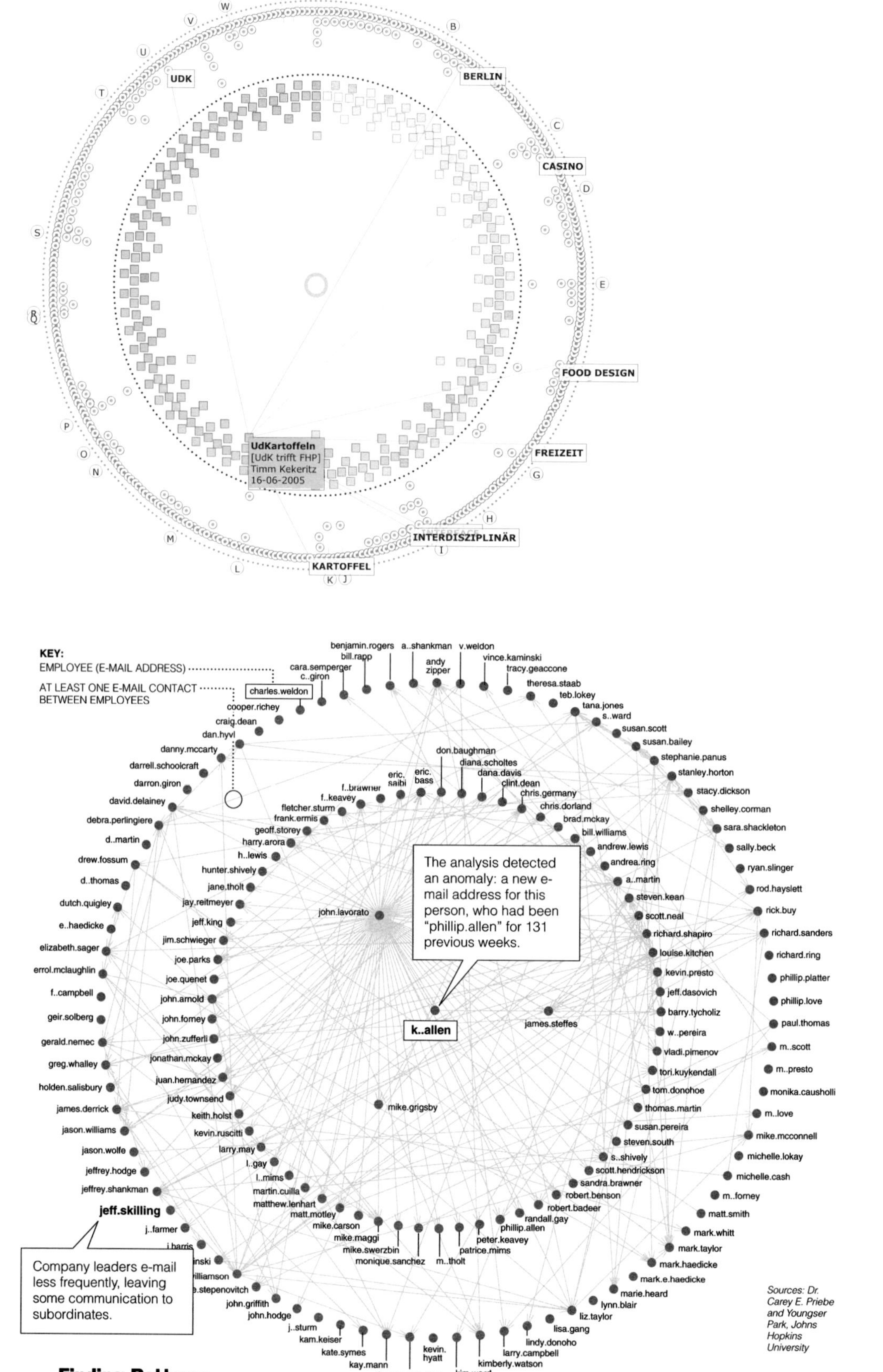

Bill Marsh for *New York Times*
Finding Patterns in Corporate Chatter
2005

A map of the email-exchange patterns of Enron employees during a single week in May 2001

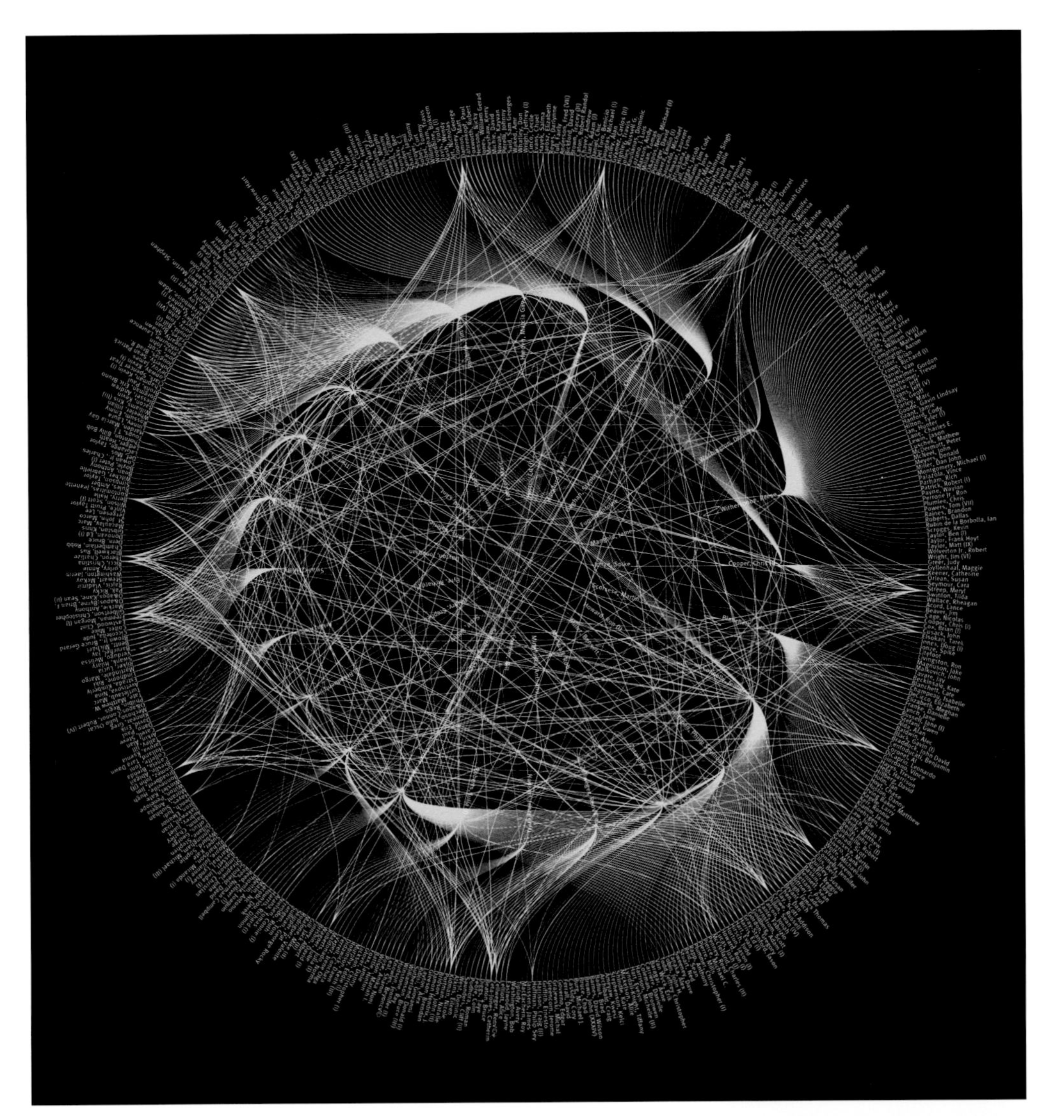

Wesley Grubbs and Nick Yahnke
Patterns in Oscar Movies
2007

A map of working relationships between directors, Oscar-winning actors, and non-Oscar-winning actors, generated using a data set of over ten thousand movie records provided by the IEEE InfoVis 2007 conference, for their annual information-visualization contest. Names of Oscar-winning actors are placed on the intermediate ring, directors they have worked with in the inner ring, and other actors they have worked with in the outer ring. Curved lines were used to draw the working relationships.

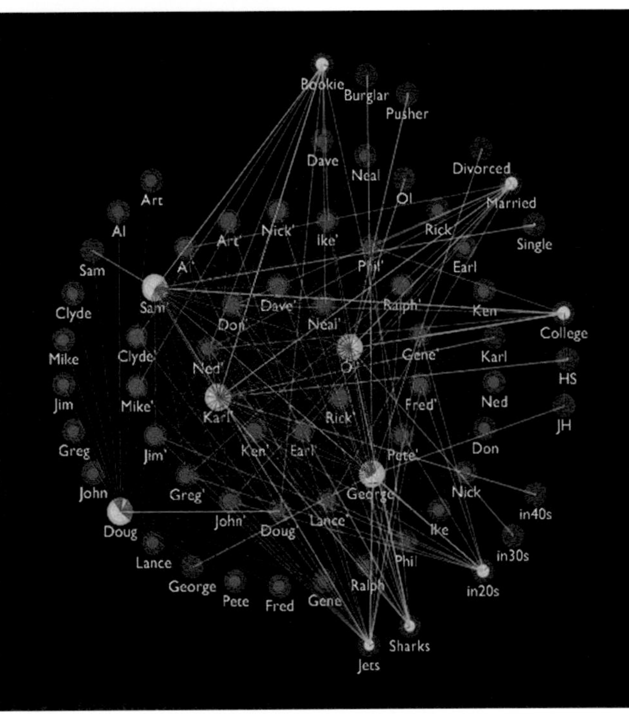

Axel Cleeremans
Interactive Activation
2004

A demonstration of an Interactive Activation and Competition network—neural networks made up of nodes or artificial neurons and used to model memory—as first described by J. L. McClelland in his paper "Retrieving General and Specific Information from Stored Knowledge of Specifics," *Proceedings of the Third Annual Conference of the Cognitive Science Society*, 1981, 170–72. Hillsdale, New Jersey.

Sphere

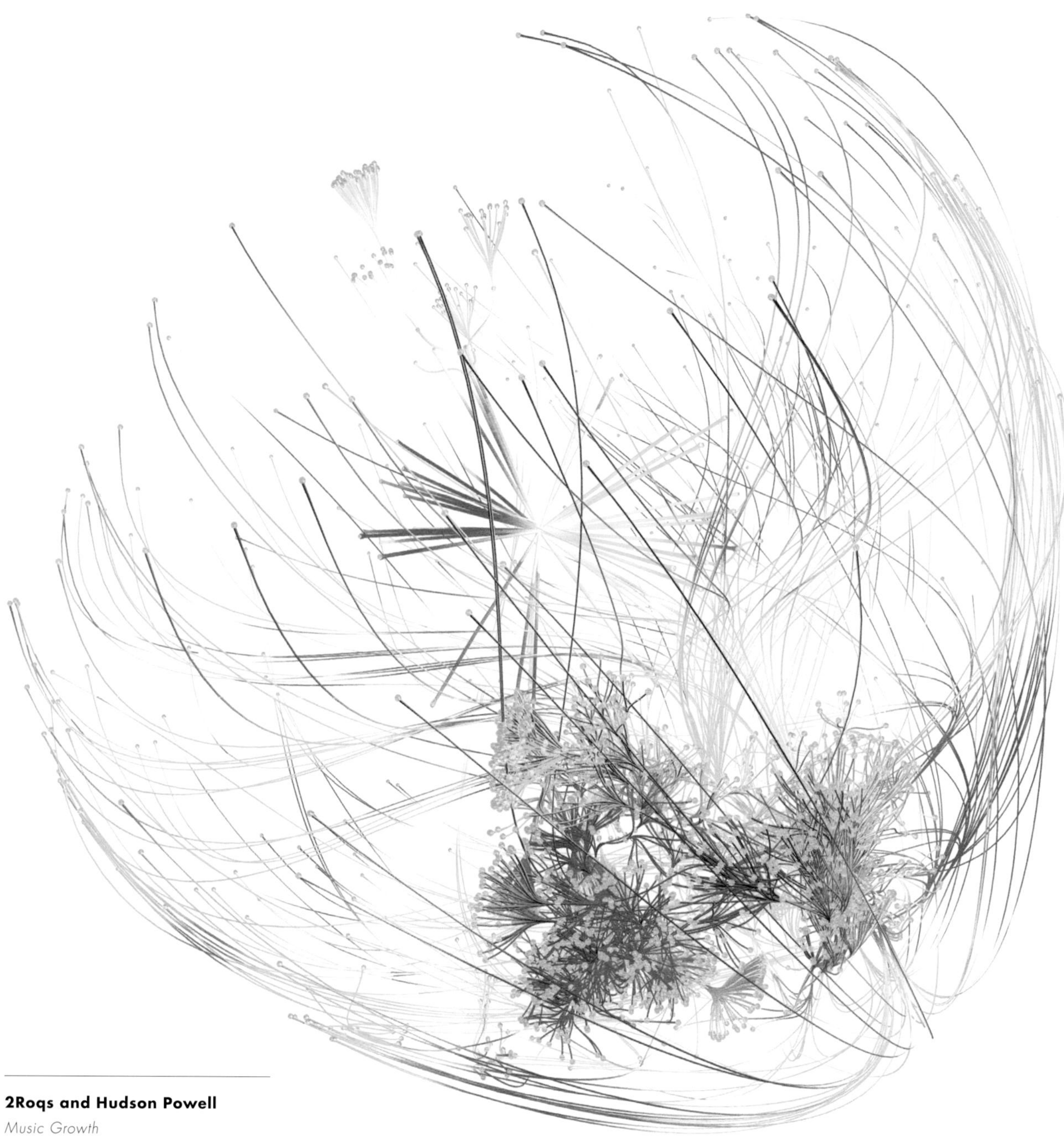

2Roqs and Hudson Powell
Music Growth
2006

A visual representation of musical pieces with audio-frequency patterns feeding the structure's growth. Made for Barbican, an art venue in London, for its *Great Performers* session booklets.

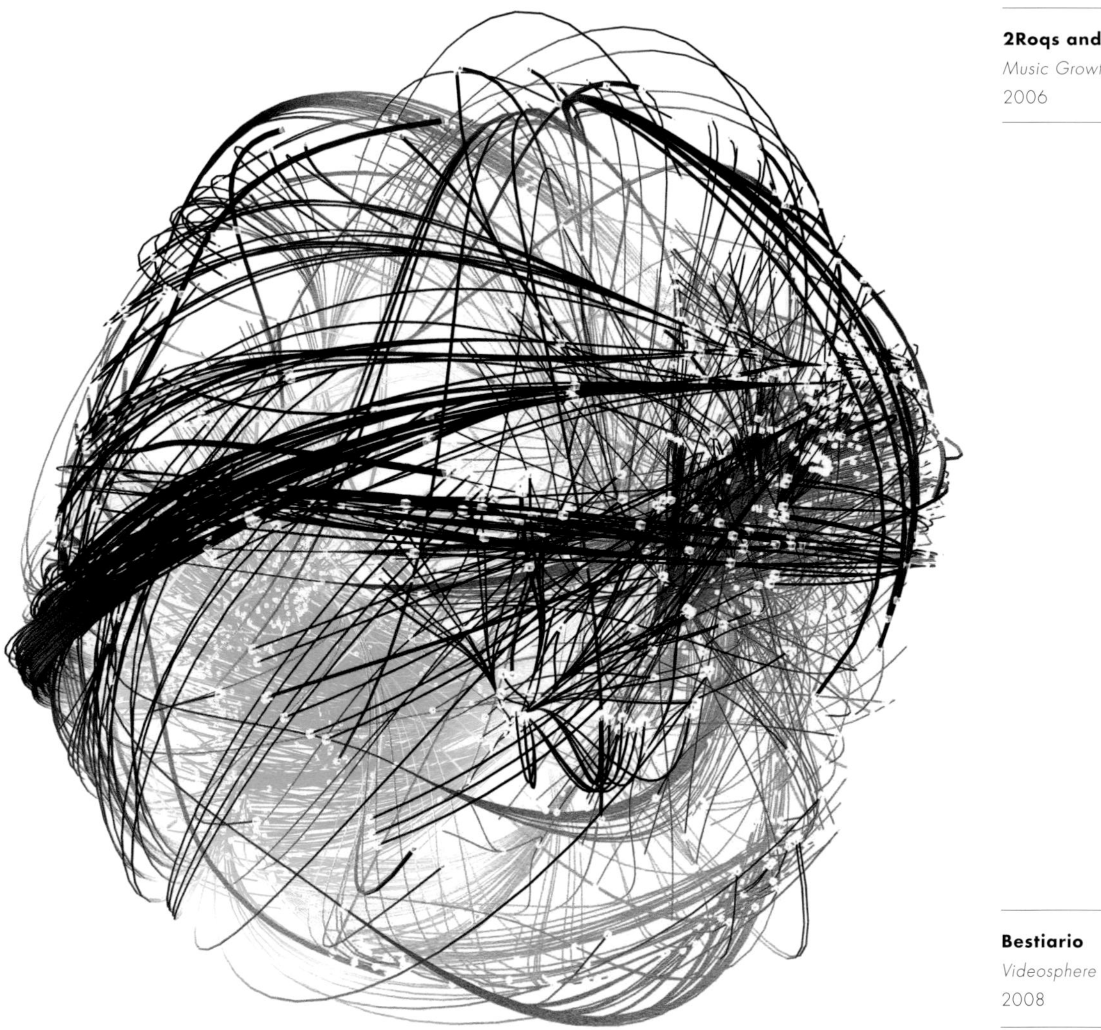

2Roqs and Hudson Powell
Music Growth
2006

Bestiario
Videosphere
2008

A three-dimensional network visualization exploring semantic overlaps between video-recorded talks at the Technology Entertainment and Design (TED) conference.

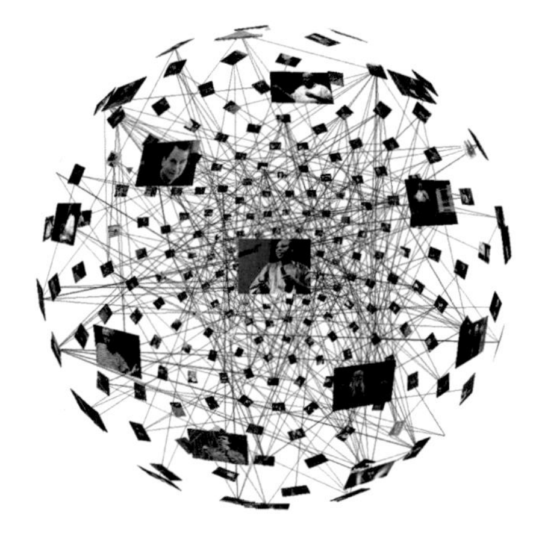

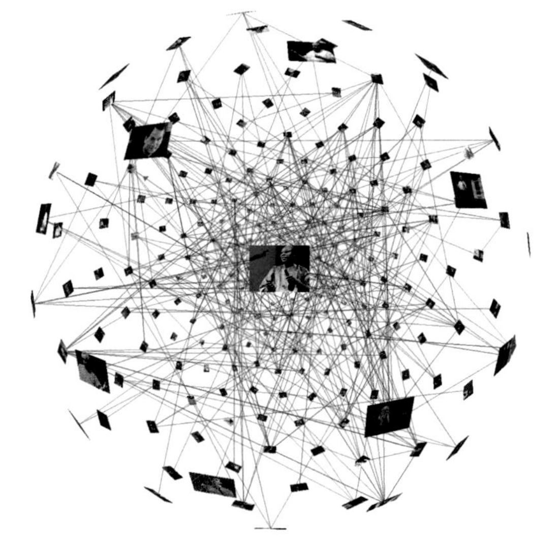

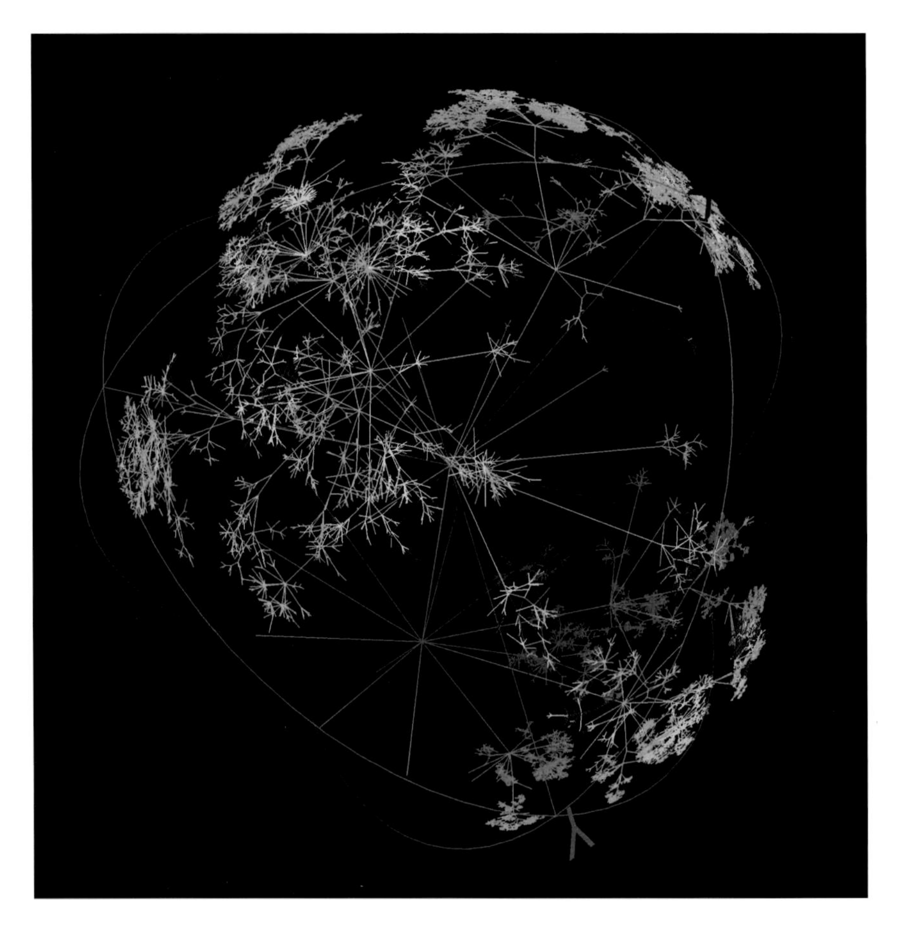

CAIDA
Walrus
2001

A conceptual, prototype visualization of abstract data outputted using Walrus, a tool for interactively visualizing large graphs in three-dimensional space. Walrus uses a common focus and context technique, allowing the user to interact with the graph by selecting a node, which smoothly moves to the center of the display and enlarges to enable the user to view fine details.

Scott Hessels and Gabriel Dunne
Celestial Mechanics
2005

A planetarium visualization revealing the location of many of the technologies hovering, flying, and drifting above us

James Spahr
Website Traffic Map
2003

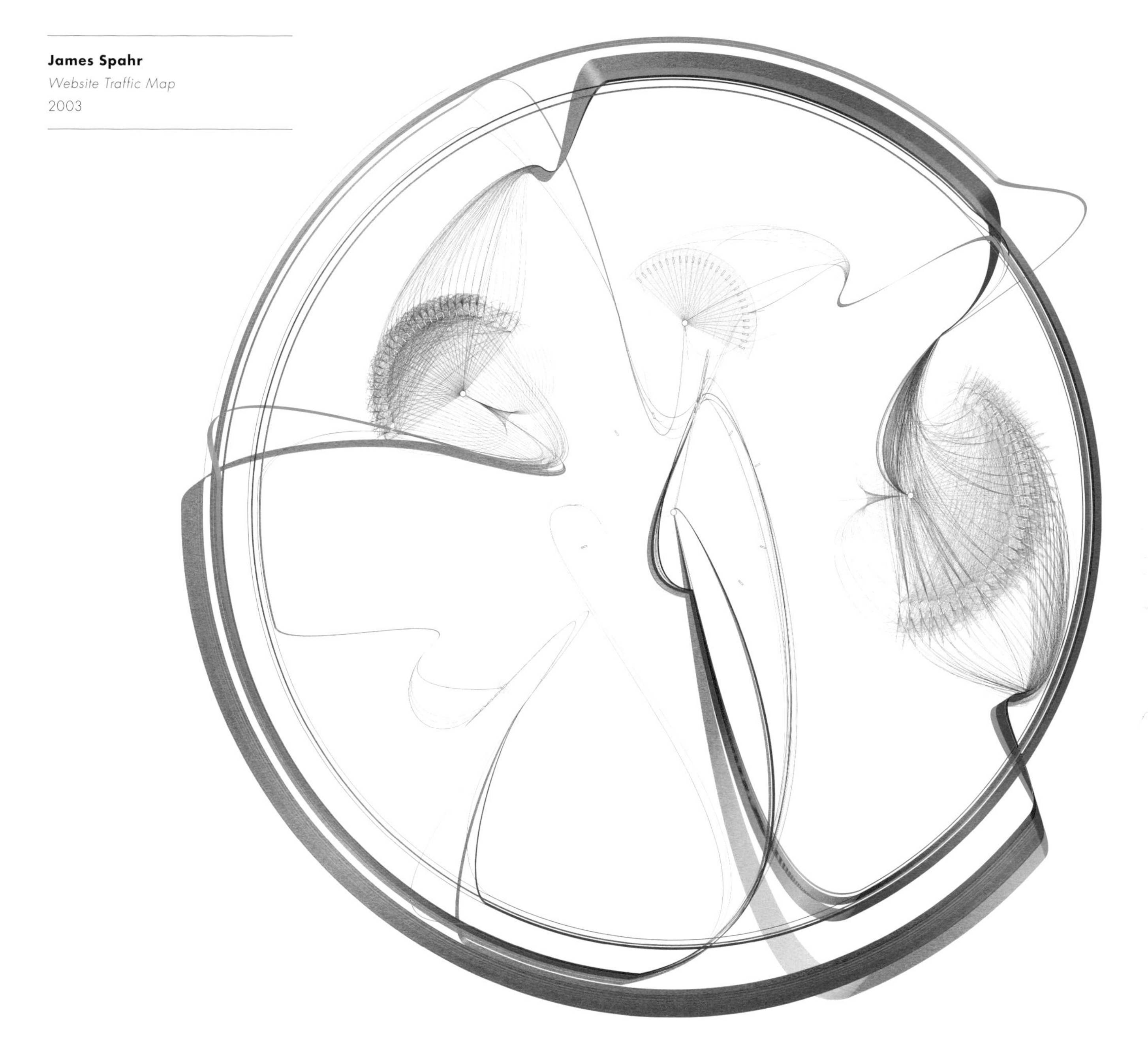

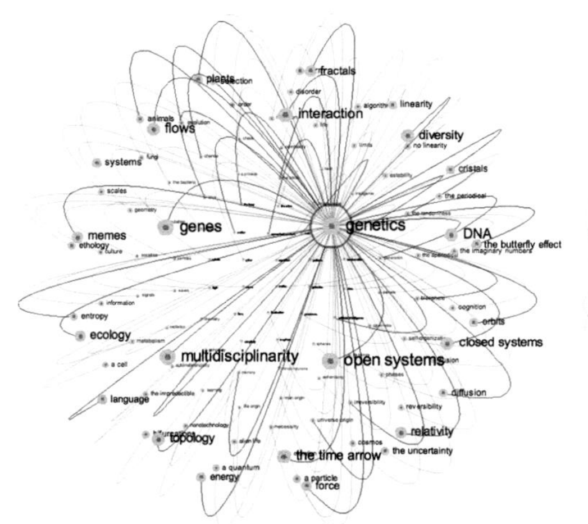

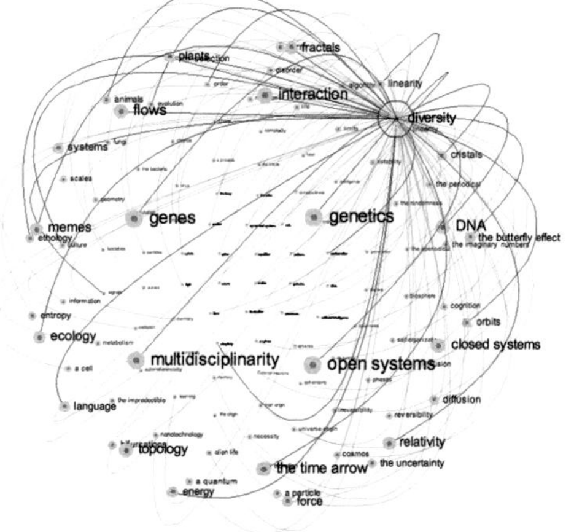

Santiago Ortiz
Spheres—Spherical Surface of Dialogue (English)
2005

The semantic network of 122 common words in the domain of complexity science

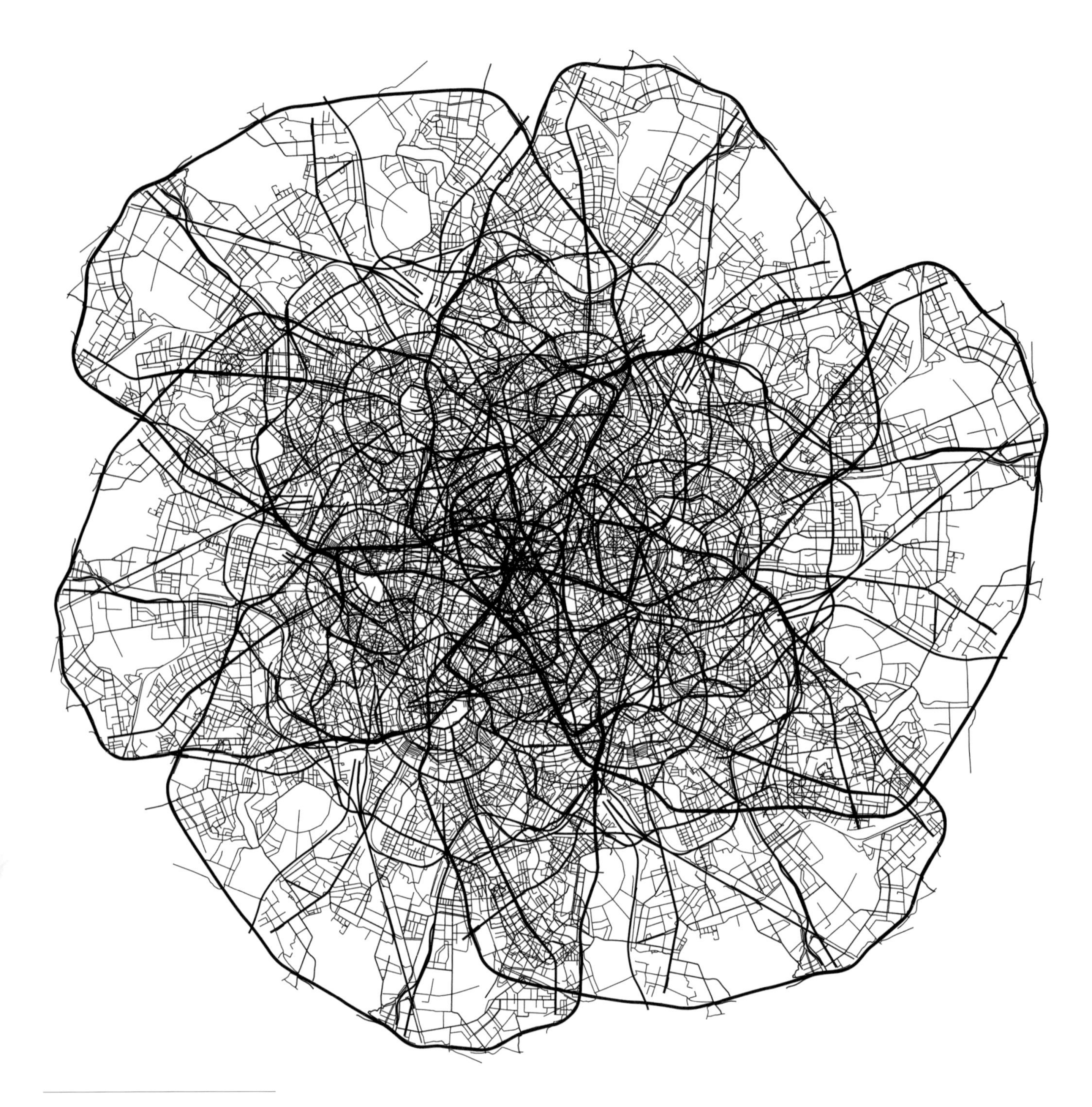

Jang Sub Lee, *Complexcity*, 2008

A graphic composition of urban-road patterns in Moscow

06 | Complex Beauty

Nature is an infinite sphere of which the center is everywhere and the circumference nowhere.

—Blaise Pascal

For the harmony of the world is manifest in Form and Number, and the heart and soul and all the poetry of Natural Philosophy are embodied in the concept of mathematical beauty.

—D'Arcy Wentworth Thompson

Complex networks are not just omnipresent, they are also intriguing, stimulating, and extremely alluring structures. Networks are not just at the center of a scientific revolution; they are also contributing to a considerable shift in our conception of society, culture, and art, expressing a new sense of beauty. As we continuously strive to decipher many of their inner workings, we are constantly bewildered by their displays of convolution, multiplicity, and interconnectedness. And the most elaborate of schemes are the ones that apparently seduce us at the deepest level. So what are the enthralling properties of complexity? Why do we feel so innately drawn to these structures? To try and answer these questions, we will investigate the various trajectories of complexity, alternating between a scientific and artistic viewpoint, before culminating in a contemporary artistic expression that truly embodies a redefined notion of complex beauty.

The two epigraphs to this chapter are drawn from Pascal, *Pensées*, 60; and Thompson, *On Growth and Form*, 326.

Holism

It is difficult—some would say impossible—to measure and understand the aesthetic appeal of art. Some artists altogether oppose the pursuit, as John D. Barrow articulates in his essay "Art and Science—Les Liaisons Dangereuses" (2003): "Most artists are very nervous of scientific analysis. They feel it destroys something about the human aspect of creativity. The fear (possibly real) of unsubtle reductionism—music is nothing but the trace of an air pressure curve—is widespread."[1] The realm of aesthetics deals, after all, with emotional values and is never a straightforward territory for investigation. Many attempts at creating a reliable framework of analysis have been derailed, but that does not mean others have stopped trying to deconstruct various elements of aesthetics.

When one looks at a painting or visual representation, one immediately perceives certain characteristics, such as composition, balance, symmetry, contrast, or color. These qualities themselves are made up of smaller building blocks—what Donis A. Dondis, professor of communication at Boston University School of Public Communication, called the "skeletal visual forces" in her eminent *A Primer of Visual Literacy* (1974). These are easily recognizable in the form of a dot, line, color, shape, direction, texture, scale, dimension, and motion. According to Dondis, these components "comprise the raw material of all visual information in selective choices and combinations, and are chosen according to the nature of what's being designed—the final aim of the piece."[2]

But when it comes to expressing particular intentions, the mere appliance of these individual elements is not enough. This is why Dondis provides a complementary inventory of visual methods, or recipes, of how to combine these ingredients—a "wide palette of means for the visual expression of content."[3] These are some of the pairings presented: balance and instability, symmetry and asymmetry, regularity and irregularity, predictability and spontaneity, subtlety and boldness, neutrality and accent, transparency and opacity, accuracy and distortion, flatness and depth, sharpness and diffusion. Dondis further describes each paired approach with a series of visual examples and short recommendations for its appropriate use. A seminal work, used in art and design classes across the globe, Dondis's study provides a set of communication-design patterns, cohesive guidelines for building the most suitable visual composition for any given intent.

It is important to understand such communication strategies by scrutinizing the different methods for reaching a particular goal. However, this reductionist effort can only tell us half of the story. When it comes to painting, literature, music, cinema, or other artistic expressions, the whole is always more than the sum of its parts.

This dominant notion of wholeness, which seems commonplace today, only matured in the early twentieth century, when a group of researchers from the Berlin School of Experimental Psychology set the foundations for what would become known as Gestalt psychology. This innovative school of thought emphasized the idea of wholeness, asserting that the operational principle of the brain was ultimately holistic. Gestaltists reinforced the belief that human experiences cannot be derived from the summation of perceptual elements, because they are ultimately irreducible. An outcome of their experiments, the Gestalt effect describes a cognitive process in which the visual recognition of shapes and forms is not based on a collection of elements—dots and lines—but on seeing them as a fully identifiable pattern: from recognizing a zebra in a grassland, to distinguishing a face in the crowd.

But Gestaltism had a much wider influence outside the walls of perceptual psychology. Its theoretical base fostered a larger shift in science and in society at large, under the name of holism. While reductionism asserts that any system can be explained by reducing it down to its most fundamental elements, holism expresses the opposite, believing that the whole is ultimately irreducible. Even though this general principle can be traced back to Aristotle, it became a fundamental school of thought in twentieth-century science. It helped to explain newly discovered natural systems, in which the whole was consistently made of interconnected and interdependent parts, and where no one unit of a system could be altered without modifying the whole.

Gestalt theory is still today an immensely valuable resource in disciplines dealing with objective visual communication, such as graphic design, interface design, and information design. The Gestalt principles of form perception, developed by Gestaltists over the years, constitute great recipes for how to better employ visual elements while also pointing to their ability to easily fool the cognitive system. However, when it comes to complex nonfigurative representations, Gestalt falls short in two key aspects. First, it is always prescriptive and never explanatory: it may describe that we are able to see a dog in a particular arrangement of dots, but it does not explain how the percept of a dog emerges in our mind. Second, Gestalt addresses the form-forming capability of our senses, the way we visually identify familiar figures and whole forms. So what happens when we are unable to identify a pattern?

Humans have a remarkable aptitude for pattern recognition, and some might argue that the idea of aesthetics relies on the satisfying reward of identifying a figure or form in a particular portrayal. The brain's propensity to look for familiar shapes is so strong that we even tend to find meaningful patterns in meaningless noise, a type of behavior usually called patternicity. Philosopher Karl Popper suggests that people are born with a certain innate proclivity for regularity, "an inborn propensity to look out for regularities, or with a *need* to *find* regularities."[4]

So what happens if that sense of order is unstable and regularity imperceptible? When complexity exceeds a "sustainable" level? Does our perceptual system become clogged, preventing us from appreciating the specific target of our attention? Architect and lecturer Richard Padovan elaborates on this delicate balance, by paraphrasing art historian Ernst Gombrich: "Delight lies somewhere between boredom and confusion. If monotony makes it difficult to attend, a surfeit of novelty will overload the system and cause us to give up; we are not tempted to analyze the crazy pavement."[5] If this is really the case, how can we explain our attraction to particularly convoluted nonfigurative subjects? Ultimately, how can we explain the alluring qualities of complexity? Well, to shed new light on this topic, we need to resort to a different type of analysis.

Complexity Encoding

When talking about intricate depictions of complex networks, there is a recurrent association with one particular art movement: abstract expressionism. As one of its main figures, Jackson Pollock was the key perpetrator of a subgenre of abstract expressionism called action painting, or gestural abstraction, which describes the act of dribbling, leaking, or splashing paint onto the canvas. His intricate and dynamic drip paintings have attracted many enthusiasts over the years and have decisively defied the mold of most traditional views on art and artistic execution. fig. 1

Pollock's drip paintings evoke large-scale views into undisclosed networked systems, where singularity is lost in the dense forest of interconnections. These intricate landscapes express a variety of dynamic patterns and organic textures that resemble those found in natural systems. fig. 2 The similarity is certainly not accidental. "My concern is with the rhythms of nature....I work inside out, like nature," Pollock declared. [6] But it is in Pollock's process that we find the closest affinities with nature, explains Richard Taylor, a physicist who has investigated many of Pollock's pieces. In the paper "Fractal Expressionism—Where Art Meets Science" (2003), Taylor describes the interesting overlap between nature and Pollock's paintings and provides five key traits of Pollock's method that reveal a unique instinctive character: (1) Cumulative layering: similar to the way leaves fall over time or erosion patterns are created, Pollock's paintings are made of a succession of strata; (2) Variation in intensity: just as weather changes over time, assuming a bright blue sky or a dark intense storm, so do Pollock's patterns, both in color and energy; (3) Horizontal plane: Pollock's canvas was the terrain—his dripping technique exploited gravity in the same way nature builds patterns; (4) The large canvas: this was seen by Pollock as an environment, which by ignoring the canvas edges resembles "Nature's expansive and unconfined patterns"; (5) Ongoing process: the cyclical routine of nature is illustrated by Pollock's habit of leaving a painting resting for a long time and occasionally revisiting it—he was a great enthusiast of the "continuous dynamic" method.[7]

In his innovative analysis of Pollock's convoluted patterns and trajectories generated by his drip paintings, Taylor goes much further and suggests the evidence of several fractal properties. Fractals are recurrent patterns found in nature that express self-similarity—any reduced part of the entity is the same in proportion to the whole—like never-ending repetitions of identical geometrical motifs. According to Taylor, Pollock's fractality appears in two visual modes: "fractal scaling" and "fractal displacement." While the former relates to a pattern that employs the same repetitious motifs at different scales or magnifications, the latter refers to the use of the same motifs at different spatial locations.

The inherent fractal features of Pollock's paintings might in part explain the captivating qualities of his work. "Nature builds its patterns using fractals as its basic building block," notes Taylor. "Having evolved surrounded by this fractal scenery, it perhaps is not surprising that humanity possesses an affinity with these fractals and an implicit recognition of their qualities. Indeed, it is possible to speculate that people possess some sort of 'fractal encoding' within the perception system of their minds."[8] fig. 3, fig. 4, fig. 5

Are we in the possession of some type of perceptual fractal encoding? And could the same encoding be extended to other natural patterns? If this holds true, it could well explain why we are so allured by depictions of complex networks. After all, it has been proven that networks are a ubiquitous topology in nature, and a type of encoding similar to fractal encoding might exist in our minds. The fact that as you read this sentence you are using your own vastly interconnected network of neurons alludes to this plausible cognitive *complexity encoding*. Perhaps we have a propensity for structures similar to our own brain—at its

fig. 3

Carden and Coast, *London GPS Tracking Map*

User-generated GPS traces of London, used as a part of the OpenStreetMap project, a large collaborative proposal aimed at creating a free, editable map of the world. (See also chapter 4, page 148.)

fig. 1

Jackson Pollock, *Autumn Rhythm: Number 30*, 1950, oil on canvas.

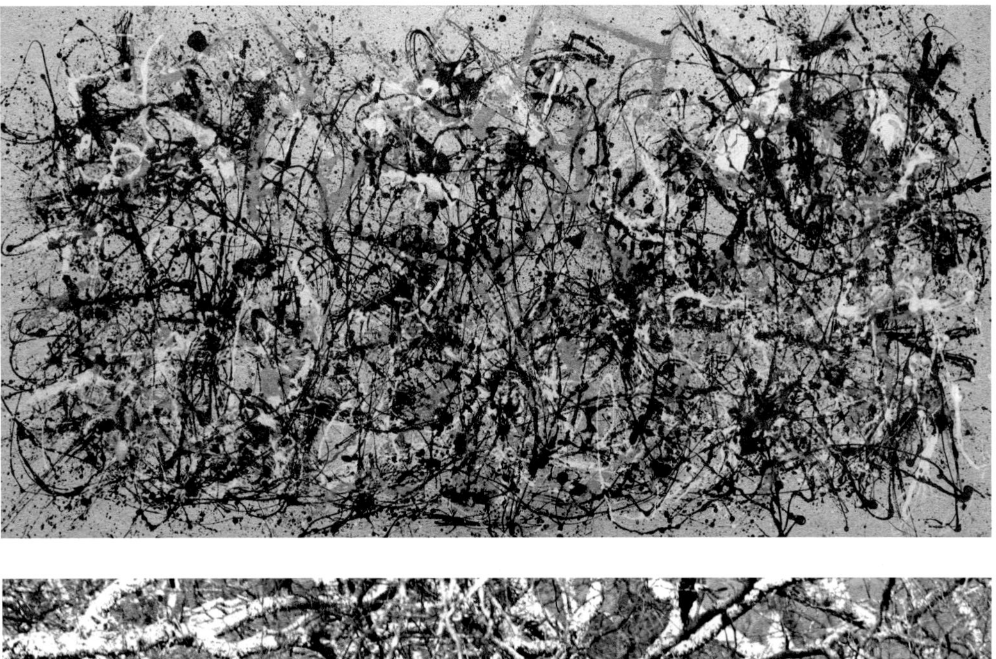

fig. 2

Graciela Blaum, *Oaks*, 2007, photograph

Oak trees tangled up in Calero Park, San Jose, California

fig. 4

Richard P. Taylor, *Pollock's Traces*, 2006

A schematic version of Jackson Pollock's poured trajectories

fig. 5

Paul De Koninck Laboratory, Dissociated culture of rat hippocampal neurons, 2005

A detailed view into a rat's neuronal network. (See also page 230, figure 13.)

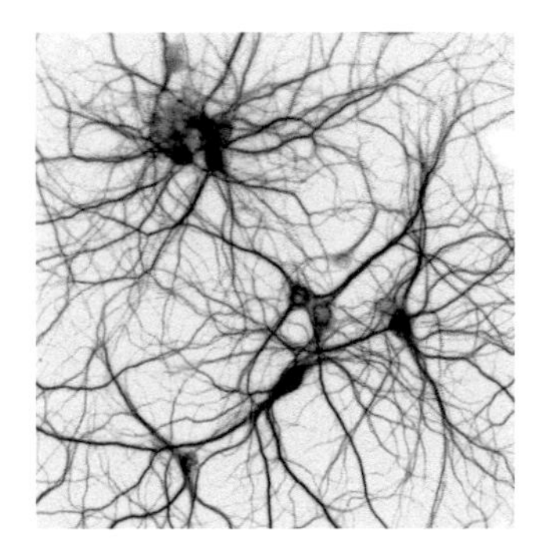

most basic cellular level—turning us into victims of a dopaminelike neurotransmitter every time we view systems that roughly resemble a neural network. fig. 6, fig. 7

Most of these queries and assertions are speculative by nature, and at this stage there is hardly a more reliable framework that can explain this appeal. As a matter of course, the outcomes of ongoing efforts may clarify this elusiveness. Aesthetic judgment has always been seen as an unempirical domain, but many researchers are striving to quantify and understand it. Whereas the growing field of computational aesthetics facilitates the analysis of aesthetic perception by means of computer-driven automated tools, the emergent subfield of empirical aesthetics, known as neuroesthetics, recognizes aesthetic perceptions at the neurological level. Even if significant discoveries arise from these innovative approaches, it is apparent that many of the obstacles preventing a new understanding of aesthetics are not related to technological or scientific limitations but to a widespread resistance to alternative concepts of beauty.

Ordered Complexity

In "Regularities and Randomness: Evolving Schemata in Science and the Arts" (2003), American physicist and Nobel Prize winner Murray Gell-Mann notes that computer scientist, psychologist, and economist Herbert Simon studied the path made by ants, which may appear complex at first, but when "we realize that the ant is following a rather simple program, into which are fed the incidental features of the landscape and the pheromone trails laid down by the other ants for the transport of food, we understand that the path is fundamentally not very complex."[9] A parallel phenomenon, known as swarm behavior, is observable in a variety of animals, from insects like bees and termites, to birds (flocking), fish (schooling), and even people (crowds). figs. 8–9

The highly synchronized movement of large groups of animals, sometimes forming perfectly aligned patterns—as in the case of bird flocking—is an intriguing occurrence, particularly since it subverts our assumptions that some type of hierarchy or centralized control is taking place. It turns out that the highly complex performance is not based on any chain of command, but on very simple local rules, followed by every animal in the group. Most swarm behaviors follow three basic directives: separation—don't crowd your neighbors (short-range repulsion); alignment—maneuver in the average direction neighbors are moving toward; cohesion—steer toward average position of neighbors (long-range attraction). This set of directives forms, in essence, a very simple recipe, manifested in various natural systems and able to create gripping complexity out of simplicity.

Our knowledge of swarming has grown considerably in the last two decades, in part due to our ability to replicate collective animal behavior in various computer simulations, initially developed by computer graphics expert Craig Reynolds in 1986. Reynolds's work was an important contribution to the field of computation evolution, which encompasses a series of computational techniques and algorithms based on evolutionary processes of biological life, but it also became a popular subject of interest for generative art.

Emerging in the 1960s, generative art refers to art created or constructed by means of computer algorithms, usually employing randomized autonomous processes. Arguably the most suitable art practice to illustrate the intricate and cumulative building of natural patterns,

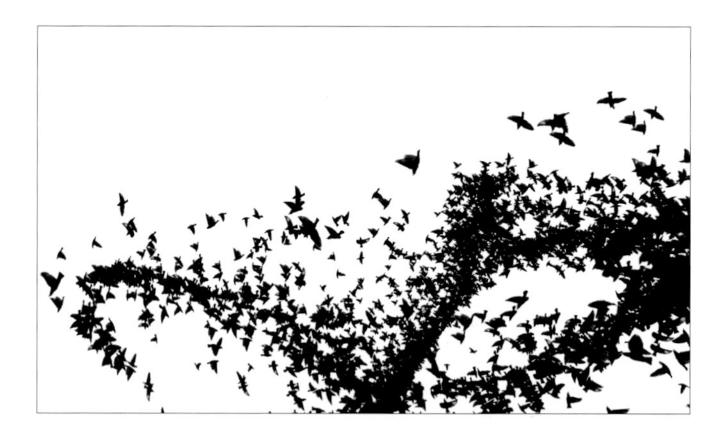

fig. 6

Jackson Pollock, *Number 5*, 1948, oil, enamel, and aluminum paint on fiberboard

fig. 7

Ecole Polytechnique Fédérale de Lausanne, *Blue Brain Project*, 2008

A computer-generated model produced with IBM's Blue Gene supercomputer shows the thirty million connections between ten thousand neurons in a single neocortical column—arguably the most complex part of a mammal's brain. The different colors indicate distinctive levels of electrical activity. (See also chapter 2, page 53.)

figs. 8–9

Robert Hodgin, *Birds*, 2007

A computer simulation of bird-flocking behavior, with more than three thousand bird silhouettes

fig. 10

Marius Watz, *Abstract 1*, 2003

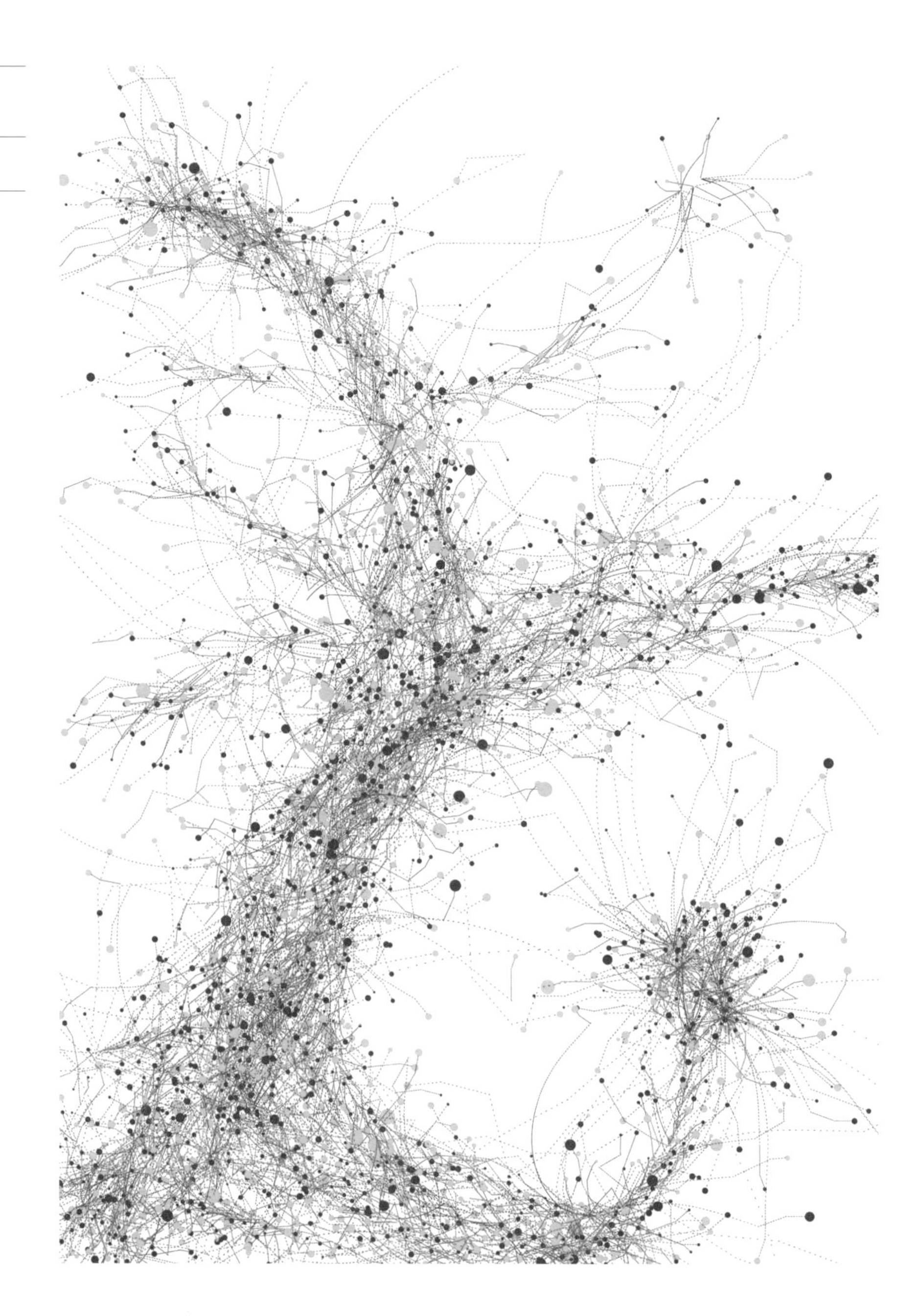

figs. 11–12

Keith Peters, *Random Lissajous Webs*, 2008

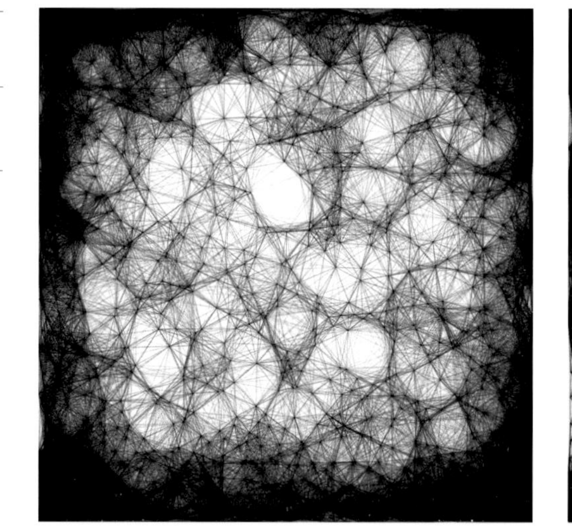

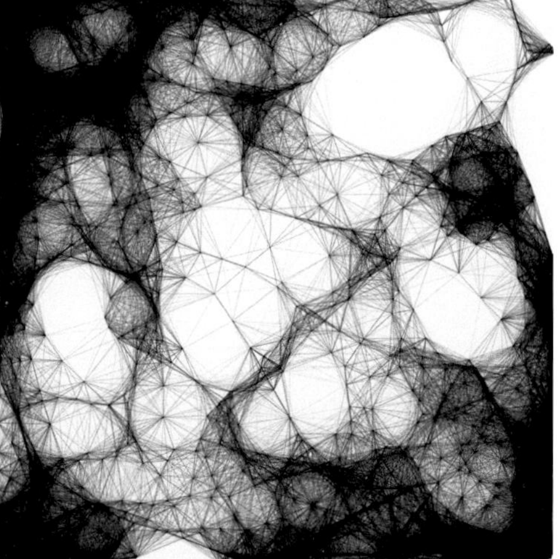

generative art has gained momentum in the past decade, with the works of artists Marius Watz, Alessandro Capozzo, Jared Tarbell, Keith Peters, Eno Henze, Casey Reas, Paul Prudence, Robert Hodgin, and Karsten Schmidt, among others, delivering riveting views on the artistic qualities of code. fig. 10, fig. 11, fig. 12

Generative art is fascinating, not only for its medium and end result, but also for its inherent building process, akin to the creation processes of nature. The associations with biology and evolution are indubitable, with many artists employing behavioral algorithms, in which coded "agents," or "virtual ants," are told to move across the screen in a series of randomized directions, progressively generating the elaborate final piece as they leave behind their colorful trails. If Pollock's affinity to natural processes helps explain the magnetism of his work, so too does generative art's nature-inspired assembly procedure explain its own magnetism. After all, the way generative art operates at a code level—as simple orders repeated over time—is remarkably similar to the way nature builds many of its convoluted patterns.

But the notion of such an ordered nature is in fact quite recent. Before the 1950s, as Taylor observes in his book *Chaos, Fractals, Nature* (2006), science showed little interest for the visual complexity that abounded in the natural environment, with the common presumption that it was no match for humanity's artificial order: "Although the individual actions of Nature were expected to be trivially simple, it was assumed that the sheer number and variety of these actions caused the combined system (the one we observe all around us) to descend into disorder. In other words, complexity was thought to exclude any hope of order."[10] But around the time of Pollock's death, in 1956, scientists began to view nature differently, understanding that even though "natural systems masqueraded as being disordered, lurking underneath was a remarkably subtle form of order."[11]

The 1960s and 1970s witnessed the growth of two groundbreaking areas of scientific inquiry, fractal and chaos theories, which in turn led to the development of the notion of *emergence*. While fractal theory looks at recursive self-similarity as a building mechanism for complexity, whereby the same geometrical motif is continuously repeated at different scales, chaos theory, famously typified by the butterfly effect, is a wide field of study that investigates the properties of complex dynamic systems sensitive to initial small changes, and covers areas like mathematics, physics, and economics. Both theories led to the awareness of emergence—the rise of complex patterns and systems from multiple simple interactions—which delivered a new understanding of natural phenomena and became an integral part of complexity science.

These theories look at dynamic systems from an entirely new perspective, in which order and complexity are not seen as opposing but as complementary elements of nature. The extraordinary work of mathematicians Alan Turing, Edward Lorenz, and Benoît Mandelbrot, which translated nature's apparent disorder into simple mathematical equations, was pivotal in this transition. Soon it became widely accepted that the newly conceived ordered complexity of natural systems relied on basic rules, coming together in a variety of intricate shapes and schemas.

In many cases it is not so much a matter of order and disorder being antagonistic as it is a matter of each prevailing at a different scale. "What seems complex in one representation may seem ordered or disordered in a representation at a different scale," say media artists and researchers Christa Sommerer and Laurent Mignonneau.[12] To elucidate this point, they provide the example of cracks in dried mud, which might seem like a flat, homogeneous surface from afar, but a disordered array when the clay particles making the mud is looked at closely. An inverse of this duality might be said to exist in the case of fractals or networks, whether biological, social, or technological. They might seem astoundingly complex from afar, but as we zoom in, we find a very defined order between individual entities and their respective ties. When we alternate between micro and macro views of the world, we can decipher different attributes of order and complexity, as if looking at a flexible universal continuum. fig. 13, fig. 14

As the idea of an ordered complexity in nature—through advances in chaos, fractal, and network theory—became accepted in the sciences, the art community likewise began to question this duality. In his seminal *Metaphysics* (ca. first century CE), Aristotle considered order, symmetry, and definiteness to be the general elements of beauty, a view that prevailed for centuries. Many other philosophers have supported and expanded on this belief. But in the beginning of the twentieth century, the notion of pure, balanced beauty was assertively challenged by vocal artists who suggested a new type of aesthetics independent of a striving for balance. Cubism and futurism helped foster this change of mind-set, gearing art toward the path of abstraction. Most people familiar with art history know what happened next: art became gradually detached from its role as a faithful representation of the real and became looser and nonliteral, often placing emphasis on the process, the emotional impact, and experimentation.

Even though art went through a major transformation in this period, the refusal to fully acknowledge complexity was carried out by a number of art critics, who later found in Pollock the perfect target for their repudiation. In 1957, art theorist and perceptual psychologist Rudolf Arnheim (1904–2007) took a stand against the latest "accidental" form of art in a harsh condemnation of contemporary art. Published in the *Journal of Aesthetics and Art Criticism*, his

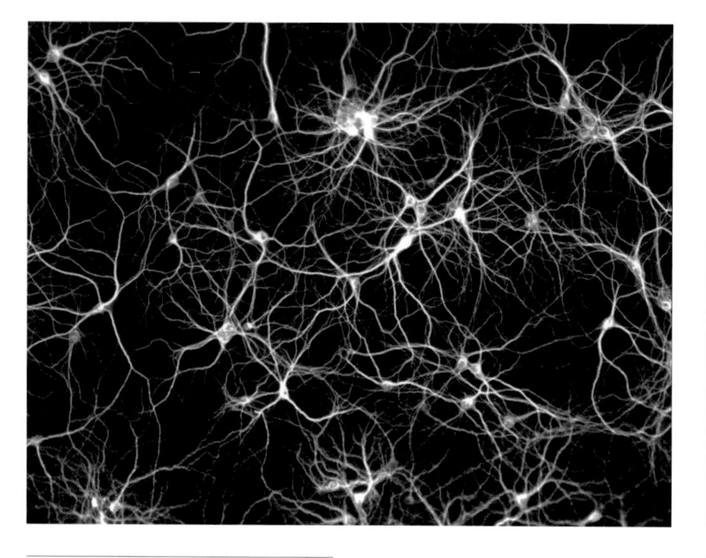

fig. 13

Paul De Koninck Laboratory, Dissociated culture of rat hippocampal neurons, 2005

A detailed view of a rat's neuronal network

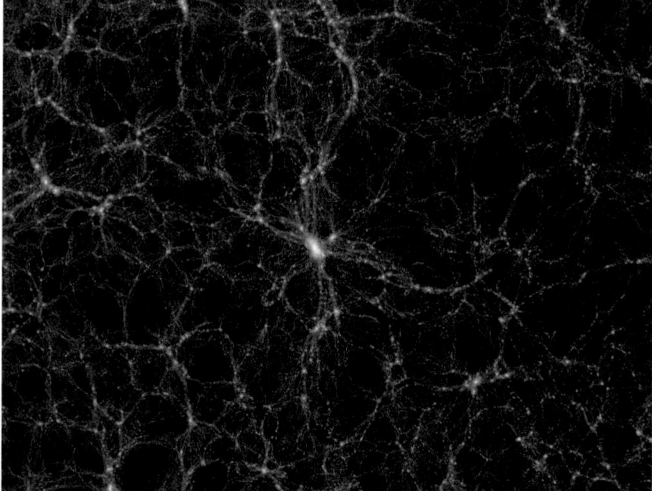

fig. 14

Volker Springel et al., *Millennium Simulation*, 2005

The largest and most realistic simulation of the formation and growth of galaxies and quasars. It re-created the evolutionary histories of approximately twenty million galaxies in twenty-five terabytes of stored data.

essay "Accident and the Necessity of Art" delivered a livid attack on modern artists, particularly Pollock, "who deliberately [rely] on accident for the production of their work."[13]

As one of the key precursors of Gestaltism, it was only natural for Arnheim to be derisive of a new form of expression lacking any obvious sign of uniformity. After all, Gestaltists believed psychological organization was facilitated by regularity, predictability, and symmetry, and if anything, order and complexity were antagonists in a struggle for constancy. Slowly, other critics started realizing that the aesthetic split between order and complexity was not clear cut, and the dismissal of Pollock's work began to fade. The influential art critic Clement Greenberg, one of the first voices of appraisal for Pollock, opposed the superficial idea of "accident" popularized by Arnheim. In the article "Inspiration, Vision, Intuitive Decision," from 1967, Greenberg writes:

> Pollock wrests aesthetic order from the look of accident—but only from the look of it.... True, this is all hard to discern at first. The seeming haphazardness of Pollock's execution, with its mazy trickling, dribbling, whipping, blotching, and staining of paint, appears to threaten to swallow up and extinguish every element of order. But this is more a matter of connotation than of actual effect. The strength of the art itself lies in the tension...between the connotations of haphazardness and the felt and actual aesthetic order, to which every detail of execution contributes. Order supervenes at the last moment, as it were, but all the more triumphantly because of that.[14]

By analyzing Pollock's work in this fashion, Greenberg became a voice of change, finally pairing order and complexity in a new type of "orderly disorder" aesthetics. This important shift, similar to the one observed in the sciences, still reverberates throughout the arts. As Padovan explains, "Order and complexity are twin poles of the same phenomenon. Neither can exist without the other, and aesthetic value is the measure of both.... Just as order needs complexity to become manifest, complexity needs order to become intelligible."[15] According to philosophy professor Ruth Lorand, this interlace leads to the formula of beauty described as "unity in diversity."[16]

We can witness this aesthetic formula in a variety of daily circumstances, simply by looking at a stirring natural landscape. It is the apparent sense of unity despite the uncountable interacting variables and inherent complexity that makes us gaze in awe when contemplating such a landscape and wonder with amazement how all these elements came to form such a striking construct. Truly, the same sense of wonder occurs when we are faced with a complex network. The dense layering of lines and interconnections might enthrall us at a deeper level, leaving us to marvel at the feeling of wholeness from disparate multiplicity. If a sense of self-awareness, driven by an innate complexity encoding, can partly explain our infatuation with networks and complexity, the notion of unity in diversity can certainly complement it. Unity in diversity is ultimately an important metaphor for a new outlook on the captivating power of networks, granting complexity greater aesthetic legitimacy than ever before.

Networkism

It is well documented that the traditional arts, such as painting and sculpture, have always been influenced by advances in science. Mathematics and art, for instance, have a long history of cross-pollination, going back to the golden ratio and the idea of symmetrical quantifiable beauty in Ancient Greece. Furthermore a multitude of artists, like Leonardo da Vinci, Piet Mondrian, M. C. Escher, and Salvador Dali, have famously incorporated mathematical themes in their work. As mathematician John L. Casti amusingly stated: "Nowadays it's almost impossible to walk into the office of a scientist or mathematician without seeing an engraving or two by the well-known Dutch artist M. C. Escher hanging on the wall."[17]

Today, more than ever, art and science are highly intertwined in a cyclical sphere of influence, and complexity science is simply a new source of inspiration. As researchers, scientists, and designers across the globe use a variety of technological tools to make sense of a wide range of complex structures, they inspire a growing number of artists infatuated with networked schema and their disclosure of hidden territories. This seductiveness is bearing fruits, with many similar art projects emerging under a new trend called networkism. Stimulated by rhizomatic properties like nonlinearity, multiplicity, or interconnectedness, and scientific advances in areas such as genetics, neuroscience, physics, molecular biology, computer systems, and sociology, networkism is a small but growing artistic trend, characterized by the portrayal of figurative graph structures—illustrations of network topologies revealing convoluted patterns of nodes and links. As a direct consequence of the recent outburst of network visualization, networkism is equally motivated by the unveiling of new knowledge domains as it is by the desire for the representation of complex systems. fig. 15, fig. 16

It is never easy to introduce an artistic trend, since most artists, and rightfully so, tend to detach themselves from definition: their work is an individual pursuit and should be analyzed as such. Nevertheless, it is still possible to establish correlations and find patterns emerging from their discrete efforts. The general motivations of networkism are best expressed in the words of artist Sharon Molloy, one of the precursors of this movement:

> My quest is to reveal how everything is interconnected. From the atom to the cell, to the body and beyond into society and the cosmos, there are underlying processes, structures and rhythms that are mirrored all around and permeate reality...one small thing leads to another and larger patterns emerge....This work embraces the multiple, the network, the paradoxical and the idea that even the smallest gesture or event has significance, and the power to change everything.[18]

In an apparent allusion to chaos theory, the words of Molloy echo many of the inherent properties of complex systems: rhythm, movement, pattern, structure, multiplicity, and interconnectedness. These are all qualities of observed networks but also intrinsic traits of the work of artists operating in this realm. As a systems art practice, networkism epitomizes the notion of ordered complexity, balancing order and disorder in a striking resemblance to both natural patterns and scientific visualizations. fig. 17, fig. 18, fig. 19, fig. 20

fig. 17 (opposite, left)

Stephen Coast, detail from *IP Mapping*, 2001

A three-dimensionally rendered network of 8,996 IP addresses with 10,334 connections

fig. 18 (opposite, right)

Sharon Molloy, detail of *Transient Structures and Unstable Networks*, 2008, oil and enamel on canvas

fig. 15

Firstborn and Digital Kitchen, detail from *Operation Smile*. A large-scale visualization of hundreds of smiling portraits of visitors to New York City's South Street Seaport. (See also chapter 5, page 202.)

fig. 16

Emma McNally, detail from *Field 4*, 2009, graphite on paper

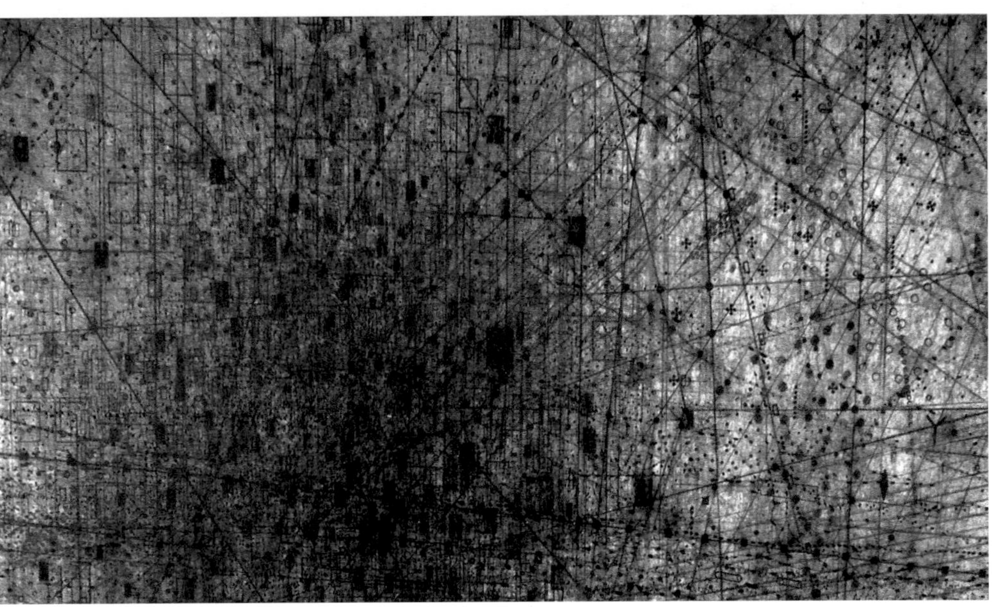

fig. 19

Sharon Molloy, *Transient Structures and Unstable Networks*

In addition to Molloy, other central players of networkism are artists Janice Caswell and Emma McNally. Although the three artists seem to be allured by the most appealing features of network formations—with imaginary landscapes of interconnected entities being the prevailing theme—their motivations are never exactly the same. While Molloy seems to be captivated by scientific discoveries and underlying structures in nature, Caswell pursues portraits of her own memories as indecipherable mental maps. McNally, on the other hand, is highly influenced by the concept of the rhizome from Deleuze and Guatarri, and her work "investigates possibilities of semiotic connections and disconnections through a visually and conceptually dense use of pencil on paper."[19]

In an article for *Art Lies*—a contemporary art quarterly—covering the exhibition *City Maps* in which Caswell took part, writer Catherine Walworth opens with the following: "Maps are gorgeous but notable artists are rarely invited to render them. They wouldn't make sense in a language of pure subjectivity... or would they?"[20] She then describes Caswell's exhibited works in detail:

> Unconstrained lines drawn in marker travel back and forth across the page creating airy, curving forms with no specific mass....Caswell's markings pause as they stroll along, conveying the slightly awkward sense that they have no intended direction. Dots of various sizes are affixed to these lines like the symbol for towns and cities on a common road map.[21]

Caswell's maps of intertwined lines and colorful pins are captivating, taking us through imaginary journeys and wanderings. The network is an ever-present topology, linking different nodes, tying different places together, in a glimpse of faded memory. Caswell has also developed a personal lexicon—which she recurrently explores in her maps—reminiscent of computer-generated network representations: "This work arises out of a desire to capture experience, an impulse to locate, arrange, and secure the past. I use a pared-down, coded language through which points, lines, and fields of color define spaces and retell narratives, making memories concrete."[22] fig. 21

McNally, whose drawings showcase remarkable landscapes of intense graphite, depicting imaginary networks, paths, and trajectories, explores a parallel cartographic conjecture. In an essay accompanying McNally's work at T1+2 Gallery, in London, curator and art historian Ana Balona de Oliveira describes McNally's studies: "Her

fig. 20

Sharon Molloy, *Transient Structure*, 2010, oil and enamel on canvas

large and small-scale drawings...offer themselves to the viewer as surfaces or sites for rhythmic relations of graphite marks disruptively connected in gatherings, collisions, swirls, and dispersals that are both geometric and chaotic. There seems to be a permanent flux of disquietingly pulsing energy achieved through these conflicting highly organized and extremely fleeting forces."[23] Oliveira explains that McNally's nodal connections show a similarity "between the processes of the radically differing micro-cosmos of the atom and macro-cosmos of the star formation" and bear a resemblance to a variety of themes, like "aerial views, battlefield maps, geological formations, oceanic charts, disease transmissions, animal migratory routes, molecule structures, black holes, etc."[24] fig. 22, fig. 23

The work of Molloy, Caswell, and McNally has a strong correspondence to the work produced in action painting, generative art, and even network visualization. So what makes networkism distinct as an independent artistic movement? Although analogous in terms of visual output, networkism is unique in a number of ways.

No tangible data. In most cases the depicted entities on canvas and their expressed linkages are fictitious and do not relate to an existing data set. This first feature of networkism puts aside apparent similarities with information design or network visualization—fields that always employ actual data sets or tangible facts.

Not entirely random. Although dealing with abstract elements and respective ties, their placement on the canvas seems to be carefully considered and planned, with a particular visual composition in mind. This differs from most generative art and other forms of algorithmic art, which, besides distinct computerized media, tend to employ randomized, autonomous processes in their work.

Nodal building. The key influence of networkism, as the name implies, is networks—widespread topologies usually represented by means of a graph. It is the specific nonlinear network configuration—defined in sets of nodes and edges—that situates networkism in a different artistic context, separate from other past and contemporary art movements, particularly action painting and other branches of abstract expressionism.

Even though the strongest manifestation of networkism is occurring in painting and illustration, it is not limited to expression in two dimensions. The work of Tomas Saraceno, particularly a piece with the fabulous title *Galaxies Forming Along Filaments, Like Droplets Along the Strands of A Spider's Web* (2008), shown at the Venice Art Biennale in 2009, is a magnificent example of what networkism is all about. In this dramatic installation, several bulbous shapes hang in the air, sustained by a dense interwoven elastic rope that stretches to the floor, walls, and ceiling. It plays with the notion of celestial space and the vast planetary landscape, like invisible strands holding groups of stars in candid emptiness; but it also alludes to smaller-scale entities, resembling at times the construct of a neuronal network. The piece is striking and theatrical, and allows users to wander freely between the multitudes of elastic strands. Art editor Kristin M. Jones describes her own experience with Saraceno's web in *Frieze Magazine*: "Bumping into one [rope] meant sending shivers throughout the skeletal cosmological web, but clumsiness worked to the viewer's advantage, providing this surrogate universe with a sense of tangible interconnectedness and mutability."[25] figs. 24–27

If Saraceno's *Galaxies Forming Along Filaments* appears to sustain the surrounding white walls in a slender assemblage, Japanese artist Chiharu Shiota's lattices

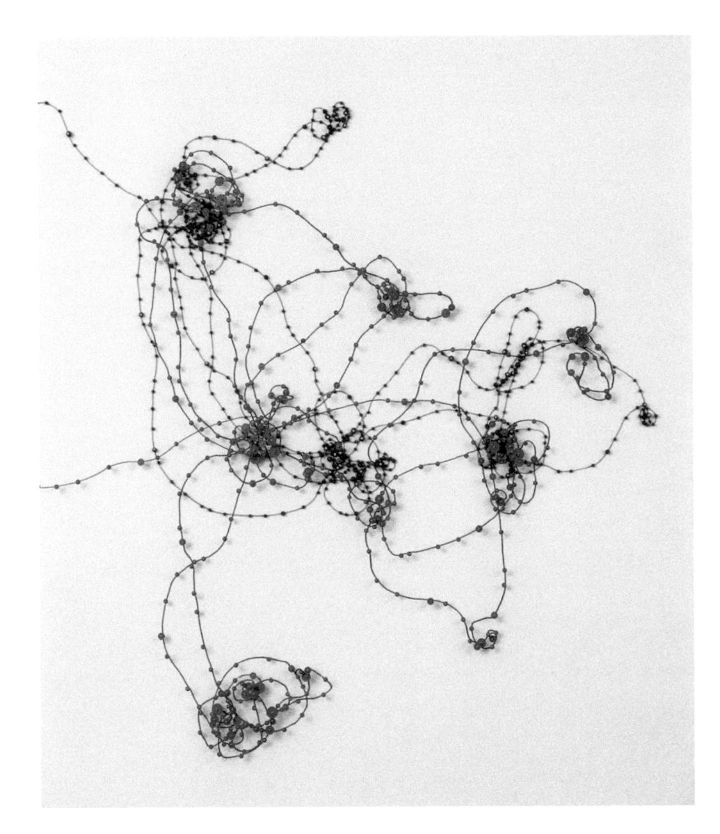

fig. 21

Janice Caswell, *The Book of Salt*, 2006, ink, paper, pins, enamel, beads on paper, mounted on aluminum-backed archival foam board

fig. 22

Emma McNally, *e1*, 2009, graphite on paper

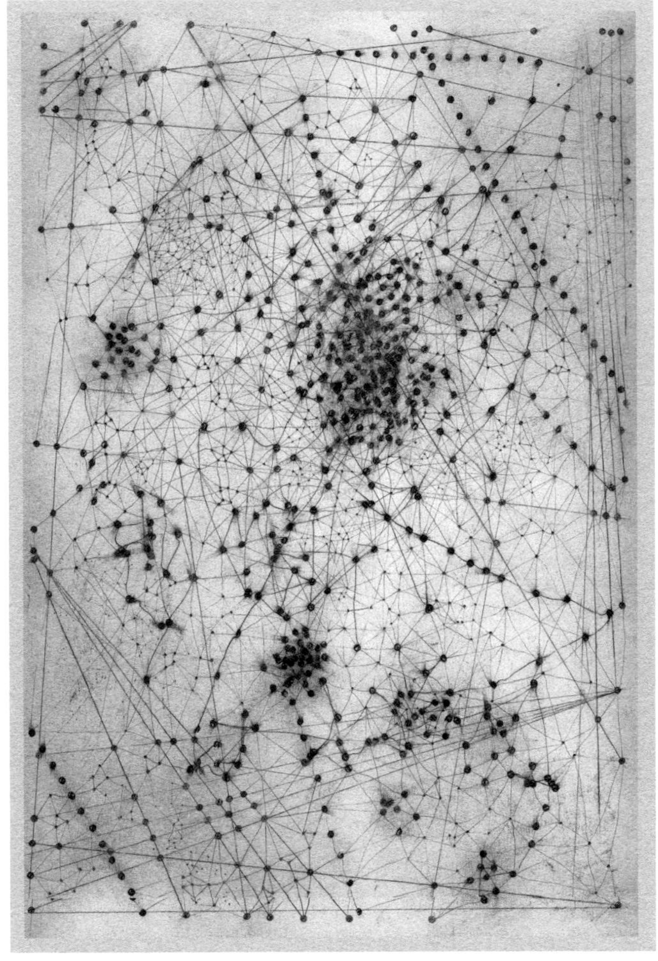

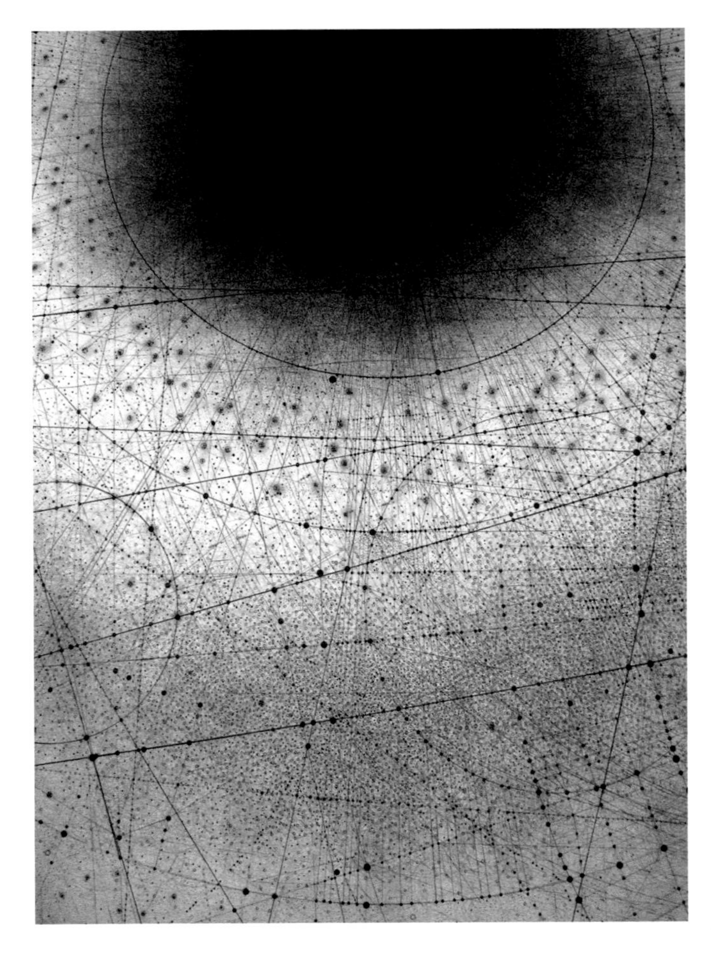

fig. 23

Emma McNally, *field 8*, 2010, graphite on paper

figs. 24–27

Tomas Saraceno, *Galaxies Forming Along Filaments, Like Droplets Along the Strands of A Spider's Web*, 2008, elastic ropes emphatically invade every corner of the room, in a gloomy swarm of dark lines. fig. 28 Her spaces are filled with hundreds of black woolen threads—dense layers that form an impenetrable cocoon—and appear to be contaminated by the intrusive web. Shiota does not have a studio, nor does she produce drawings or notes beforehand. She works only on location and relies solely on recollection. This explains why her captivating installations resemble dreamy scenarios, invoking the passage of time or the erosion of memory.

Bosnian artist Dalibor Nikolic has also been exploring the meanderings of the network, but instead of using elastic bands or woolen threads, Nikolic uses plastic pipes and wires to produce many of his convoluted shapes in a remarkable effort of systematization. figs. 29–30 His constructions are made by the continuous replication of basic patterns, always with a simple, or the simplest possible, assembly process. Accepting the premise that everything is made up of atoms, Nikolic finds great inspiration in the repetition of uncomplicated patterns, as a universal mode

fig. 28

Chiharu Shiota, *In Silence*, 2008, elastic ropes and black woolen thread

figs. 29–30

Dalibor Nikolic, *Network*, 2007

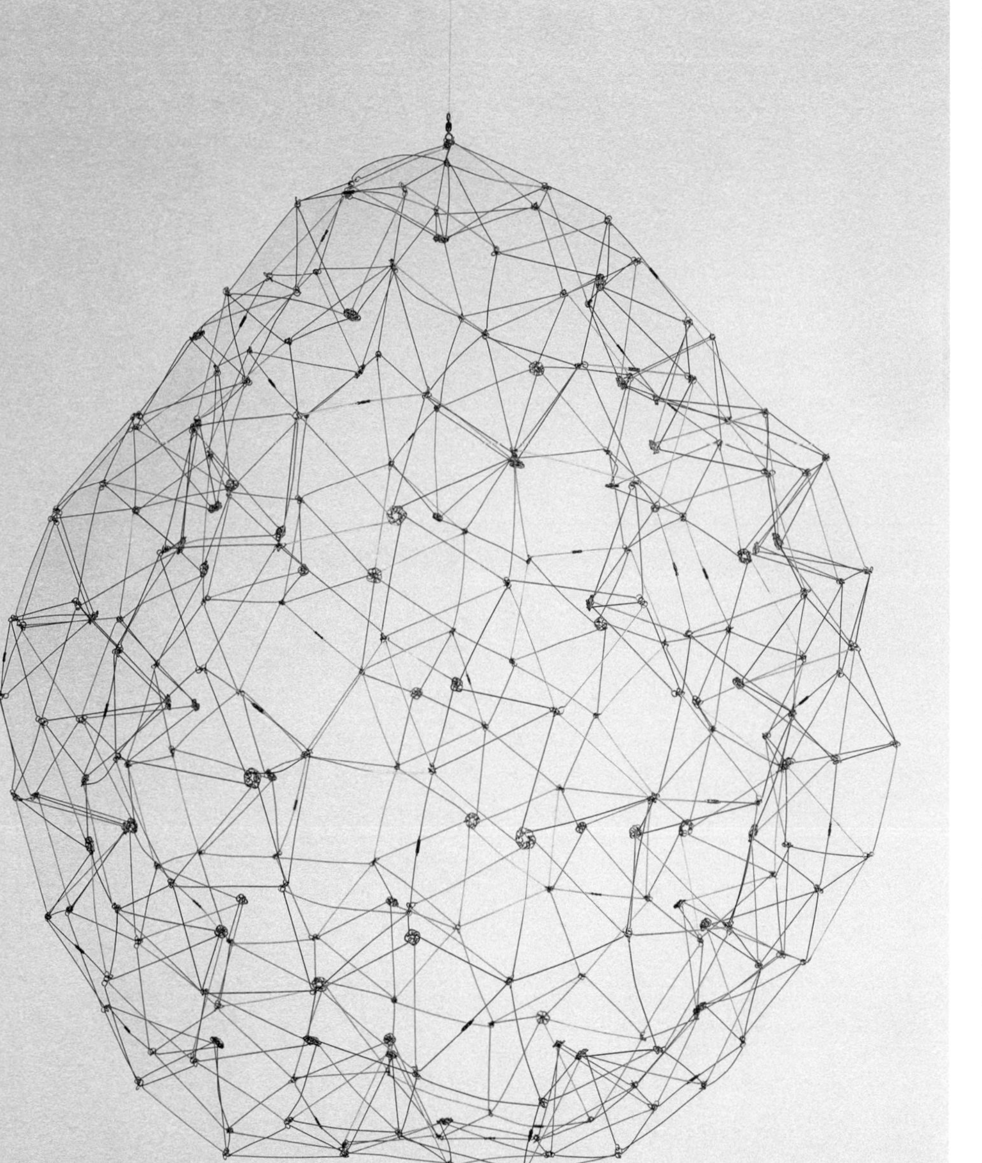

fig. 31

Gertrud Goldschmidt (Gego), *Esfera* (Sphere), 1976, stainless-steel wire

of production, and the duality of absoluteness and nothingness. In one of his pieces, appropriately titled *Network* (2007), Nikolic builds a dense globe of interconnections of intersecting pipes and assembled joints, resembling visions of a noosphere, the sphere of human thought.

A visionary precursor to the works of Saraceno, Shiota, and Nikolic comes from the hands of the notorious artist Gertrude Goldschmidt, known as Gego. A true predecessor of networkism, Gego was born and raised in Germany. In 1939, at the age of twenty-seven, she moved to Venezuela and lived there the rest of her life, until passing away in 1994. Her artistic vocabulary was unique and constantly changing. In her drawings, prints, and sculptures, she showcased an unconventional and independent viewpoint, apparently immune to trends or styles. fig. 31

Gego's *Reticulárea* is the most popular and striking piece in her diverse body of work. figs. 32–33 First exhibited at the Museo de Bellas Artes de Caracas in 1969, several meshes of aluminum and steel were tied together and dispersed irregularly in the confined space of a room. In a June 2003 article for *Art in America*, critic and curator Robert Storr described *Reticulárea* as "an astonishing tessellation of suspended, interlocking stainless-steel wire elements that fills a large white room whose corners have been rounded so that viewers can more easily lose themselves and their sense of scale in the triangulated, volumetric webs that surround them, webs through which they move like planes navigating the gaps in a cloud bank."[26]

The web of influence between Gego, Saraceno, Shiota, and Nikolic is evident not only in the similarity of their resulting structures but also in the ingrained principles that tie their work. In *Questioning the Line: Gego in Context* (2003), curators Mari Carmen Ramírez and Theresa Papanikolas draw a comparison between Gego's *Reticulárea* and Deleuze and Guattari's rhizome:

> As in the rhizome, two principles rule instead of a master plan: a principle of "connection" and a principle of "heterogeneity".... By connecting any one point with any other, the Reticulárea, like all rhizomes, "makes multiple" without adding a "superior dimension"; it neither begins nor ends, but is instead always in the middle, "in the midst of things," and therefore always lacking a "culminating point."[27]

The rhizome, discussed in chapter two, is not only a strong metaphor for Gego's work but for all artists within the sphere of networkism. Rhizome showcases an entirely new conception of aesthetic quality—opposed to our obsession for order, tidiness, and linear narrative—that relies on multiplicity and interconnectedness to express the inner construct of the world and its striking invisible beauty. "The net has no center, no orbits, no certainty," writes technology editor and writer Kevin Kelly. "It is an indefinite web of causes."[28] If complexity science is networkism's scientific mentor, then rhizome is its philosophical counterpart.

Networkism typifies a new conception of art, stretching as far as our scientific eye can take us and embracing all scales of human understanding, from atoms, genes, and neurons to ecosystems, the planet, and the universe. A seeming consequence of the complex connectedness of modern life, networkism follows a revised idea of metanarrative, or grand narrative, introduced by French philosopher Jean-François Lyotard in the 1970s, in this case pertaining to the omniscience of science. The network is at the center of

figs. 32–33

Gertrud Goldschmidt (Gego), *Reticulárea* (Reticula + area), 1969. Museo de Bellas Artes, Caracas, Venezuela.

this belief, embodying a transcendent and universal truth, an archetype that represents "all circuits, all intelligence, all interdependence, all things economic, social, or ecological, all communications, all democracy, all families, all large systems, almost all that we find interesting and important."[29] Ultimately, networkism is an absorbing testimony of the network's widening influence.

As we recognize its ubiquity—not as a superficial model but as a structural dynamic force—the network will continue to challenge any conventional notion of beauty. The awareness of this widespread topology is driving a considerable perceptual shift, replacing many of complexity's dubious qualities with new, evocative ones. Networks show us that there is order in disorder, that there is unity in diversity, and above all, that complexity is astonishingly beautiful.

Notes

1 Casti and Karlqvist, *Art and Complexity*, 1.
2 Dondis, *Primer of Visual Literacy*, 39.
3 Ibid., 110.
4 Padovan, *Proportion*, 41.
5 Ibid.
6 Emmerling, *Jackson Pollock*, 48.
7 Taylor, *Chaos, Fractals, Nature*, 56.
8 Casti and Karlqvist, *Art and Complexity*, 142.
9 Ibid., 49.
10 Taylor, *Chaos, Fractals, Nature*, 11.
11 Ibid.
12 Casti and Karlqvist, *Art and Complexity*, 94.
13 Arnheim, *Toward a Psychology of Art*, 162.
14 Taylor, *Chaos, Fractals, Nature*, 132.
15 Padovan, *Proportion*, 41, 42.
16 Lorand, *Aesthetic Order*, 10.
17 Casti and Karlqvist, *Art and Complexity*, 25.
18 Molloy, "Sharon Molloy: Artist Statement."
19 Oliveira, "Emma McNally."
20 Walworth, "City Maps."
21 Ibid.
22 Caswell, "Artist Statement."
23 Oliveira, "Emma McNally."
24 Ibid.
25 Jones, "Tomas Saraceno."
26 Storr, "Gego's Galaxies," 108–13.
27 Ramírez and Papanikolas, *Questioning the Line*, 93.
28 Kelly, *New Rules for the New Economy*, 9.
29 Ibid.

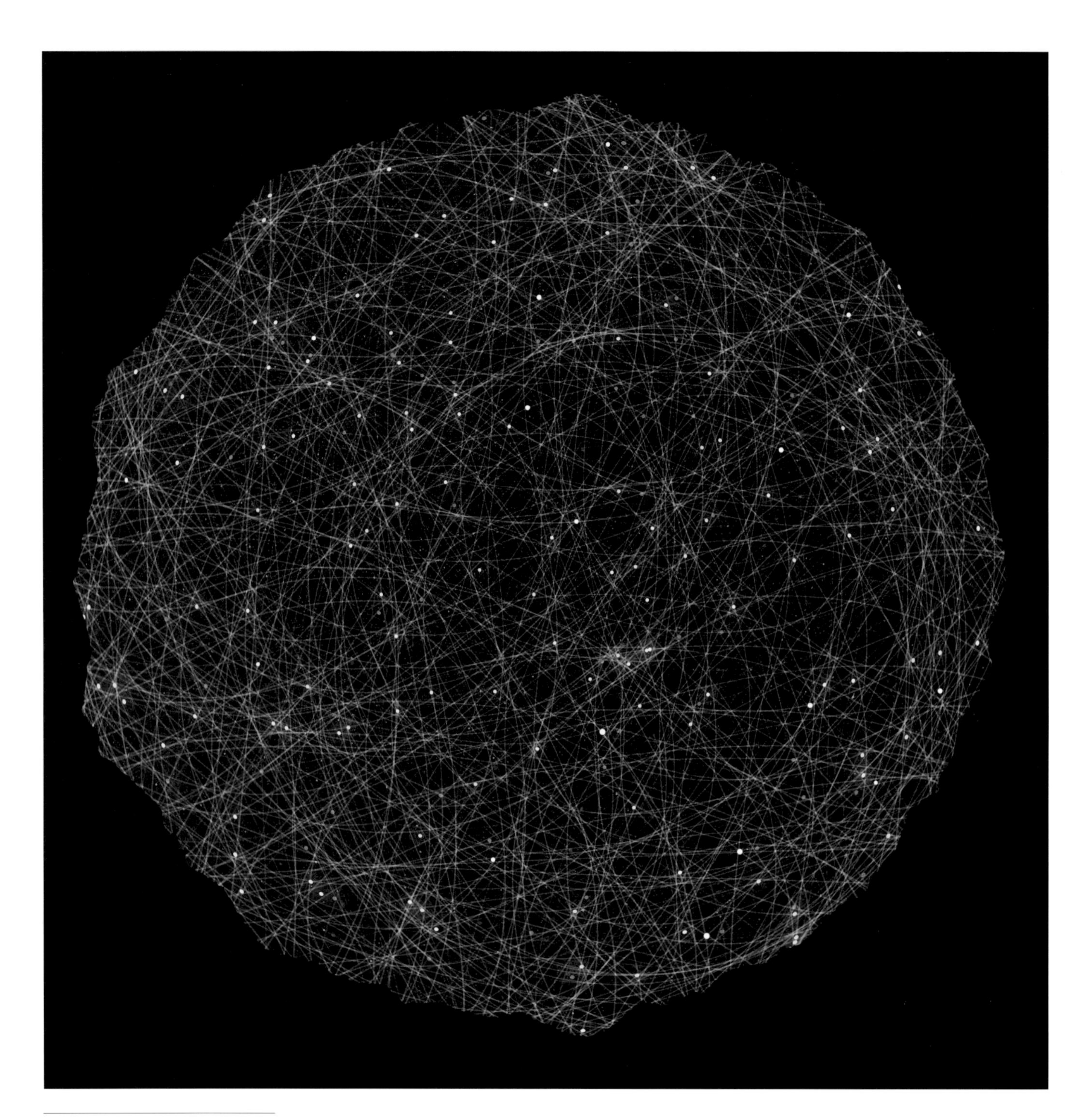

Scott Hessels and Gabriel Dunne,
Celestial Mechanics, 2005

A still image from a visual animation depicting the paths of several satellites drifting above our planet. It is part of a larger planetarium-based installation that visualizes the statistics, data, and protocols of man-made aerial technologies.

07 | Looking Ahead

Today we live invested with an electric information environment that is quite as imperceptible to us as water is to a fish.

—Marshall McLuhan

This quantity over quality shift in our culture has created an even deeper need for truly informing experiences—for insight, the most precious form of information.

—Nathan Shedroff

"Information gently but relentlessly drizzles down on us in an invisible, impalpable electric rain."[1] This is how physicist Hans Christian von Baeyer starts his engaging book *Information: The New Language of Science* (2005). The statement is not only an intriguing metaphor for our contemporary data-inundated world but an acute look into the future. If visualization has until now played a major role as a substantiated filter of relevance, disclosing imperceptible patterns and hidden connections in von Baeyer's electric rain, it will simply become indispensable as raindrops swiftly turn into a drenching downpour. Visualization will become imperative not solely as a response to the growing surge of data, but also as a supporting mechanism to the various political, economic, cultural, sociological, and technological advances shaping the coming years.

In the spirit of network diversity and multiplicity, this chapter will embrace a number of different views on the future of visualization, covering an array of trends and technologies that will shape the progress of the field. Some of these directions have just begun to show the first signs, while others have been maturing for years. The essays cover topics ranging from ubiquitous visualization to social-data collection, neo-cartography, ambient visualization, collective intelligence, mass digitization, and sensory networks. Through these discussions, their authors forecast an engaging and malleable use of visualization as a fundamental tool of discovery in an increasingly interconnected and complex world.

The two epigraphs to this chapter are drawn from McLuhan, *Counterblast*, 5; and Wurman, Sume, and Leifer, *Information Anxiety 2*, 16.

1 von Baeyer, *Information*, 3.

Seeing the World in Data

by Nathan Yau

In the early 1900s, an anthropologist, a poet, and a filmmaker started a project in the United Kingdom called Mass-Observation, in which the goal was to gain a better understanding of their community. They asked participants to keep diaries documenting their daily lives. Sometimes participants were given specific objectives, like count how many people in a pub wore hats. Other times, the subject of documentation was open-ended with very little to no structure. The collective result was a micro view of the United Kingdom, made possible by thousands of individuals working toward a common goal. These journals were meaningful to the individuals who kept them but also provided something useful as a collection. This, of course, was before computers.

Current technological developments offer the opportunity to collect data in the same vein as Mass-Observation, at an even more detailed, and a much larger, scale. We can use advancing technology, like mobile phones and the internet, to collect information about our surroundings and ourselves. An individual can collect thousands of data points during a single day without even batting an eye or picking up a pencil and a notepad. Hundreds of thousands of people are a part of this fast-paced digital world.

With these advances come a number of possible applications. One area is citizen science. People can take active roles in their community by collecting data about what is around them, contributing to a common database that experts can in turn analyze to find solutions to local problems. For example, individuals could report traffic levels, which in turn could help others find the best route home or to work in real time. Citizens can collect pollution levels in their area, which could collectively provide a view of a city's air quality and provide a clear direction for public policy. Less serious matters can also be measured, like helping people find a fun place to hang out.

It is easy to see the potential in such an idea; however, we are still very much at the beginning of social-data collection, or Participatory Sensing. Before we hit any milestones and really make use of these new flows of data, there are three main areas that we have to work on: collection, analysis, and interaction.

Data Collection

With the huge growth and adoption of mobile technology, data collection is easier than ever. We can record our location every few seconds with GPS devices, take digital pictures on a whim, and send text messages wherever there is connectivity, which is just about everywhere nowadays. Some data flows are self-updating and automatic. Others are manual and involve more active collection procedures. Either way, one of the keys to data collection is to make the process easy and to intertwine it with daily activities.

Millions of people around the world own mobile phones and subscribe to services so that they are connected almost everywhere they go. These phones could be repurposed into data-collection devices with software that tasks the phone to retrieve data a few times per minute so that people can collect data almost everywhere they go. Programs could be created to allow users to take pictures and annotate them with location and metadata.

It is, of course, not as easy as it sounds. Like with any experiment, there are many kinks to resolve before moving on to the next step. Connectivity, for example, will inevitably be spotty in some areas. How do we account for a loss of connection? How should we store the data? We can cache data on a phone's memory, but what happens when it runs out of space? Along with connectivity and storage come synchronization issues with phone and server.

Then there are concerns with data privacy. Who sees your data when you upload it to the server? How safe is your data and how long will it stay in memory? If the data is used for research, is the data properly "anonymized"? For example, some might not mind sharing what they eat, but most people are not comfortable revealing their location every single minute of the day.

Ultimately, these are issues to address while retaining transparency for the user. When the user, who is not necessarily professionally trained to deal with data, becomes analyst, it is important he or she knows what is going on.

Analysis

Once there are streams of data to work with, we have to figure out what to do with it. There is an inclination to show all the data at once, which may be appropriate at times, but what happens when there is too much data to fit on a single screen? In these cases, which continue to grow in number, analysis occurs between data collection and the end use. Algorithms and traditional statistical techniques help locate the useful points in the data, which are then visualized.

For example, imagine a camera phone programmed to take pictures several times a minute, perhaps with the intent of examining whom we interact with or the foods that we eat during a single day. Without proper algorithms to distill the data, the hundreds of thousands of images are difficult to process. Our brains are great at finding patterns, but when we have gigabytes or terabytes of information, it is easy to see how important details could be missed. Not only does analysis help find the interesting points in our data, but it also filters out the outliers and corrupted files, and automates tedious tasks like classification and correlation.

Interaction

Once the information is in the database and properly vetted, we can finally see our world in data through visualization. With the growing amount of data, many tools—some artistic, some analytical, and others in between—have been applied to provide unique views of our surroundings. And the web has made it much easier for these ideas to propagate.

Dynamic online mapping, or so-called neocartography, has brought intuitive interfaces in a familiar geographic setting to all users to access specific information about their county or city. What if you were able to see what was going on in your neighborhood from a data standpoint? Is your block due for a road repair because there are too many potholes? Which street lamps are flickering on and off at night? Are the noise levels too high in the middle of the night? People can easily access this information, which not just satisfies curiosity but also provides quantitative evidence that can be used in public-council meetings or provided to policy makers. Participants can also collect and aggregate information about their neighborhoods themselves. So while people interact with the data through a computer, they are actually able to improve their communities.

Although not everyone who "analyzes" this data will have a background in the proper techniques, a certain level of data literacy must be developed. Visualization will be essential in making the data more accessible.

Looking Forward

In the end, it all comes down to the data. It comes down to the individual taking an interest in his or her surroundings. Visualization is only as good as the data that creates it, and if there is no data, there is nothing to analyze—no new understanding of the world.

If no one had sent in journals for Mass-Observation, there would be no localized narrative of Britain in the archives. We have come a long way since the early 1900s. Data collection is much easier today, and we have the opportunity to engage nonprofessionals with visualization and analysis. In some ways, this is already happening with microblogging on social applications like Twitter and Facebook, or with citizen reporting on popular news sites, such as CNN, MSNBC, and ABC. The next step is to add structure and tools that take advantage of these open applications, and when that happens, we gain micro views into our neighborhoods. But most importantly, we will start to see undiscovered relationships between neighborhoods, communities, states, countries, and continents. We will see how we, as individuals, interact with our surroundings and how we can use that information for the better. Computers are often thought of as technology that takes us away from the physical world and social interactions. Quite the opposite. Computers, through data, visualization, and interfaces, can bring us closer—and that is certainly something to look forward to.

The Fall and Rise of Ambient Visualization

by Andrew Vande Moere

Released back in 2002, the Ambient Orb is a cute light-emitting globe able to change its external color depending on various data, such as the forecasted temperature, current stock-market performance, or sports statistics. It is a perfect example of an "ambient display," a new approach to visualization that combines two intrinsically contradicting goals: to constantly inform viewers about useful, time-varying information and to do so without disturbing them. Or, in academic terms, ambient visualization communicates information through the "periphery of user attention" so that one can remain committed to other, more important, tasks at hand.[1] The significant contradiction lies in the fact that our visual sense tends to be so sensitive that the slightest perceivable change within our visual range will inevitably obstruct any concurrent activity. But why is there a need for a display of data, when one does not want to be disturbed by it anyway?

The idea of ambient visualization is thought to have been conceived by a group of researchers at the MIT Media Lab, who investigated the idea to discover a way to convey information by means beyond the traditional computer screen.[2] They proposed the dynamic alteration of environmental qualities according to changes in data, such as slight changes in sound or subtle adaptations of light, so that people would be continuously provided with information like live traffic conditions, trends in airline prices to favorite holiday destinations, or the online status of friends and family. Many examples of ambient visualization have been developed, although most have not appeared outside of the bubble of academia. For instance, in 2000 researchers in Sweden proposed the concept of informative art, an attempt to introduce dynamic data visualizations in public space by exploiting the social acceptance of art paintings.[3] An electronic version of a Mondrian-style painting slowly changed its visual composition according to the world's weather conditions or the real-time locations of public-transportation buses. Passersby would initially be puzzled, often unaware that the wall-mounted Mondrian replica was a slow-changing computer art work, even when informed citizens revealed that the painting actually portrayed meaningful information.

The Downfall

The wish to present information without being noticed demonstrates the contradicting requirements of ambient visualization, a discipline that surfaced when information visualization was introduced to the field of ubiquitous computing (ubicomp), which investigates how computing technology can disappear into the fabric of everyday life.[4] The issues facing ambient visualization are manifold, including a widespread belief that people do not appreciate textual interfaces. For instance, displaying the forecasted outside temperature in an ambient way is often more difficult to comprehend than simply reading a numerical value: while the relatively small difference between 20 and 23 degrees Celsius could lead to a fashion adjustment, it might just be a slight, almost indistinguishable color hue shift on the Ambient Orb. In addition, the urge for creating visualizations that have more decorative than informative qualities has led to the fact that most laypeople cannot understand the need to dedicate their money or space for something that requires deciphering for it to become functional. There has also been a severe shortage of interesting data sets that go beyond simple weather reports or stock-market quotes and establish a personal relationship with the user. Who really cares about the weather when one can simply look outside the window? And are people who really care about the real-time performance of their stock-market portfolio not much better off with more precise tools than noticing an odd twist in the Mondrianesque pattern?

The Current

During the last few years, some of these issues have dramatically changed. Data visualization is becoming a communication medium in its own right, able to both inform and engage people. There is also the emerging trend of self-surveillance and Participatory Sensing, as people voluntarily record their personal behaviors for self-analysis or social-media sharing. (See Nathan Yau's essay "Seeing the World in Data," on pages 246–48.) Finally, we have become aware of the urgent need to use modern technology for persuasive goals—for instance, information feedback displays that are specifically designed to change people's behavior, attitude, or opinions.[5] In short, the unique combination of these three phenomena will enable ambient visualization to rise out of its ashes and forge a purpose that goes beyond the triviality of representing environmental conditions through changes too subtle to be really noticed or even comprehended.

The Rise

Slowly but surely, we are moving toward a world that urgently requires us to become more aware of the environmental consequences of our actions—perhaps by introducing some sort of visual feedback that reveals their true cost or impact. As most of our actions are already electronically traced and stored, most of the required data already exists somewhere in today's huge corporate databases—for example, our day-to-day energy usage, water consumption, or food-spending habits. However, there is still no convincing interface that reveals the consequences of such consumer actions, right when such information would be needed the most. Where are the visceral multimedia experiences that could encourage us to keep up even minor behavioral changes, show our progress in relation to set goals, or benchmark our habits with those of similar people?

Ambient visualization research should explore how such socially relevant data depictions could become integrated within our physical experience, potentially using the plethora of electronic displays of today for more socially relevant goals. People tend to make more considerate decisions when directly confronted with pertinent and trustworthy information about the consequences of our intended actions. These situated data depictions should be present when and where it really counts, namely, when we make decisions.

Moreover, the physical location of such data representations could contextualize the information beyond highlighting data patterns or trends that lie beyond our own personal experience. Instead, situated-data depictions that immediately relate human activities to their impact could demonstrate the underlying principles that actually drive the data. They could even give meaning to the physical environment by making invisible but place-specific data streams that define the local identity perceivable to the local community. Ambient feedback in a public context could further emphasize the current movement toward the democratization of data access, to empower the people whose very data is being represented.

Abstract versus Realistic

The use of visualization for persuasive goals has actually been attempted before. However, the majority of such initiatives have focused on photorealistic simulations of potential outcomes, such as conveying the implications of climate change or the negative consequences of unhealthy eating habits. Visualizations that simulate reality, however, tend to emphasize the visual aspects of change, which may exacerbate people's existing tendency to relate their decision-making solely to what they can see. Instead, a well-designed ambient visualization could shift the focus to the hidden, but often more meaningful, insights that actually should be considered, such as causal factors that relate our actions to their impact. The actual shape and form of such situated visualizations can be manifold, reaching beyond the normal flat displays of today. Ranging from shape-changing clothing and reactive-wall textures to multitouch walls and kinetic urban monuments, we should attempt to integrate information displays within our everyday experience by exploiting our natural capability to understand our environment through its natural affordances.

The Future

The future role of visualization in an ambient context is potentially much more versatile than just providing feedback to encourage behavioral change. A well-designed ambient visualization should have the unique power to also help shape our identity as well as our experience of a place. In the information society of today, we have lost a sense of presence, of ourselves, of our actions, and of other people that surround us. However, a sense of presence still persists through the continuous creation of data in our society. To capture the true nature of our existence, we should look for it through the lens of socially relevant data—in the form of real-time digital traces—and consider their qualitative impact on our lives, through which many characteristics shaping our unique identity can be distinguished. By tracking this data that is continuously generated all around us and sharing it through the medium of visualization, a meaningful sense of existence, as well as responsibility, could be instated. When mass-produced consumer weather station displays can inform us of the impact of our actions, in addition to the outside conditions we have no control over, and when architectural media facades finally display information that is meaningful and socially relevant to their immediate surroundings, ambient visualization might finally enjoy the success it deserves.

Notes

1 Weiser and Brown, "Designing Calm Technology."
2 Wisneski, Ishii, Dahley, Gorbet, Brave, Ullmer, and Yarin, "Ambient Displays," 22–32.
3 Redström, Skog, and Hallnäs, "Informative Art," 103–114.
4 Weiser, "The Computer for the 21st Century," 94–104.
5 Fogg, *Persuasive Technology.*

Cybernetics Revisited: Toward a Collective Intelligence

by Christopher Kirwan

Living in our time poses a formidable challenge that will require the collective intelligence of mankind to reach a new dimension of cooperation to harness and apply the vast amount of technological resources we have at our disposal. If we can learn to overcome the real issues that are preventing progress (geopolitical division, economic disparity, and intellectual property domination), we can then begin to effectively work as a global network to monitor and manage the state of entropy that mother earth is experiencing. In this way we can achieve what Norbert Weiner, the founder of the term *cybernetics*, described as resistance: a way to slow the planet's inevitable systemic decline.

Today, more than a half a century after the publication of Weiner's groundbreaking book *The Human Use of Human Beings: Cybernetics and Society* (1950), in which he describes the general theory of organizational and control relationships in systems and the cooperation between human and machines, we need to revisit the underlying concepts of cybernetics and the ways they have influenced the rapidly evolving world of human-computer interface. Successful application of this has the potential to play a critical role in the efforts to preserve our ecosphere and create a more peaceful, equitable world community.

Weiner and his colleagues, building on the work of other pioneers in mathematics and computation, developed cybernetics as a new discipline linking the fields of electrical and mechanical engineering, logic theory, biology, and neurophysiology. Coined from the Greek word *kybernetes*, meaning "steersman, pilot, governor," *cybernetics* traces back to Plato, who used the concept in his essay on self-governance. Cybernetics examines the structure of regulatory systems, investigating and defining those systems that have goals and operate within circular causal loops, and in which action, information input (feedback), and response all interact, causing the system to adapt and change its behavior. It is notable that the military was the first employer of cybernetics, often in ways dangerous to humans and our environment—quite different from what we expect from the intelligent systems described herein.

The early applications, known as first-order cybernetics, utilized relatively simple applications of cybernetic principles for the control and regulation of mechanical systems. While being very effective in solving problems of limited complexity, they did not factor in the influence of the observer in the process. To overcome this problem, new approaches called second-order cybernetics evolved to incorporate a broader and more complex set of circumstances, representative of a more holistic approach. These include the user point of view, the environmental context, and the models and methods being applied. Second-order cybernetics has been central to rapid development of a myriad of scientific and technological fields, including biology, medicine, demography, robotics, semiotics, management, artificial intelligence, and a number of important analytical disciplines, such as systems theory, decision theory, and information theory.

One of these new applications of cybernetics, emergence theory, is concerned with complex, open systems and the way order can be created without central command and control, made possible by open systems, which have the capability of extracting information directly from the environment. This is the basis for the popularized term *rhizome*, which describes a nonhierarchical system, exemplified by the World Wide Web and its self-generating, exponential growth. (See also chapter 2, page 44.) Another finding indicates that when a large number of complex actions/reactions occur simultaneously, it may lead to the creation of new phenomena and new constructs. Weiner documented the unique ability of information systems, because they operate within organizational structures, to perform negentropically (to resist entropy). If these facts are true, then the fusion of information systems and organic life may become part of our solution to arrest global decline.

Many eminent scientists, in their later years, incorporated new philosophical, historical, and cultural insights into their foundational theories. Weiner was one of those prescient thinkers, like Albert Einstein and others, who eventually came to understand the critical need for our species to accept a greater sense of responsibility for the stewardship of planet Earth and the cosmological system with which we are inextricably linked. In *The Human Use of Human Beings*, Weiner compared Earth's precarious state to the biological fact that all organisms reach a peak in their life cycle when their cells stop reproducing. He termed this natural process *resistance* and used the idea to reinforce the need for man-made systems to develop feedback mechanisms to assist in the control of entropy.

We have just begun to recognize the importance of his predictions, with Barack Obama, Al Gore, and other world leaders and activists striving frantically to find new solutions to our global dilemmas. Scientific studies from many credible sources inform us that our biological system is reaching its tipping point and is in imminent danger of failure. In order to counteract this seemingly inexorable trajectory toward global disorder, we must devise an imaginative new cybernetic process to ensure the viability of our life-sustaining environmental systems. A vital component in developing such a process will be our capacity to plan, organize, and manage complex systems and entities with the capacity to link both organic life and digital networks and to allow us to monitor larger patterns of complex systemic forces at work. Importantly, the resulting systems must become self-generating to be sustainable.

This new cybernetic process will rely on our exponential increase in computational power to process and filter the massive data required to provide a real-time documentation of global activity and resulting behavioral patterns. Each element within our collective biosphere will need to be carefully tracked, interrelating multiple factors such as geologic and climatic conditions, predatory traits, and life cycles. In this complex and multidimensional data environment, the role of visualization will be key in providing the capacity to recognize the emergent patterns and processes of these phenomena. Visualization will itself become organic, as it will need to adapt to simulate information from a wide spectrum of sources, ranging from micro/organic to macro/planetary states. The role of artificial intelligence will be critical in creating this new cybernetic form of resistance, revealing abnormal trends and anomalies and giving us the ability to utilize resources more effectively and to prevent major catastrophes before they occur.

A new process I am calling regenerative networks could be the next phase of cybernetics, a real-time living system of communication, capable of dealing with highly complex interactions. In early-stage electronics, a simple one-way/two-way communication—a computer responding to a query—was considered a breakthrough. But soon we will have the capability to integrate electromechanical devices with organic structures, permitting us to monitor and regulate the flow of information from multiple data sources and creating a macroscale feedback system designed to evolve gradually and to link more and more networks until there is a virtual reproduction of organic states. This new regenerative process would have the capability to monitor and control both environmental and man-made systems. The Smart City program, currently underway in a number of large cities, integrates smart electronics into both building systems and urban infrastructure. These are examples of the potential of the new systems and tools already at our command.

The prognostications of Ray Kurzweil in his book *The Age of Spiritual Machines* (1999) is a reflection of our efforts to create a productive, sustainable future—one requiring local and global cooperation. Kurzweil is confident that by 2020, computers will approximate the memory capacity and computational power of the human brain. Yet he argues that raw computational power alone will not supplant the human ability to foresee events and chart new courses. We must now employ what he refers to as "the software of intelligence" to construct a global "open network," which utilizes the collective capacities of both man and machine. He notes, however, that while our advanced technologies and systems offer great benefits to mankind, they also present serious obstacles to progress, due to our innate human weaknesses—fear, hate, greed, and the lust for power.

The threat of imminent disaster to our planet may well be the catalyst for leaders in government, industry, and religion to reach across ancient and contemporary boundaries to create innovative new partnerships and networks as well as a set of shared ethical and moral principles to move toward what might be called a state of collective intelligence. Within the resulting collaborative framework, diverse cultures, environments, and life forms could be harnessed together as a technological extension of both natural processes and human consciousness. On an organic level, such a new system would allow unions of collective intelligence to form a holistic web that would no longer be constrained by the divisions and boundaries of human creation.

In our effort to create this collective intelligence, we will need to work together to find equitable ways of sharing economic opportunities, create new public-private partnerships, incorporate new technologies rather than allow the protectionist interests of large corporations to inhibit their use by the public, and move away from our current fossil-fuel-based economy toward a new era of bio-organic and artificial intelligence. The liberators of the future will be those from diverse industries, fields of knowledge, and cultural backgrounds who seek to facilitate this critically needed and daunting integration process, to sustain our planet, and to realize what Weiner called "the human use of human beings."

Reflexive Ecologies: Visualizing Priorities

by David McConville

Who are we? Within this simple question is contained the essence of what it means to be human: our capacity to reflect on our own consciousness. This reflexive impulse is so central to our character that we call our subspecies *homo sapien sapiens*, identifying sapience—the wisdom to act with appropriate judgment—as the primary trait that distinguishes us from other animals.

Our collective quest to know ourselves begins with imagining the world and our place in it. The success of our species is largely attributable to our ability to imagine and map abstract concepts, which help us to study, communicate, and synchronize with our local environments. We create and imbue imagery with symbolism derived from interactions with our surroundings, often accompanied by stories, artifacts, and practices that give clues to their meaning. These culturally constructed modes of communication enable us to share our experiences and cultivate knowledge across generations, providing important contextual understanding that helps us to situate ourselves in the world and the cosmos.

The power of imagery to profoundly affect our sense of place has been exquisitely demonstrated by a few key examples in the history of science. Nicolaus Copernicus's sixteenth-century illustration of a sun-centered solar system has been widely credited as the primary factor in the precipitation of the scientific revolution. In one fell swoop, he redrew the cosmic order and by extension, much of the Western world's understanding of humanity's relationship to the heavens. Almost five hundred years later, the Apollo 8 *Earthrise* image recontextualized perceptions of humanity's place in the cosmos for much of the world, with its first photographic view of Earth from outer space. This image is often attributed with instigating an environmental-paradigm shift, inducing numerous commentaries concerning the fragility of our home planet and the interconnectedness of the global community. Designers Charles and Ray Eames further pushed the reorienting potential of imagery to new heights (and depths) with their seminal 1972 short film *Powers of Ten*. It took viewers on an impossible journey across many orders of magnitude from quarks to quasars, pioneering the dynamic long-zoom camera technique that illustrates how strongly our concepts of reality are shaped by sensory experiences.

Today our self-reflective search has expanded into new dimensions. While these earlier examples shifted spatial awareness, we are increasingly able to measure and represent temporal, spectral, and relational characteristics of our environs. A latticework of satellites, telescopes, and other measuring instruments are perpetually scanning and providing voluminous amounts of data about our surroundings. Time-lapse and hyperspectral photographs shed new light on atmospheric, biospheric, and cosmic processes. GPS and radio-frequency identification devices (RFID) track interactions between people, products, and processes around the globe. And with the ever-expanding integration of internet-connected gadgets into our daily lives, data about our activities, interests, and movements are generated across physical, social, and virtual domains.

As many of us attempt to make sense of the gestalt of information being generated by and about us, it is little surprise that interest in computer visualization is exploding. Mass digitization yields endless territories to map, while increased accessibility of graphics software enables widespread experimentation with novel representation techniques. Geospatial visualizations provide instant access to worldwide atlases of information, now ubiquitously available through GPS, web maps, and digital globes. Scientific visualizations are widely used to visually simulate phenomena at various scales, appearing regularly in news reports, exhibitions, websites, and mobile applications. Information visualizations are used to reveal hidden patterns within interdisciplinary networks of large-scale data collections. A new generation of information cartographers has taken up the challenge of exploring the aesthetic possibilities of these databases, and these ongoing investigations are yielding intriguing—and occasionally useful—renderings to disclose previously imperceptible relationships.

Burgeoning interest in these visualizations suggests that they may also prove useful for illuminating the most complex and important network of all: Earth's biosphere. Composed of all of the ecosystems on the planet, the biosphere regulates the countless vital interactions that are essential for supporting life as we know it. These include not only the biological networks that sustain us but also the generation of the "natural resources" that feed consumer society's global production and distribution networks. While most of these ecological processes have been made invisible as externalities with modern economic systems, it is apparent that their healthy functioning can no longer be taken for granted.

The sensory networks that monitor our home planet have brought to light some alarming trends in recent decades. Industrial societies have been consuming resources much faster than the planet can regenerate them, resulting in the destabilization of environmental conditions upon which human civilizations have been dependent for millennia. Specific planetary boundaries have now been identified as the "safe operating space for humanity," within which we must stay to avoid disastrous consequences. Since 1968 (ironically, the same year the *Earthrise* photograph helped to birth the environmental movement), we have been slipping further into "ecological debt" as we rapidly expand our global footprint and exceed the "safe operating space."[1] As a result, we are facing a convergence of interconnected environmental crises, including ocean acidification, mass species extinction, overfishing, peak oil, peak water, land degradation, deforestation, and plastics pollution—not to mention climate change.

Developing appropriate responses to these urgent issues requires more effective tools for reflexively examining humanity's relationship with global ecological systems. Derived from the Greek root *oikos* and *logos*, *ecology* appropriately means the "study of relations" and is used to describe many studies of interactions between organisms and their environments. Practitioners in the field of complex network visualization are well positioned to apply their artistic and technical experiences to focus much-needed attention on these essential interconnections.

A number of nascent efforts are already exploring how aesthetic approaches to visualization and mapping can provide new perspectives on critical ecological interactions. NASA's Scientific Visualization Studio incorporates satellite data with 3-D animations to demonstrate a wide range of scientifically measured phenomena.[2] Designers Tyler Lang and Elsa Chaves have illustrated that interconnected global systems and events can influence each other with *Connecting Distant Dots* (2009). (See chapter 5, page 194.) Media artist Tiffany Holmes creates and curates artworks devised to reveal the processes of consumption under the rubric of *eco-visualization*, which she defines as the "creative practice of converting real-time ecological data into image and sound for the purpose of promoting environmental awareness and resource conservation."[3] Photographer Chris Jordan has created a series of sobering images chronicling the unimaginable scale of mass consumption with the series Running the Numbers: An American Self-Portrait (2006–9).[4] The *Sourcemap* project from MIT uses geospatial data and information visualization to reveal the global impact of product supply chains.[5]

But these efforts are only the beginning. The accelerating environmental challenges faced by modern civilization are necessitating that we reimagine our relationships to the natural world. Meeting the needs of global society does not require infinite economic growth but an understanding of and respect for the regenerative limits of the biosphere. As accelerating global changes force us to find innovative ways of enhancing the integrity of local and global ecosystems, visualizations will play an essential role in making our connections to these ecological processes explicit. We will likely find that our species' unique ability to creatively imagine and map our place in the world will once again be key to adapting to changing environments.

Notes

1 "Global Footprint Network"; Rockstrom et al., "A Safe Operating Space," 472–75.
2 "NASA Goddard Scientific Visualization Studio."
3 Holmes, "Ecoviz.org."
4 Jordan, "Running the Numbers."
5 "Sourcemap—Open Supply Chains."

Bibliography

Adamic, Lada A., and Natalie Glance. "The Political Blogosphere and the 2004 U.S. Election: Divided They Blog." In *Proceedings of the 3rd International Workshop on Link Discovery.* Chicago: ACM, 2005. http://doi.acm.org/10.1145/1134271.1134277.

Alexander, Christopher. "A City is Not a Tree." Pts. 1 and 2. *Architectural Forum* 122 (April 1965): 58–62; 122, (May 1965): 58–62. http://www.rudi.net/node/317.

———. *Notes on the Synthesis of Form*, Harvard Paperbacks. Cambridge, MA: Harvard University Press, 1964.

———. *A Pattern Language: Towns, Buildings, Construction.* Center for Environmental Structure Series. New York: Oxford University Press, 1977.

Anklam, Patti. *Net Work: A Practical Guide to Creating and Sustaining Networks at Work and in the World.* Burlington, MA: Butterworth-Heinemann, 2007.

Arnheim, Rudolf. *Toward a Psychology of Art: Collected Essays.* 1966. Reprint, Berkeley and Los Angeles: University of California Press, 1972.

———. *Visual Thinking.* 1969. Reprint, Berkeley and Los Angeles, CA: University of California Press, 2004.

Artz, Lee, and Yahya R. Kamalipour, eds. *The Globalization of Corporate Media Hegemony.* Albany, NY: State University of New York Press, 2003.

Bacon, Francis. *Francis Bacon: The Major Works.* Oxford World's Classics. Edited by Brian Vickers. 2002. Reprint, New York: Oxford University Press, 2008.

Barabási, Albert-László. *Linked: How Everything Is Connected to Everything Else and What It Means for Business, Science, and Everyday Life.* New York: Plume, 2003.

Barabási, Albert-László, and Eric Bonabeau. "Scale-Free Networks." *Scientific American* 288 (May 2003): 60–69. http://www.scientificamerican.com/article.cfm?id=scale-free-networks.

Baran, Paul. "On Distributed Communications: Introduction to Distributed Communications Networks." Rand Report RM-3420-PR. Santa Monica, CA: Rand Corporation, 1964.

Bates, Brian. *The Real Middle Earth: Exploring the Magic and Mystery of the Middle Ages, J.R.R. Tolkien, and "The Lord of the Rings.* 2003. Reprint, New York: Palgrave Macmillan, 2004. First published 2002 by Sidgwick & Jackson.

Berners-Lee, Tim. "Tim Berners-Lee on the next Web." Audiovisual file of lecture. TED, February 2009. http://www.ted.com/talks/tim_berners_lee_on_the_next_web.html.

———. "Tim Berners-Lee: The Year Open Data Went Worldwide." Audiovisual file of lecture. TED, February 2010. http://www.ted.com/talks/tim_berners_lee_the_year_open_data_went_worldwide.html.

Bertin, Jacques. *Semiology of Graphics: Diagrams, Networks, and Maps.* Translated by William J. Berg. Madison, WI: The University of Wisconsin Press, 1984.

Best, Steven, and Douglas Kellner. *Postmodern Theory: Critical Interrogations.* New York: Guilford Publications, 1991.

Biggs, Norman L., E. Keith Lloyd, and Robin J. Wilson. *Graph Theory 1736–1936.* 1976. Reprint, New York: Oxford University Press, 1998.

Brafman, Ori, and Rod A. Beckstrom. *The Starfish and the Spider: the Unstoppable Power of Leaderless Organizations.* New York: Portfolio, 2006.

Brinton, Willard C. *Graphic Methods for Presenting Facts.* New York: The Engineering Magazine Company, 1914.

Brinton, Willard C. *Graphic Presentation.* New York: Brinton Associates, 1939.

Brown, Lloyd A. *The Story of Maps.* New York: Dover Publications, 1980. First published 1949 by Boston, Little, Brown.

Butler, Jill, Kritina Holden, and Will Lidwell. *Universal Principles of Design: 100 Ways to Enhance Usability, Influence Perception, Increase Appeal, Make Better Design Decisions, and Teach Through Design.* 2003. Reprint, Beverly, MA: Rockport Publishers Inc., 2007.

Card, Stuart K., Jock D. Mackinlay, and Ben Shneiderman. *Readings in Information Visualization: Using Vision to Think.* San Francisco: Morgan Kaufmann, 1999.

Casti, J., and A. Karlqvist, eds. *Art and Complexity.* Amsterdam: Elsevier Science B.V., 2003.

Caswell, Janice. "Artist Statement." Accessed March 15, 2010. http://janicecaswell.com/statement.html.

Chambers, Ephraim. *Cyclopædia, or, An universal dictionary of arts and sciences: containing the definitions of the terms, and accounts of the things signify'd thereby, in the several arts, both liberal and mechanical, and the several sciences, human and divine: the figures, kinds, properties, productions, preparations, and uses, of things natural and artificial: the rise, progress, and state of things ecclesiastical, civil, military, and commercial: with the several systems, sects, opinions, etc: among philosophers, divines, mathematicians, physicians, antiquaries, criticks, etc: the whole intended as a course of antient and modern learning.* London: J. and J. Knapton, 1728. http://digital.library.wisc.edu/1711.dl/HistSciTech.Cyclopaedia01.

Chen, Chaomei. *Information Visualization: Beyond the Horizon.* 2nd ed. London: Springer, 2006.

Chi, Ed H. *A Framework for Visualizing Information.* Dordrecht, Netherlands: Kluwer Academic Publishers, 2002.

Christ, Karl. *The Handbook of Medieval Library History.* Edited and translated by Theophil M. Otto. Rev. ed. Metuchen, NJ: The Scarecrow Press, Inc., 1984.

Crosby, Alfred W. *The Measure of Reality: Quantification in Western Europe, 1250–1600.* Cambridge: Cambridge University Press, 1997.

Darwin, Charles. *The Origin of Species.* New York: Gramercy Books, 1995. First published 1859 by John Murray.

Darwin Correspondence Project Database, letter no. 2465. Accessed February 2, 2010. http://www.darwinproject.ac.uk/entry-2465/.

Deleuze, Gilles, and Felix Guattari. *A Thousand Plateaus: Capitalism and Schizophrenia.* Minneapolis, MN: University of Minnesota Press, 1987.

Diderot, Denis. "Encyclopedia." In *The Encyclopedia of Diderot & d'Alembert Collaborative Translation Project.* Translated by Philip Stewart. Ann Arbor, MI: Scholarly Publishing Office at the University of Michigan Library, 2002. Originally published as "Encyclopédie." In *Encyclopédie ou Dictionnaire raisonné des sciences, des arts et des métiers* (Paris: Briasson, 1751). http://hdl.handle.net/2027/spo.did2222.0000.004.

Dondis, Donis A. *A Primer of Visual Literacy.* Cambridge, MA: MIT Press, 1973.

Edson, Evelyn. *Mapping Time and Space: How Medieval Mapmakers Viewed Their World.* The British Library Studies in Map History, vol. 1. London: British Library Board, 1998.

Emmerling, Leonhard. *Jackson Pollock: 1912–1956.* Köln, Germany: Taschen, 2003.

Fellmann, Emil A. *Leonhard Euler.* Translated by Erika Gautschi and Walter Gautschi. Basel, Switzerland: Birkhäuser Basel, 2007.

Fildes, Jonathan. "Artificial brain '10 Years Away,'" *BBC*, July 22, 2009. http://news.bbc.co.uk/2/hi/8164060.stm.

Fogg, B. J. *Persuasive Technology: Using Computers to Change What We Think and Do.* San Mateo, CA: Morgan Kauffmann, 2002.

Foskett, Douglas John. *Classification and Indexing in the Social Sciences.* Washington DC: Butterworths, 1963.

Freeman, Linton C. *The Development of Social Network Analysis: A Study in the Sociology of Science.* Vancouver, BC: Empirical Press, 2004.

———. "Visualizing Social Networks." *Journal of Social Structure* 1, no. 1 (2000). http://www.cmu.edu/joss/content/articles/volume1/Freeman.html.

Friendly, Michael. "Milestones in the history of thematic cartography, statistical graphics, and data visualization." Accessed on August 24, 2009. http://www.math.yorku.ca/SCS/Gallery/milestone/milestone.pdf.

Fry, Ben. "Computational Information Design." PhD diss., Massachusetts Institute of Technology, 2004.

———. *Visualizing Data.* Sebastopol, CA: O'Reilly Media, Inc., 2008.

Gerli, E. Michael, ed. *Medieval Iberia: An Encyclopedia.* New York: Routledge, 2002.

"Global Footprint Network." Accessed June 26, 2010. http://www.footprintnetwork.org.

Gogarten, Peter. "Horizontal Gene Transfer: A New Paradigm for Biology." Lecture presented at the Evolutionary Theory: An Esalen Invitational Conference, Big Sur, California, November 5–10, 2000. http://www.esalenctr.org/display/confpage.cfm?confid=10&pageid=105&pgtype=1.

Greenfield, Susan A. *The Human Brain: A Guided Tour.* Science Masters. London: Phoenix, 1998.

Hageneder, Fred. *The Living Wisdom of Trees: Natural History, Folkore, Symbolism, Healing.* London: Duncan Baird, 2005.

Harris, Michael H. *History of Libraries of the Western World.* Lanham, MD: The Scarecrow Press, Inc., 1999.

Heart, F., A. McKenzie, J. McQuillan, and D. Walden. *ARPANET Completion Report.* Burlington, MA: Bolt, Beranek and Newman, 1978.

Heyworth, P. L., ed. *Medieval Studies for J. A. W. Bennett: Aetatis Suae LXX.* New York: Oxford University Press, 1981.

Hobbs, Robert. *Mark Lombardi: Global Networks.* New York: Independent Curators International, 2003.

Holmes, Tiffany. "Ecoviz.org: Reviews of Pro-environmental Art and Design Projects." Accessed June 26, 2010. http://ecoviz.org/.

Huberman, Bernardo A. *The Laws of the Web: Patterns in the Ecology of Information.* Cambridge, MA: MIT Press, 2001.

Ifantidou, Elly. *Evidentials and Relevance.* Amsterdam: John Benjamins Publishing Company, 2001.

Jacobs, Jane. *The Death and Life of Great American Cities.* New York: Vintage Books, 1992.

Jacobson, Robert. *Information Design.* Cambridge, MA: MIT Press, 2000.

James, E. O. *The Tree of Life: An Archaeological Study.* Leiden, The Netherlands: E. J. Brill, 1966.

Jencks, Charles. *The Architecture of the Jumping Universe—A Polemic: How Complexity Science is Changing Architecture and Culture.* West Sussex, UK: Academy Editions, 1997.

Johnson, Steven. *Emergence: The Connected Lives of Ants, Brains, Cities, and Software.* New York: Scribner, 2002.

Jones, Kristin M. "Tomas Saraceno." *Frieze Magazine* 116 (June–August 2008). http://www.frieze.com/issue/review/tomas_saraceno1.

Jordan, Chris. "Running the Numbers: An American Self-Portrait (2006–2009)." Accessed June 26, 2010. http://www.chrisjordan.com/gallery/rtn/.

Kelly, Kevin. *New Rules for the New Economy: 10 Radical Strategies for a Connected World.* New York: Penguin Books, 1999.

Kerren, Andreas, John Stasko, Jean-Daniel Fekete, and Chris North, eds. *Information Visualization: Human-Centered Issues and Perspectives.* New York: Springer, 2008.

Klapisch-Zuber, Christiane. *L'ombre des ancêtres.* Paris: Fayard, 2000.

Kohler, Wolfgang. *Gestalt Psychology: An Introduction to New Concepts in Modern Psychology.* New York: Liveright Publishing Corporation, 1992.

Korzybski, Alfred. "A Non-Aristotelian System and its Necessity for Rigour in Mathematics and Physics." Paper presented at the American Association for the Advancement of Science, New Orleans, LA, December 28, 1931.

Kosara, Robert. "Visualization Criticism—The Missing Link Between Information Visualization and Art." In *Proceedings of the 11th International Conference on Information Visualization.* Washington, DC: IEEE CS Press, 2007, 631–36.

Kruja, Eriola, Joe Marks, Ann Blair, and Richard Waters. "A Short Note on the History of Graph Drawing." Lecture notes for Computer Science 2265 of the International Symposium on Graph Drawing, Vienna, Austria, September 23–26, 2001. http://www.merl.com/reports/docs/TR2001-49.pdf.

Kuhn, Thomas S. *The Structure of Scientific Revolutions.* 3rd ed. Chicago: The University of Chicago Press, 1996.

Kunin, Victor, Leon Goldovsky, Nikos Darzentas, and Christos A. Ouzounis. "The Net of Life: Reconstructing the Microbial Phylogenetic Network." *Genome Research* 15, no. 7 (2005): 954–59. http://genome.cshlp.org/content/15/7/954.full.

Kurzweil, Ray. *The Age of Spiritual Machines: When Computers Exceed Human Intelligence.* New York: Viking Press, 1999.

Lecointre, Guillaume, and Hervé Le Guyader. *The Tree of Life: A Phylogenetic Classification.* Translated by Karen McCoy. Cambridge, MA: Belknap Press, 2007.

Lehrer, Jonah. "Out of the Blue." *Seed Magazine,* March 3, 2008. http://seedmagazine.com/content/article/out_of_the_blue/.

Lewis, Peter. *Maps and Statistics.* London: Methuen, 1977.

Livio, Mario. *The Golden Ratio: The Story of Phi, the World's Most Astonishing Number.* London: Headline Review, 2003.

Lorand, Ruth. *Aesthetic Order: A Philosophy of Order, Beauty and Art.* Routledge Studies in Twentieth Century Philosophy. New York: Routledge, 2000.

Lovink, Geert. *The Principle of Notworking: Concepts in Critical Internet Culture.* Amsterdam: HVA Publicaties, 2005. Accessed February 12, 2010. http://www.hva.nl/lectoraten/documenten/ol09-050224-lovink.pdf.

Marineau, René F. *Jacob Levy Moreno 1889–1974: Father of Psychodrama, Sociometry, and Group Psychotherapy.* New York: Routledge, 1989.

Massumi, Brian. *A Shock to Thought: Expressions After Deleuze and Guattari.* New York: Routledge, 2002.

Mau, Bruce, Jennifer Leonard, and Institute Without Boundaries. *Massive Change.* London: Phaidon Press, 2004.

Mazza, Riccardo. *Introduction to Information Visualization.* London: Springer, 2009.

McLuhan, Marshall. *Counterblast.* London: Rapp & Whiting, 1970.

Meisenheimer, Peter. "Seals, Cod, Ecology and Mythology." Technical Briefing no. 95.1, International Marine Mammal Association, Guelph, Ontario, Canada, January 1995.
Miller, Reese P., and Valentine Rodger Miller, trans. *René Descartes: Principles of Philosophy: Translation with Explanatory Notes.* Dordrecht, The Netherlands: Kluwer Academic Publishers, 1991.

Moles, Abraham A. *Information Theory and Esthetic Perception.* Champaign: University of Illinois Press, 1969.

Molloy, Sharon. "Sharon Molloy: Artist Statement." White Columns. April 17, 2009. http://registry.whitecolumns.org/view_artist.php?artist=526.

Moreno, Jacob Levy. *Who Shall Survive? A New Approach to the Problem of Human Interrelations.* Washington, DC: Nervous and Mental Disease Publishing Co., 1934.

Morville, Peter. *Ambient Findability.* Sebastopol, CA: O'Reilly Media, Inc., 2005.

Murdoch, John E. *Antiquity and the Middle Ages.* Album of Science, vol. 5. New York: Scribner, 1984.

"NASA Goddard Scientific Visualization Studio." Accessed June 26, 2010. http://svs.gsfc.nasa.gov/.

Nastase, Adrian. "Utilizing Mind Maps as a Structure for Mining the Semantic Web." Master's diss., University of Liverpool, 2009. http://www.scribd.com/doc/16612282/.

Newman, Mark, Albert-László Barabási, and Duncan J. Watts. *The Structure and Dynamics of Networks.* Princeton, NJ: Princeton University Press, 2006.

New York Times, "Emotions Mapped by New Geography: Charts Seek to Portray the Psychological Currents of Human Relationships." April 3, 1933.

Norman, Donald A. *Things That Make Us Smart: Defending Human Attributes in the Age of the Machine.* New York: Perseus Books, 1993.

Northway, Mary L. *A Primer of Sociometry.* Toronto, ON: University of Toronto Press, 1953.

Oliveira, Ana Balona. "Emma McNally: Fields, Charts, Soundings." T1+2 Gallery. Accessed November 30, 2009. http://www.t12artspace.com/artists0/emma-mcnally/essay/.

Padovan, Richard. *Proportion: Science, Philosophy, Architecture.* London: Taylor & Francis, 1999.

Pascal, Blaise. *Pensées.* London: Penguin Classics, 1995.

Philpot, J. H. *The Sacred Tree in Religion and Myth.* New York: Dover Publications, 2004. Originally published as *The Sacred Tree* (London; New York: Macmillan, 1897).

Playfair, William. *The Commercial and Political Atlas and Statistical Breviary.* Cambridge: Cambridge University Press, 2005.
Pollack, Rachel. *The Kabbalah Tree: A Journey of Balance & Growth.* Woodbury, MN: Llewellyn Worldwide, 2004.

Pombo, Olga. "Combinatória e Enciclopédia em Rámon Lull." Accessed October 16, 2009. http://www.educ.fc.ul.pt/hyper/enciclopedia/cap3p2/combinatoria.htm.

———. *Unidade da Ciência: Programas, Figuras e Metáforas* (Unity of science: programs, figures, and metaphors). Lisbon: Edições Duarte Reis, 2006.

Popper, Karl. *The Poverty of Historicism.* Routledge Classics. New York: Routledge, 2002. First published 1957 by Routledge & Kegan Paul.

Porteous, Alexander. *The Forest in Folklore and Mythology.* New York: Dover Publications, 2002. Originally published as *Forest Folklore, Mythology, and Romance* (New York: Macmillan, 1928).

Preus, Anthony, and John P. Anton, eds. *Essays in Ancient Greek Philosophy V: Aristotle's Ontology.* Albany: State University of New York Press, 1992.

Ramírez, Mari Carmen, and Theresa Papanikolas. *Questioning the Line: Gego in Context.* Houston, TX: The Museum of Fine Arts, 2003.

Redström, Johan, Tobias Skog, and Lars Hallnäs, "Informative Art: Using Amplified Artworks as Information Displays." In *Proceedings of DARE 2000 on Designing Augmented Reality Environments.* Elsinore, Denmark: ACM, 2000. http://www.johan.redstrom.se/thesis/pdf/infoart.pdf

Robinson, Arthur H. *Early Thematic Mapping in the History of Cartography.* Chicago: The University of Chicago Press, 1982.

Rockström et al. "A Safe Operating Space for Humanity." *Nature* 461 (September 2009): 472–75. http://www.nature.com/nature/journal/v461/n7263/full/461472a.html.

Rossi, Paolo. *Logic and the Art of Memory: The Quest for a Universal Language.* Translated by Stephen Clucas. London: Continuum, 2006. Originally published as *Clavis Universalis: Arti Della Memoria E Logica Combinatoria Da Lullo A Leibniz* (Bologna, Italy: Societa editrice il Mulino, 1983).

Sandifer, C. Edward. *The Early Mathematics of Leonhard Euler.* Washington DC: The Mathematical Association of America, 2007.

Simon, Herbert A. *The Sciences of the Artificial.* 3rd ed. Cambridge, MA: MIT Press, 1996.

Sivartha, Alesha. *The Book of Life: The Spiritual and Physical Constitution of Man 1912.* Whitefish, MT: Kessinger Publishing, LLC, 2004.

Smith, W. H. *Graphic Statistics in Management.* New York: McGraw-Hill, 1924.

"Sourcemap—Open Supply Chains & Carbon Footprint." Accessed June 26, 2010. http://www.sourcemap.org/.

Sowa, John F. "Building, Sharing, and Merging Ontologies." Accessed October 12, 2009. http://www.jfsowa.com/ontology/ontoshar.htm.

Sperber, Dan, and Deirdre Wilson. *Relevance: Communication & Cognition.* 2nd ed. Oxford: Blackwell Publishers, 1995.

Staley, David J. *Computers, Visualization, and History: How New Technology Will Transform Our Understanding of the Past.* Armonk, NY: M. E. Sharpe, 2002.

Storr, Robert. "Gego's Galaxies: Setting Free the Line." *Art in America* 91 (June 2003): 108–13.

Studtmann, Paul. "Aristotle's Categories." In *Stanford Encyclopedia of Philosophy.* Stanford University, 1997–. Articled published September 7, 2007. http://plato.stanford.edu/archives/fall2008/entries/aristotle-categories/.

Tapscott, Don, and Anthony D. Williams. *Wikinomics: How Mass Collaboration Changes Everything.* New York: Portfolio, 2006.

Taylor, Richard. *Chaos, Fractals, Nature: A New Look at Jackson Pollock.* Eugene, OR: Fractals Research Laboratory, 2006.

Thackara, John. *Design After Modernism: Beyond the Object.* London: Thames & Hudson, 1989.

Thalimair, Franz. "Technological Mimesis: In Conversation with the Artist Marius Watz." CONT3XT.NET. Accessed March 14, 2010. http://cont3xt.net/blog/?p=253.

Thomas, James J., and Kristin A. Cook. *Illuminating the Path: The Research and Development Agenda for Visual Analytics.* Richland, WA: National Visualization and Analytics Center, 2005.

Thompson, D'Arcy Wentworth. *On Growth and Form.* Canto series. Cambridge: Cambridge University Press, 1992.

Thrupp, Sylvia Lettice, ed. *Change in Medieval Society: Europe North of the Alps, 1050–1500.* London: Owen, 1965.

Tidwell, Jenifer. *Designing Interfaces: Patterns for Effective Interaction Design.* Sebastopol, CA: O'Reilly Media, Inc., 2005.

Tufte, Edward R. *The Visual Display of Quantitative Information.* 2nd ed. Cheshire, CT: Graphics Press, 2001.

———. *Beautiful Evidence.* Cheshire, CT: Graphics Press, 2006.

Van Doren, Charles. *A History of Knowledge: Past, Present, and Future.* New York: Ballantine Books, 1992. First published 1991 by Carol Publishing Group.

Varnelis, Kazys. *Networked Publics.* Cambridge, MA: MIT Press, 2008.

Viegas, Fernanda, and Martin Wattenberg. "Tag Clouds and the Case for Vernacular Visualization." *interactions* 15 (July–August 2008): 49–52.

von Baeyer, Hans Christian. *Information: The New Language of Science.* Cambridge, MA: Harvard University Press, 2005. First published 2003 by Weidenfeld & Nicolson.

Walworth, Catherine. "City Maps: ArtPace—A Foundation For Contemporary Art." *Art Lies: A Contemporary Art Quarterly* 42 (Spring 2004). http://www.artlies.org/article.php?id=143&issue=42&s=1.

Ware, Colin. *Information Visualization: Perception for Design.* 2nd ed. San Francisco: Morgan Kaufmann, 2004.

Wasserman, Stanley, and Katherine Faust. *Social Network Analysis: Methods and Applications.* Cambridge: Cambridge University Press, 1994.

Weaver, Warren. "Science and Complexity." *American Scientist* 36 (1948): 536–44. http://www.ceptualinstitute.com/genre/weaver/weaver-1947b.htm.

Weigel, Sigrid. "Genealogy: On the Iconography and Rhetorics of An Epistemological Topos." Accessed October 1, 2009. http://www.educ.fc.ul.pt/hyper/resources/sweigel/.

Weinberger, David. "Taxonomies and Tags: From Trees to Piles of Leaves." January 20, 2006. http://www.hyperorg.com/blogger/misc/taxonomies_and_tags.html.

Weiner, Norbert. *The Human Use of Human Beings: Cybernetics and Society,* 1950. Reprint, Cambridge, MA: Da Capo Press, 1988.

Weiser, Mark. "The Computer for the 21st Century." *Scientific American* 265 (September 1991): 94–104. http://www.ubiq.com/hypertext/weiser/SciAmDraft3.html.

Weiser, Mark, and John Seely Brown. "Designing Calm Technology," December 21, 1995. http://www.ubiq.com/hypertext/weiser/calmtech/calmtech.htm.

Wisneski, Craig, Hiroshi Ishii, Andrew Dahley, Matt Gorbet, Scott Brave, Brygg Ullmer, and Paul Yarin, "Ambient Displays: Turning Architectural Space into an Interface between People and Digital Information." In *Proceedings of the First International Workshop on Cooperative Buildings.* New York: Springer, 1998, 22–32.

Woolman, Matt. *Digital Information Graphics.* New York: Watson-Guptill Publications, 2002.

Wurman, Richard Saul, David Sume, and Loring Leifer. *Information Anxiety 2.* Indianapolis, IA: Que, 2001.

Contributor Biographies

Christopher Grant Kirwan is a multidisciplinary consultant and educator with expertise in urban planning, architecture, and new media. He grew up in a large, artistic Unitarian Universalist family in Cambridge, Massachusetts; graduated from RISD with a bachelor of architecture; attended MIT Center for Advanced Visual Studies; has been an adjunct faculty at Parsons The New School for Design since 1996 and a visiting lecturer at Harvard GSD; worked and lived throughout Europe, the Middle East, and Asia; and cofounded the firm Newwork, with offices in New York and Beijing.
http://kirwandesign.com

David McConville is a media artist and theorist, whose investigations focus on how visualization environments can transform perspectives on the world. He is cofounder of The Elumenati, a design and engineering firm with clients ranging from art festivals to space agencies. He is also on the board of directors of the Buckminster Fuller Institute, through which he helps to develop Idea Index 1.0, a network of projects designed to apply ecological principles to human endeavors.
http://elumenati.com

Andrew Vande Moere is a senior lecturer at the University of Sydney and specializes in the research and teaching of data visualization, interaction design, and media architecture. He is also the sole author of the blog Information Aesthetics, which collects contemporary examples of the blending of design, art, and visualization.
http://infosthetics.com

Nathan Yau is a PhD candidate in Statistics at UCLA and has a background in computer science and graphic design. He focuses on visualization and making data more accessible to those without scientific training, which he covers regularly on his blog FlowingData.
http://nathanyau.com

Image Credits

Alex Adai 139, 140; Lada Adamic 101b; Eytan Adar 166b; Christopher Adjei 188–89; Syed Reza Ali 211b; José Ignacio Alvarez-Hamelin 120, 202t; Kunal Anand 63b, 106 bl, 106br; Aran Anderson 178b; Bergamini Andrea 191t; Burak Arikan 152t; © 2010 Artists Rights Society (ARS), New York / VG Bild-Kunst, Bonn 239; Christopher Paul Baker 115; Michael Balzer 164; Marian Bantjes 181t; Jeff Baumes 117t; Augusto Becciu 151; Tom Beddard 20; Skye Bender-deMoll 85b, 110; Biblioteca Pública del Estado en Palma de Mallorca 32; © Graciela Blaum (http://bit.ly/9Nvmcn) 225c; Marco Borgna 113t; Dave Bowker 135; Trina Brady 213; Stefan Brautigam 136–37; Heath Bunting 142–43; Lee Byron 128; © 2005 Tom Carden 85r; © 2005 Tom Carden and Steve Coast, map data CC-BY-SA, OpenStreetMap.org contributors. 148 top, 224; © 2006 Tom Carden 149tl; Janice Caswell 237tl; Pascal Chirol 182t; Nicholas Christakis 171b; Marshall Clemens. © shiftN 194b; Axel Cleeremans 215b; Stephen Coast 121b, 233bl; Colección Fundación Museos Nacionales-Galería de Arte Nacional. Archivo Fundación Gego. Photo by Paolo Gasparini. 242; Dan Collier. http://dancollier.co.uk. 127; Peter Crnokrak 129; Pedro Miguel Cruz 147; Franck Cuny. © 2009 Linkfluence, Gephi. 60; Ian Dapot 3; Maurits de Bruijn 180; Paul De Koninck, Laval University (http://www.greenspine.ca) 225br, 230l; Gerhard Dirmoser 181b; Reproduced by permission from Matthias Dittrich 114; Matthias Dittrich 130l; Martin Dittus 161b, 162b; Douglas H. Gordon Collection, Special Collections, University of Virginia Library 35; Gabriel Dunne 218b, 244; Reprinted by permission of the publisher from Francesco Rao and Amedeo Caflisch, "The Protein Folding Network," *Journal of Molecular Biology* 342, no. 1 (September 3, 2004): 299–306. 141; © 2004, with permission from Elsevier 141; Andrew Coulter Enright 90; © FAS.research 203; Jean-Daniel Fekete 105; Firstborn 202b, 233t; Courtesy of Eric Fischer. Base map © OpenStreetMap, CC-BY-SA. 1; FMS Advanced Systems Group 144b; Francesco Franchi 132; © Eric Gaba and user Bamse for Wikimedia Commons 89l; Julien Gachadoat. http://www.2roqs.com I http://www.v3ga.net I http://www.hudson-powell.com 216; Christoph Gerstle 175b; © Govcom.org 200; Ramesh Govindan 170t; Wesley Grubbs 111, 112t, 215t; Baris Gumustas 199t; Sameer Halai 168; Chris Harrison 118, 155, 156–57, 161t; Marcus Hauer 211t; Dan Haught 145; Jeffrey Heer 96, 144t, 165b; Felix Heinen 162t; Sebastian Heycke 214t; Robert Hodgin 226, 227b; Danny Holten 198t; J. D. Hooge 209t; Ben Hosken 109; Matthew Hurst 99; © 1996 IEEE 177br, 179 bottom row; International Center for Joachimist Studies 29, 30, 31; Hawoong Jeong. Reprinted by permission from Macmillan Publishers Ltd: Nature, © 2001. 170br; Greg Judelman 185t; Chris King 177bl; Robert King 192; Joris Klerkx 108b; Josh Knowles 116; Andreas Koberle 199b; Andreas Koller. http://similardiversity.net/ 124–25; Victor Kunin 68b; Martin Krzywinski 197; Anthony Kyriazis 130b; Reproduced by permission from David M. Lavigne, "Marine Mammals and Fisheries: The Role of Science in the Culling Debate," in *Marine Mammals: Fisheries, Tourism and Management Issues*, eds. N. Gales, M. Hindell and R. Kirkwood (Collingwood, Australia: CSIRO Publishing, 2003). 70; Jang Sub Lee 220; Manuel Lima 28, 45, 63c, 66t, 75, 91; © 2009 Linkfluence, Gephi, INIST, CNRS 165t; Donald Lombardi and Pierogi Gallery 72; Patent Pending & Copyright © Lumeta Corporation 2010. All Rights Reserved. 121t; Barrett Lyon 119; Ludovico Magnocavallo 98; Brandon Martin-Anderson 47t; Sean McDonald 113b, 174b; Daniel McLaren 153b; Emma McNally 233c, 237r, 237b; © 2010 Microsoft Corporation, © NAVTEQ. 93; Ernesto Mislej 166t; © MIT SENSEable City Lab. Courtesy of Aaron Koblin. 176; Sharon Molloy 232br, 234, 235; © motiroti 2006. Helen Mitchell. 172; Boris Muller 212; Muckety.com. 185b; © 2008 National Academy of Sciences, U.S.A. 102, 170bl; © 2007 National Academy of Sciences, U.S.A. Courtesy of Jukka-Pekka Onnela. 206t; Copyright © 2005 *New York Times*. Reprinted by Permission. 214b; Copyright © 2007 *New York Times*. Courtesy of Jonathan Corum. 198b; Sandra Niedersberg 193; Dalibor Nikolic 240tl, 240tr; Inan Olcer 106t; Josh On 112b; Jose Luis Ortega 169; Mariona Ortiz 131; Santiago Ortiz 87 top row, 177t, 196, 217, 219b; W. Bradford Paley 42, 85tl, 104, 123, 184, 186–87; Andrew Pavlo 210b; Daniel Peltz 195; Keith Peters 228 bottom row; Doantam Phan 191b; Sebastien Pierre 173b; © 2010 The Pollock-Krasner Foundation / Artists Rights Society (ARS), New York. 225t, 227tl; Mario Porpora 190; Stefanie Posavec 122, 205; Private Collection. Reproduced with permission from Fundación Gego. 240b; Marco Quaggiotto 130r, 133, 167; Reproduced by permission of Rand Corporation via Copyright Clearance Centre 55t; Jacob Ratkiewicz 152b; From J. Ratkiewicz, M. Conover, M. Meiss, B. Gonçalves, S. Patil, A. Flammini, and F. Menczer, "Detecting and tracking the spread of astroturf memes in microblog streams." CoRR eprint arXiv:1011.3768, 2010. http://arxiv.org/abs/1011.3768. Courtesy of Jacob Ratkiewicz. 152b; © 2002 The Regents of the University of California. All Rights Reserved. Used by permission. 218t; Donato Ricci 210t; Antonin Rohmer. © Linkfluence. 59t, 100b; Antonin Rohmer. © AFCR, Linkfluence, La Netscouade. 101t; Andrey Rzhetsky 175t; Marcel Salathé 206b; Nikos Salingaros 47b; Tomas Saraceno 238; Michael Schmuhl 207; Felix Schürmann © BBP/EPFL 53, 227tr; Eduardo Sciammarella 182b; seedmagazine.com. Included here by permission. 194t; Aaron Siegel 107; Jørgen Skogmo, shiftcontrol.dk 178t; James Spahr 219t; Jürgen Späth 3, 126; Stamen 148b; Moritz Stefaner 2, 86–87, 95b, 103, 173t; Lisa Strausfeld 142l; Richard Taylor 225bl; © TeleGeography (www.telegeography.com) 89r; © TesisDG. Juan Pablo de Gregorio. 208; Kris Temmerman. 153t; Jer Thorp 134; Jer Thorp 150; Ian Timourian 108t, 162t; Christophe Tricot 94t; Makoto Uchida 100t; Peter Uetz 171t; University of Chicago: ARTFL Encyclopédie Project 38, 40; University of Wisconsin Digital Collections 37; Frederic Vavrille 209b; Volker Springel, Max-Planck-Institute for Astrophysics, Germany 230r; Patrick Vuarnoz 183; Martin Wattenberg 160; Marius Watz 4, 228t; Marcos Weskamp 117b, 174t; Yose Widjaja 138; Roland Wiese 204; Richard Wolton (http://wolton.net) 179 top row; Jeremy Wood 146, 149tr, 149b

Index

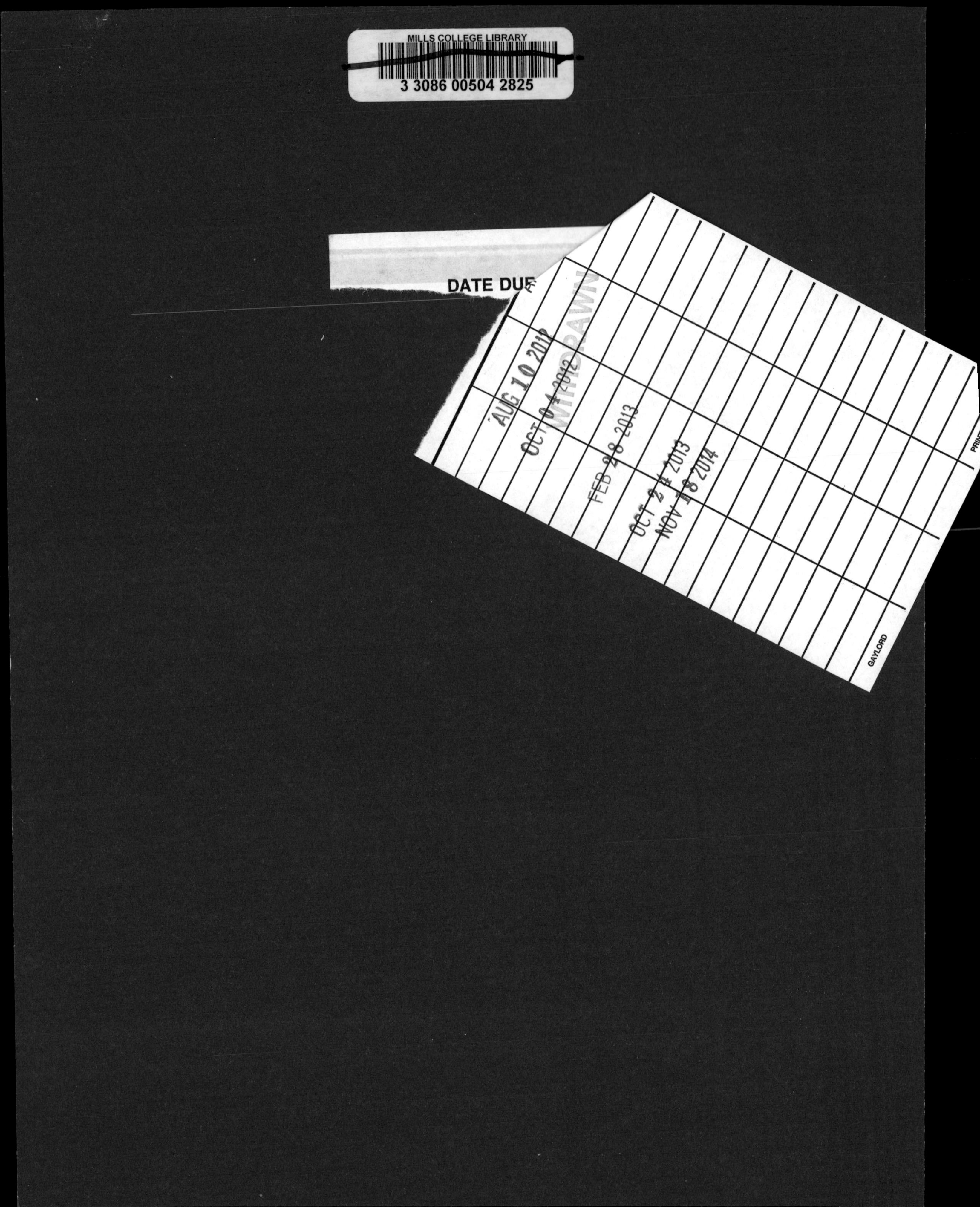
MILLS COLLEGE LIBRARY
3 3086 00504 2825
DATE DUE
AUG 10 2012
OCT 04 2012
WITHDRAWN
FEB 28 2013
OCT 24 2013
NOV 18 2014
GAYLORD